# An Introduction to
# Beam Physics

# Series in High Energy Physics, Cosmology, and Gravitation

Series Editors:  **Brian Foster**, *Oxford University, UK*
**Edward W Kolb**, *Fermi National Accelerator Laboratory, USA*

This series of books covers all aspects of theoretical and experimental high energy physics, cosmology and gravitation and the interface between them. In recent years the fields of particle physics and astrophysics have become increasingly interdependent and the aim of this series is to provide a library of books to meet the needs of students and researchers in these fields.

*Other recent books in the series:*

**The Standard Model and Beyond**
P. Langacker

**Particle and Astroparticle Physics**
U. Sakar

**Joint Evolution of Black Holes and Galaxies**
M. Colpi, V. Gorini, F. Haardt, and U. Moschella *(Eds)*

**Gravitation: From the Hubble Length to the Planck Length**
I. Ciufolini, E. Coccia, V. Gorini, R. Peron, and N. Vittorio *(Eds)*

**Neutrino Physics**
K. Zuber

**The Galactic Black Hole: Lectures on General Relativity and Astrophysics**
H. Falcke and F. Hehl *(Eds)*

**The Mathematical Theory of Cosmic Strings: Cosmic Strings in the Wire Approximation**
M. R. Anderson

**Geometry and Physics of Branes**
U. Bruzzo, V. Gorini, and U. Moschella *(Eds)*

**Modern Cosmology**
S. Bonometto, V. Gorini, and U. Moschella *(Eds)*

**Gravitation and Gauge Symmetries**
M. Blagojevic

**Gravitational Waves**
I. Ciufolini, V. Gorini, U. Moschella, and P. Fré *(Eds)*

**Neutrino Physics, Second Edition**
K. Zuber

Series in High Energy Physics, Cosmology, and Gravitation

# An Introduction to Beam Physics

**Martin Berz**
*Michigan State University*
*East Lansing, Michigan, USA*

**Kyoko Makino**
*Michigan State University*
*East Lansing, Michigan, USA*

**Weishi Wan**
*Lawrence Berkeley National Laboratory*
*Berkeley, California, USA*

CRC Press
Taylor & Francis Group
Boca Raton   London   New York

CRC Press is an imprint of the
Taylor & Francis Group, an **informa** business

A TAYLOR & FRANCIS BOOK

CRC Press
Taylor & Francis Group
6000 Broken Sound Parkway NW, Suite 300
Boca Raton, FL 33487-2742

First issued in paperback 2016

© 2015 by Taylor & Francis Group, LLC
CRC Press is an imprint of Taylor & Francis Group, an Informa business

No claim to original U.S. Government works

Version Date: 20141015

ISBN 13: 978-1-138-19890-6 (pbk)
ISBN 13: 978-0-7503-0263-0 (hbk)

**Visit the Taylor & Francis Web site at**
**http://www.taylorandfrancis.com**

**and the CRC Press Web site at**
**http://www.crcpress.com**

# Contents

# List of Figures

# *Preface*

This volume provides an introduction to the physics of beams. This field touches many other areas of physics, engineering and the sciences, and in turn benefits from numerous techniques also used in other disciplines. In general terms, beams describe ensembles of particles with initial conditions similar enough to be treated together as a group, so that the motion is a weakly nonlinear perturbation of that of a chosen reference particle.

Applications of particle beams are very wide, including electron microscopes, particle spectrometers, medical irradiation facilities, powerful light sources, astrophysics – to name a few – and reach all the way to the largest scientific instruments built by man, namely, large colliders like LHC at CERN.

The text is based on lectures given at Michigan State University's Department of Physics and Astronomy, the online VUBeam program, the US Particle Accelerator School, the CERN Academic Training Programme, and various other venues. Selected additional material is included to round out the presentation and cover other significant topics.

The material is at a level to be accessible to students of physics, mathematics and engineering at the beginning graduate or upper division undergraduate level and can be viewed as an introductory companion to the more advanced *Modern Map Methods in Particle Beam Physics* by M. B., published by Academic Press. Emphasis has been placed on showing major concepts in their original incarnations and through historic figures. Finally, some of the sections and chapters that contain more advanced material are marked by a * symbol and can be omitted in a first reading.

Many organizations and individuals have helped directly and indirectly at various stages in the development of this book. MSU's Physics and Astronomy Department provided an environment of support for this and other books, the VUBeam program, as well as many of our other activities.

For two decades of continuous financial support that were instrumental to the success of the book, the VUBeam program, and indeed much of our research, we are grateful to the US Department of Energy, and in particular to Dr. Dave Sutter, the long-term coordinator of their beam physics activities.

K. M. would like to thank her daughter Kazuko for her own great interest in physics and science and much encouragement during the finalization of this text.

W. W. would like to thank Dr. D. Robin for his encouragement, Dr. E. Forest for stimulating discussions on various aspects of beam dynamics such as normal form theory, and his wife Juxiang Teng for her unwavering support throughout this project.

All of us want to thank Béla Erdélyi, Gabi Weizman, Pavel Snopok and He Zhang for thoughtful comments about the material. We also are thankful to many authors, national laboratories and publishers allowing us to reproduce published figures. The details are described in the corresponding figure captions.

Last but not least, we are very grateful to the entire staff of Taylor & Francis for their continuous support, in particular to Francesca McGowan for her great interest and productive comments.

Martin Berz
Kyoko Makino
Weishi Wan

# Chapter 1

## Beams and Beam Physics

In this chapter we will lay the foundations of basic concepts about beam physics, and discuss various important mechanisms of production and acceleration of beams. Because of the breadth of the material and the multitude of existing devices for each of the mechanisms, we will focus only on key concepts, and introduce them through the eyes of their inventors by using their original historical drawings, with only minor adjustments for uniformity of style and technical clarity.

## 1.1 What Is Beam Physics?

The field of beam physics deals with motion of **ensembles of particles** (usually charged) in electromagnetic **fields**. It is called beam physics due to the fact that, in most cases, those particles have **similar coordinates**, which is the rough definition of a beam. In many cases, the **positions** and **momenta** of the particles are sufficient to describe their motion. In this case, the particles are described by a state vector consisting of positions and momenta

$$\vec{Z} = (x, p_x, y, p_y, z, p_z).$$

In other cases, additional coordinates may be needed; typical examples include the **mass**, sometimes the **charge**, or the **spin** vector and the related **magnetic moment** and possibly **electric moment** of the particle.

An ensemble of particles with such similar coordinates is called a **beam** (see Fig. 1.1), and the sub-fields concerned with the study of such beams is called **beam physics**. There are other fields of physics that can be described in very similar terms and language, some of the most notable examples being **light optics** and **astrodynamics**. There are also other sub-fields of physics dealing with the study of the motion of such ensembles of particles; important examples are **plasma physics** and the **dynamics of galaxies**. These fields are different from beam physics in that in their cases, the particles usually do not have rather similar coordinates but occupy larger regions.

The space of state vectors $\vec{Z}$ is often called **phase space**, and a coordinate system showing $\vec{Z}$ is often called a **phase space diagram**. The **volume** of

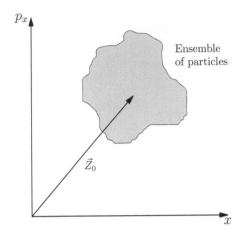

**FIGURE 1.1:** A beam — an ensemble of particles in the vicinity of a reference particle with phase space coordinate $\vec{Z}_0$.

the cloud of particles in phase space has a special name. It is called **emittance**. As we shall see later, in many systems the emittance is conserved and hence plays a special role.

Because all particles are close together, it is often useful to pick one of these particles, typically one that is somewhere in the middle, and describe the motion of the others **relative** to this **reference particle**. So if the reference particle has coordinates $\vec{Z}_0$, then the motion of the particles would be described in the relative coordinates $\Delta\vec{Z} = \vec{Z} - \vec{Z}_0$.

In many cases, the density of particles is so low that their **interaction** can be **neglected** or expressed by simple collective models. In other cases, it is necessary to include the study of the self-interaction, i.e., we have to take into account the fields due to the **space charge**.

If the fields are electromagnetic, then the motion is described by the **Lorentz force law**, which in SI units is

$$\frac{d\vec{p}}{dt} = q\left(\vec{E} + \vec{v} \times \vec{B}\right). \tag{1.1}$$

Here $\vec{E}$ and $\vec{B}$ are the electric and magnetic fields, respectively. These fields are connected to the scalar potential $V$ and the vector potential $\vec{A}$ via the relations

$$\vec{B} = \vec{\nabla} \times \vec{A}, \quad \vec{E} = -\frac{\partial\vec{A}}{\partial t} - \vec{\nabla}V.$$

Although this may not be directly relevant in this book, we want to note here for the sake of completeness that the equations of motion in the form of the

Lorentz force law can also be obtained from the **Lagrangian**

$$L = -mc^2\sqrt{1 - \frac{\vec{v}^2}{c^2}} + q\vec{v} \cdot \vec{A} - qV; \qquad (1.2)$$

refer to eqs. (1.85), (1.145) in [5], and, for example, [29]. From this Lagrangian, one can also obtain a **Hamiltonian** of the motion in a procedure that is standard for all Lagrangian systems. One begins by defining the canonical momentum as:

$$\vec{p}_{can} = \frac{\partial L}{\partial \vec{v}},$$

which here has the form

$$\vec{p}_{can} = \gamma m\vec{v} + q\vec{A} = \vec{p}_{dyn} + q\vec{A}, \qquad (1.3)$$

where

$$\gamma = \frac{1}{\sqrt{1 - \vec{v}^2/c^2}},$$

and the canonical momentum $\vec{p}_{can}$ is different from the relativistic dynamical momentum

$$\vec{p}_{dyn} = \gamma m\vec{v}. \qquad (1.4)$$

The Hamiltonian of the motion can then be found as

$$H = \vec{p}_{can} \cdot \vec{v} - L.$$

This expression initially contains both $\vec{p}_{can}$ and $\vec{v}$, and it is necessary to eliminate $\vec{v}$ and express it in terms of $\vec{p}_{can}$. Because $\vec{p}_{dyn} = \gamma m\vec{v}$ and $\vec{p}_{dyn} = \vec{p}_{can} - q\vec{A}$ from eq. (1.3), $\gamma m\vec{v} = m\vec{v}/\sqrt{1 - \vec{v}^2/c^2} = \vec{p}_{can} - q\vec{A}$, leading to $m^2\vec{v}^2/(1 - \vec{v}^2/c^2) = (\vec{p}_{can} - q\vec{A})^2$, so we find

$$\vec{v} = c \cdot \frac{\vec{p}_{can} - q\vec{A}}{\sqrt{\left(\vec{p}_{can} - q\vec{A}\right)^2 + m^2c^2}} = c \cdot \frac{\vec{p}_{dyn}}{\sqrt{\vec{p}_{dyn}^2 + m^2c^2}}, \qquad (1.5)$$

where the expression in terms of $\vec{p}_{dyn}$ is listed as well. Using this, $\sqrt{1 - \vec{v}^2/c^2}$ in the Lagrangian $L$ in eq. (1.2) is expressed in terms of $\vec{p}_{can}$, also in terms of $\vec{p}_{dyn}$, as

$$\frac{1}{\gamma} = \sqrt{1 - \frac{\vec{v}^2}{c^2}} = \frac{mc}{\sqrt{\left(\vec{p}_{can} - q\vec{A}\right)^2 + m^2c^2}} = \frac{mc}{\sqrt{\vec{p}_{dyn}^2 + m^2c^2}}. \qquad (1.6)$$

Thus, we obtain for the Hamiltonian

$$H = c \cdot \sqrt{\left(\vec{p}_{can} - q\vec{A}\right)^2 + m^2c^2} + qV;$$

refer to eq. (1.149) in [5] for details of the derivation.

When studying the evolution of the beam from the time it is born until it is used, there are usually four steps involved. First, there must be a way for the **production** of the beam, and for the sake of efficiency if possible in such a way that its emittance is small. Second, in most cases the energy of the beam has to be increased; there has to be a mechanism of **acceleration**. Because of the outstanding importance of this process, the whole field is often called **accelerator physics**. Then it is necessary to **transport** the beam to where it is being used. And finally, there is often a need for **storage** of the beam for use at a later time or reuse. Lastly, often there is a need for **analysis** of the beam, in particular after the beam has been used for its purpose, which frequently is the facilitation of certain nuclear or high energy reactions. The field of beam physics spans all these steps, and each of the steps has it own unique problems to be solved.

## 1.2    Production of Beams

The mechanisms used for the production of the beam depend very much on the particular kind of particles and the characteristics of the beam that is needed, and they include mechanisms from a variety of different fields including thermal, electrical, atomic, nuclear, and even high energy physics processes. Common beams consist of electrons, protons, or $H^-$, and some of the beams produced through nuclear and high energy physics processes include positrons, antiprotons, pions, kaons and radioactive nuclei. Overall, due to the diversity of the species of the particles and the required properties of the beam, there are dozens of different ways of producing various beams. We here restrict ourselves to some of the source types that are most commonly used in particle accelerators and electron microscopes.

### 1.2.1    Electron Sources

Electrons exist in abundance in metals, and forming them into beams requires their extraction from the metal, called the **cathode**. For this the electrons need to overcome the potential barrier, i.e., the **work function**, at the boundary between the metal and the environment. The work function usually ranges from a fraction of an electron Volt (eV) to a few electron Volts; for comparison, the average kinetic energy of gas molecules at room temperature amounts to approximately 1/40 eV. This can be achieved by either supplying additional energy to the electrons so that they can leave the material, or by lowering the work function. In the following we discuss some common approaches based on these methods.

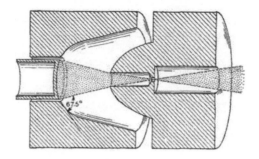

**FIGURE 1.2**: Sketch of an early thermionic emission electron source. (Reprinted with permission from J. R. Pierce, *J. of Appl. Phys.*, 11:548, 1940 [57]. Copyright 1940, AIP Publishing LLC.)

The first of these processes is **thermionic emission**. By heating a piece of metal to temperatures exceeding around 1000°C, a small fraction of the electrons will achieve energies exceeding the work function and can thus leave the metal. This type of source is usually called the **thermionic gun**. Once outside the metal, the electrons can be pulled away further by the application of strong electric fields, the distribution of which is adjusted to achieve high gradient and optimal focusing. An example of such a device is shown Fig. 1.2. Here the number of electrons available is determined by the temperature of the donor metal or **cathode**, and only those electrons in the tail of the Fermi-Dirac distribution above the work function can be extracted. This process is quantitatively described by the **Richardson-Dushman equation**

$$J = \left(\frac{4\pi e m k_B^2}{h^3}\right) T^2 e^{-W/k_B T}, \tag{1.7}$$

where $J$ is the current density, $e$ is the charge, $m$ is the mass, $k_B$ is the Boltzmann constant, $h$ is the Planck constant and $T$ is the temperature of the cathode. It is obtained from the third law of thermodynamics and characterizes an idealized situation of a sufficiently large piece of cathode material to avoid quantum mechanical influences, and the absence of electric fields influencing extraction.

In practice the extracted current is also greatly affected by any electric field applied to the cathode. This is the result of the Coulomb repulsion among the extracted electrons, where electrons extracted earlier can push those extracted later back into the cathode. When the electric field at the surface vanishes, no more electrons will be extracted. The relation between the maximum current density and the applied electric field, for a parallel flat cathode and a matching anode, is the **Child-Langmuir Law**

$$J = \frac{4}{9}\epsilon_0 \left(\frac{2e}{m}\right)^{1/2} \frac{V_0^{3/2}}{d^2}, \tag{1.8}$$

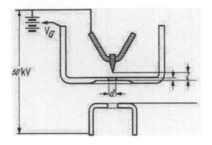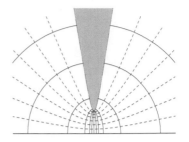

**FIGURE 1.3**:  Left: Sketch of one of the earliest electron sources using point cathodes. (From Y. Sasaki and S. Maruse, Über die Arbeitsweise und die elektronenoptischen Eigenschaften der Spitzenkathode, in G. Möllenstedt, H. Niehrs, and E. Ruska, eds., *Physikalisch-Technischer Teil*, 1:9, Springer-Verlag, 1960, © Springer-Verlag Berlin Heidelberg 1960 [61]. Abb. 3, "Zwei Anordnungen der Elektrodensysteme für die Spitzenkathode." With kind permission from Springer Science and Business Media.) Right: The potential (dashed) and the field (solid) distribution near the cathode tip.

where $J$ is the current density, $\epsilon_0$ is the dielectric constant in vacuum, $e$ is the charge, $m$ is the mass, $V_0$ is the applied voltage between the cathode and the anode, and $d$ is the distance between the cathode and the anode. In practice the situation is more involved, and the maximum current density is usually the smaller of the two quantities. For flat thermionic cathodes, usually eq. (1.7) sets the limit for the extracted current.

Due to the high operating temperature, the energy spread of the extracted electron beam is relatively large. Nonetheless, the thermionic gun is simple and reliable, and hence is still widely used as the source for many devices where the large energy spread of the electrons at the cathode is not limiting the performance of the machine. One significant example are circular electron accelerators, where the ultimate energy spread in the beam is dominated by other processes including synchrotron radiation discussed later.

The second process to produce electrons is **field emission**. In this mechanism, a **sharp needle** is brought into strong external electric fields. This type of source is usually called the **field emission gun**. Because the needle is a conductor, it acts as an equipotential surface, and thus produces very strong electric fields near its tip. In practice the radius of curvature at the tip often ranges from below 1 nm to nearly the range of single atoms to 1 $\mu$m, and one locally obtains a very strong field ranging from 1 to 3 GV/m. These strong electric fields acting near the surface lower the work function and simultaneously reduce the width of the potential barrier, which allows electrons to escape through the well through quantum tunneling. All these electrons emerge from a small area. Furthermore, due to the fact that the electrons are diverging near the surface of the needle, the actual source is quite a bit smaller than the emitting area. Meanwhile, for low current, their spread in momenta is rather small as well. Fig. 1.3 shows the basic principle.

The small emitting area imposes a severe limit on the total current extracted, but it is the standard electron source for transmission electron microscopes discussed in more detail below. Here the current need is low, but the origination from a small area and with small energy spread amounting to what is called high **brightness** is very useful. The use of sharp needles (point filaments as they were historically called) was pioneered in the 1950s in order to increase brightness of the electron beam in microscopes.

Particularly fruitful is combining needle geometries, resulting in an effective lowering of the work function due to stronger electric field at surface, with heating, which results in an increase of electrons of higher energy than the work function that can thus traverse it. This kind of thermionic emission with significant external field is called **Schottky emission**. In the 1960s, a new generation of cathodes (mainly ZrO/W, which is a tungsten tip covered with a thin layer of ZrO) were developed with stronger field at the surface, where field emission plays a significant role and complements Schottky emission. This regime is called the **extended Schottky emission**. This kind of emitters are the main sources of electrons for electron microscopes. The emitters that produce electrons through only the field emission process, called **cold field emission gun (CFEG)**, have been studied since the 1970s, and recently have been able to produce high brightness beams.

The third process to produce electrons is **photoemission** where electrons are produced via the photo effect. Exposing a surface to a large flux of photons leads to some of the photons being absorbed by electrons within the material, which consequently increase their kinetic energy. This additional kinetic energy acts very similar to intense local heating in the thermionic gun and leads to electrons with energies exceeding the work function and which can consequentially escape. In certain cases photo-energized electrons can also leave the material directly without colliding with other electrons, in a ballistic process. In practice, photons are supplied through laser pulses, of which the intensity, spot size and duration are relatively easy to adjust. This leads to the **photocathode gun**. This approach has greatly facilitated the advance of **free electron lasers** (FELs) in the last few decades and is an important component in efforts of time-resolved spectroscopy and microscopy.

Again the extracted current is limited by the Child-Langmuir law (1.8), despite the fact that an intense laser pulse can often produce large numbers of electrons. Different classes of materials have been developed for the cathode, including GaAs that has been cesiated, i.e., covered by less than a mono-layer of cesium. This has allowed the production of electrons with energy spreads down to fractions of $10^{-1}$ eV, which is the range of thermodynamic energies encountered at room temperature.

Other materials such as copper are able to withstanding the harsh environment of a radio frequency (RF) gun, where very high extraction fields can be produced that significantly exceed those of the electrostatic case. Fig. 1.4 shows the layout of an RF gun and the field distribution. It consists of roughly

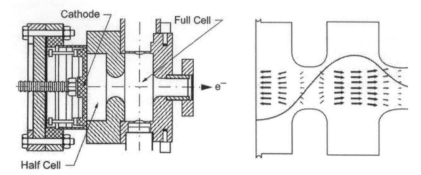

**FIGURE 1.4**: Layout (left) and field distribution (right) of the RF gun at the Linac Coherent Light Source (LCLS). (From J. Arthur, et al., SLAC-R-593, 2002 [2]. Courtesy SLAC National Accelerator Laboratory.)

1.5 cells (1.6 for this example) with the field in the two cells in opposite phase to ensure that electrons emitted into high extracting field end up with high energy and low emittance at the exit. In addition, materials such as cesiated GaAs with strained lattice can also produce polarized electrons, which has been used at SLAC National Accelerator Laboratory, California, USA, for high energy physics experiments on the one hand and in spin polarized low energy electron microscope (**LEEM**) on the other.

## 1.2.2 Proton Sources

In most proton accelerators, the protons are produced through **stripping** the two electrons in a negative hydrogen ion $H^-$ at the entrance, mostly by passing the ion through a thin carbon foil, although lasers have also been experimented with recently. In some applications, such as medical accelerators, $H_2^+$ ions instead are produced and accelerated.

Although the stripping process makes the technology a little more complicated, it has one distinct advantage over injecting protons directly. For **injection** into circular accelerators over many turns, the protons generated through stripping can easily be aligned with earlier produced protons already in the accelerator, because before stripping, both types of particles follow different paths due to the differing charge. As a result, the density of protons in the beam can increase more and more over an extended injection period. On the other hand, protons injected directly require much more care and can only be injected with positions and directions that are different from those of other protons already in the ring, since because of time reversal, all protons with the same position and direction must have followed the same earlier trajectory.

The $H^-$ ions can be produced by a variety of different methods. Here, only the **surface plasma source** of the **magnetron** type is discussed. It obtained its name due to the fact that it is similar in configuration to the ubiquitous

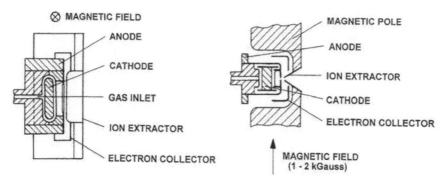

**FIGURE 1.5**:　Drawing of a surface plasma source of the magnetron geometry. (From J. Ishikawa, Negative ion sources, in I. G. Brown, ed., *The Physics and Technology of Ion Sources*, 2nd ed. [31]. Copyright (c) 2005 Wiley-VCH. With permission.)

microwave source called the magnetron, which is used in common appliances such as the microwave oven. The main process of producing the negative hydrogen ions is through electron capture of neutral hydrogen atoms on the cathode surface, where electrons penetrate the potential barrier through quantum tunneling. The presence of the electric and magnetic field creates a dense plasma near the surface of the cathode where ions, most of which are positive, are produced. Those positive ions and neutral particles bombard the cathode partially covered with cesium and $H^-$ ions are produced. The presence of cesium lowers the work function and greatly increases the probability of barrier penetration and hence $H^-$ production.

Some of the produced $H^-$ ions are neutralized shortly after production. An electric field between the cathode and the anode is used to accelerate the remaining $H^-$ ions towards the exit. However, electrons are accelerated towards the anode as well. But since electrons are much lighter than the $H^-$ ions, they are bent much more easily than the $H^-$ ions, and are absorbed by the electron collector (Fig. 1.5).

The magnetron type of negative hydrogen ion source was first developed in the former Soviet Union in the early 1970s and quickly spread to Europe and the United States. It has become the main choice for high energy proton accelerators.

### 1.2.3　Ion Sources

There are a large variety of different sources for ions which found applications in many fields. Presenting even a brief overview is already beyond the scope of this book; for more comprehensive reviews, see [77, 8]. As in the previous subsection, we limit our discussion to one important kind, which is the **electron cyclotron resonance (ECR) ion source**. Fig. 1.6 shows the mechanism schematically. The chamber on the left holds the plasma where ions are produced. First, the gas of the element of interest is injected to the chamber. Collisions among the atom generates electrons and ions. The magnetic field (see top part of Fig. 1.6) forms a magnetic mirror that confines the ions and the electrons. For the example shown, this time is around 100 $\mu$s, and for more modern and advance versions for around 10 ms. High frequency (2.45 to 28 GHz) microwaves are injected into the chamber and electrons with rotation frequency matching that of the microwaves are accelerated to between 1 and 20 keV. This process of heating up the electron gas is called **electron cyclotron resonance heating**. The resulting hot electrons collide with ions and neutral atoms and generate more ions. Furthermore, through step-by-step ionization, even multiply charged ions (e.g., $Xe^{38+}$) can be produced.

In order to produce sufficient quantities of multiply charged ions, the ion confinement has to relatively long ($\sim 10$ ms). The major improvement in this aspect is the addition of a sextupole magnet, which ensures that the magnetic field at the center of the chamber is at the minimum, which prevents the ions from drifting to the side wall. Another consequence of this configuration of the magnetic field is that the surface on which electron cyclotron resonance heating takes place is now closed, which significantly reduces hot electron loss. This configuration of the magnetic field also makes ion production more efficient, since electrons are confined longer and can collide with ions and be reheated many times. The fact that the ECR ion source does not use a cathode to generate electrons makes it a much more reliable source compared to other varieties.

The ECR ion source was first developed in the mid-1960s and, by the mid-1970s, many had been built around the world. It is probably the best source to produce multiply charged ions and has become the main choice of ion sources for nuclear physics facilities.

---

## 1.3   Acceleration of Beams

We now assume that an ensemble of particles occupying a small volume of phase space has been created, and we thus have what is called a **beam**. In many if not most of the practical cases, the energy that the beam has after being produced by the source is not sufficient for the purpose it is to be used for, which frequently amounts to furnishing the energy necessary for atomic, nuclear, or particle processes of interest.

In most cases, the motion is best studied by first considering the motion of the **reference particle**, and once this motion is understood satisfactorily, to study the **relative** motion of the other particles. For a simple analysis

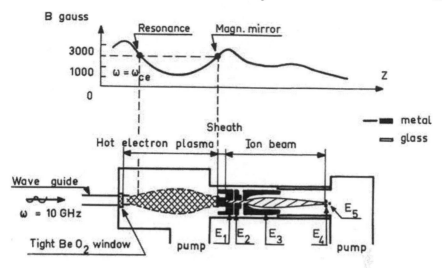

**FIGURE 1.6**: Layout of the first electron cyclotron resonance (ECR) ion source that produced multiple charged ions. (Reprinted with permission from R. Geller, *Appl. Phys. Lett.*, 16:401, 1970 [28]. Copyright 1970, AIP Publishing LLC.)

of the relative motion, often a linear approximation with all the resulting simplifications is possible, but frequently a full understanding of the motion can only be achieved by considering the nonlinear effects.

Considering the special shape of the Lorentz force law (see eq. (1.1)), since $\vec{v} \times \vec{B}$ is perpendicular to the velocity $\vec{v}$, it is apparent that magnetic fields cannot be used for purposes of acceleration, which requires forces in the direction of the particle. Thus any **acceleration** has to be **provided by electric fields**. However, as we shall see, also magnetic fields have very good use in particle accelerators, as they can be employed to guide the beam to where it is needed. In particular, in the process of acceleration they are often used to guide the beam through the same region of electric field repeatedly and thus allow the device to maximize the use of the electric fields. Indeed, for this purpose of guiding the beam, magnetic fields are usually even better suited than electric fields. This is because for the high velocities that beams usually have after even modest acceleration, the forces that can be attained with technologically available magnetic fields far exceed those that can be achieved with the respective electric fields.

Very generally, the amount of energy $K$ a particle gains while traveling from time $t_1$ to time $t_2$ in an electric field $\vec{E}(\vec{r}, t)$ that depends on position and time is given by the path integral

$$K = q \cdot \int_{t_1}^{t_2} \vec{E}(\vec{r}(t), t) \cdot \vec{v}(t)\, dt,$$

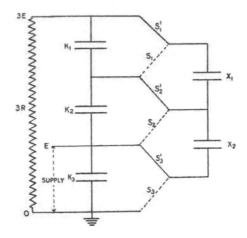

**FIGURE 1.7**:   The general principle of the Cockcroft-Walton generator. (From J. D. Cockcroft and E. T. S. Walton, *Proc. Royal Soc. London, A*, 136:619, 1932 [14]. With permission.)

where $\vec{r}(t)$ is the particle's position as a function of time and $\vec{v}(t)$ its velocity. In the special case that $\vec{E}$ is **time independent** and hence can be written in terms of a potential via $\vec{E} = -\vec{\nabla}V$, this path integral reduces in a natural way to the difference in potential as

$$K = q \cdot (V(\vec{r}_1) - V(\vec{r}_2)).$$

This simple fact implies a very important consequence for the design of electric accelerating fields: if there is to be any chance to utilize the same electric field **repeatedly** for the purpose of acceleration, then the electric field has to be **time dependent**, because otherwise repeated passing just results in a periodic increase and decrease of energy. In fact, the attempt to build an accelerator trying to increase energy repeatedly by flying through the same time independent field is tantamount to the attempt to build a perpetual motion machine.

### 1.3.1   Electrostatic Accelerators

The first class of important accelerators are those based on static electric fields. The kind that grew directly out of the area of electric circuit is the voltage multiplier that is now known under the name **Cockcroft-Walton** generator. It consists of a simple but clever circuit made of diodes and capacitors, forming the voltage multiplier ladder. Each time a base voltage is applied, the first capacitor is filled with charge. Each time the voltage is removed, the first capacitor passes on some of its charge to the second capacitor, which in turn feeds the third capacitor, and so on. Depending on the number

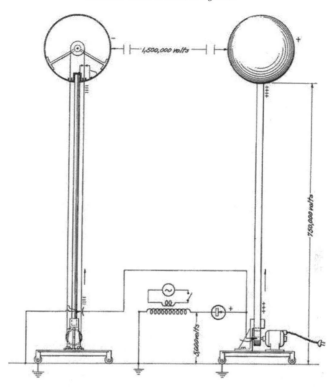

**FIGURE 1.8**: Design sketch of the Van de Graaff high voltage generator. (From R. J. Van de Graaff, US Patent 1,991,236, 1931 [18].)

of capacitors and the number of cycles applied, quite high voltages can be obtained very easily.

For small applications it is possible to simply apply alternating current (AC) at the feeding end. For larger applications it may require a longer time to transfer the charges from the lower to the higher capacitors, and the change in input voltage is achieved through mechanical switches. Fig. 1.7 illustrates "the general principle underlying the method adopted," and is from the original paper on the matter [14]. Many high energy proton accelerators use Cockcroft-Walton generators as the first stage of acceleration.

Another method to obtain high voltages for electrostatic accelerators is the Van de Graaff accelerator [20] and several similar devices derived from it are the main representatives of the class of accelerators utilizing time independent fields. The voltage difference that the particles travel through is obtained with a **Van de Graaff generator**, which consists of an endless non conducting belt onto which charge is sprayed from a tip via field emission, and which is then transported to the inside of a hollow metal sphere where it is deposited.

Since any charge on a conducting object accumulate on the outside and

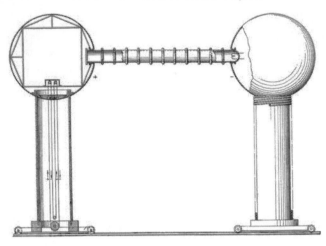

**FIGURE 1.9**:   Design sketch of the use of the Van de Graaff generator as a particle accelerator. (From R. J. Van de Graaff, US Patent 1,991,236, 1931 [18].)

create a field-free interior, a new charge can be brought in from the belt on the inside of the sphere without experiencing any opposing fields, and thus large amount of charges can be accumulated on the sphere, resulting in very high potentials. The mechanism of the Van de Graaff generator is shown in the left part of Fig. 1.8 and the complete machine is shown in Fig. 1.9. The sketches are from the original patent on the device [18]; see also [20].

In passing it is worthwhile to remark that while the newly added charge does not experience a field when moving from the belt to the inside of the sphere, it certainly experiences a field while approaching the sphere and being attached to the belt. Thus the potential energy contained on the charged sphere does not come for free. It is generated through the mechanical work that is necessary to move the belt and the attached charges toward the sphere.

The charged sphere is connected to a metal enclosure containing the ion source, thus elevating the source to the potential of the charged sphere, which can then be utilized for the acceleration of the particles.

The main practical limitation of the Van de Graaff accelerator is the necessity to prevent **sparks**. This is achieved on the one hand by sheer **size**, because at the same potential difference, larger size means less electric field strength. On the other hand, it is important to **inhibit** the **spark formation** process. Microscopically, sparks form in a gas when small numbers of charged particles have a mean free path length that is long enough so they can attain energies sufficient to ionize other particles upon collision, resulting in an avalanche. This can be avoided by choosing **inert gases** like He (helium) or $SF_6$ (sulfur hexafluoride), and on the other hand applying high pressure to reduce the mean free path length.

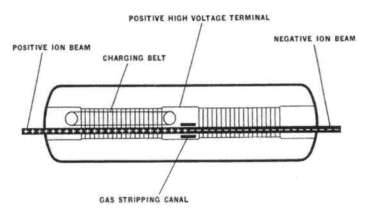

**FIGURE 1.10**: The principle of the tandem Van de Graaff accelerator. (Reprinted from *Nucl. Instrum. Methods*, v. 8, R. J. Van de Graaff, Tandem electrostatic accelerators, p. 195–202, Copyright (1960), with permission from Elsevier [19].)

The Van de Graaff accelerator has several desirable features; for example, it can produce a fully continuous beam (often denoted by the term "**cw**" for continuous wave) and at high beam current. Its main limitation is the relatively low energies that it can produce, which seldom exceed about 20 MeV.

The tandem Van de Graaff is an efficient modification of the Van de Graaff concept, in which both the source and the target are kept at ground potential and which can efficiently **increase the energy** that can be obtained. For this purpose, a source is chosen that produces negatively charged ions, which are then sent through a regular Van de Graaff. At the end of the accelerating section, the ions are sent through a thin foil, in which many of them are **stripped** of some of their electrons, resulting in positive ions. Because the particles already have substantial energy when hitting the foil, often much higher charge states can be produced than in the ion source itself. These positive ions are then sent through a second stage Van de Graaff, which is essentially a reversion of the first stage. At the location of the target, depending on their charge state after stripping, their energy is increased by a factor of two or more. The mechanism of the tandem Van de Graaff accelerator is shown in Fig. 1.10. Having very similar characteristics to the original Van de Graaff, the energies that can be achieved in this way are in the range of up to 60 MeV.

## 1.3.2 Linear Accelerators

It is an important observation that the field strength that can be obtained **in quickly oscillating** (radio frequency, RF) **electric fields** can be **substantially higher** than what can be obtained statically in devices of similar

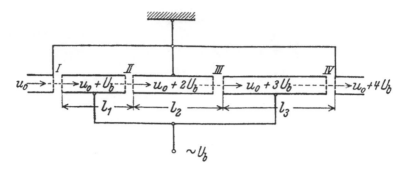

**FIGURE 1.11**: Sketch of the principle of the linear accelerator of the Wideröe type. (From R. Wideröe, Über ein neues Prinzip zur Herstellung hoher Spannungen, *Archiv für Elektrotechnik*, 21:387, 1928, © Springer-Verlag Berlin Heidelberg 1928 [73]. Bild 1, "Prinzip der Spannungstransformation mit Potentialfeldern." With kind permission from Springer Science and Business Media.)

size. This is mostly due to reduced presence of spark formation, because the formation of an avalanche of charged particles requires time scales that are usually larger than the time the field is in one phase.

The use of an oscillating field, however, immediately entails that only half of the cycle can be used for acceleration, and thus is different from static accelerators, as the resulting beams always have a temporal **micro-structure**, also called **bunched**. In practical use, usually several RF resonators are used sequentially, each one of which accelerating the particles, and it is very important that the phase relationship between the individual accelerating sections is correct. This is usually achieved by applying the fields between the edges of adjacent conducting tubes. This kind of device is called the **linear accelerator** or **linac**.

The concept of the earliest linacs is schematically shown in Figs. 1.11 and 1.12. From the wiring scheme, it is clear that the electric field in adjacent gaps points to opposite directions. In order to ensure that charged particles are accelerated in every gap, the lengths of the tubes are chosen in such a way that the time the particles require to fly through them equals one half of the RF period. So the length $L_i$ of the $i$th tube has to be chosen so that it satisfies

$$L_i = \frac{1}{2} v_i T_{\mathrm{rf}},$$

where $T_{\mathrm{rf}}$ is the period of the RF frequency. Apparently this leads to a system of tubes of increasing length, i.e., $L_1 < L_2 < L_3 < \ldots$. The exact lengths $L_i$, of course, depend on the relationship between the kinds of particles and the values of the accelerating voltages, and so often these designs are rather customized geometries. Since metal wires are used to connect the tubes, the

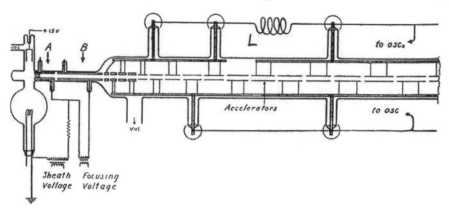

**FIGURE 1.12**: Illustration of a linear accelerator, designed by E. O. Lawrence and H. D. Sloan. (Reprinted the middle picture of Fig. 1 with permission from [64] as follows: D. H. Sloan and E. O. Lawrence, *Phys. Rev.*, 38, 2021, 1931. Copyright (1931) by the American Physical Society.)

frequency of the oscillating field is limited to below 100 MHz due to increased radiation at high frequency. This in turn imposes an upper limit on the velocity of the particles ($\beta = v/c < 0.03$) due to practical limit of the length of the tubes. Meanwhile, use of wires at low frequency can significantly reduce the size of the accelerating structure compared to a closed structure, called **RF cavity**.

Another type of linac, called the **Alvarez linac**, developed in the late 1940s is shown in Fig. 1.13. It uses closed structures, the RF cavities, to increase the oscillating frequency and reduce the length of the tubes. Another difference is that the field in adjacent gaps points to the same direction. As a result, the length $L_i$ of the $i$th tube has to be chosen so that it satisfies

$$L_i = v_i T_{\rm rf}.$$

Apparently, the frequency has to be at least twice of the Wideröe type to reduce the tube length. In practice, the frequency of the Alvarez linac is roughly an order of magnitude higher than that of the Wideröe type.

Fig. 1.13 is remarkably informative of the physics of the accelerator. The following sentences are excerpts from the original US patent [1]. The left top picture is "a diagram showing a normal cylindrical wave guide and the axial electric field distribution therein." The left bottom picture is "a diagram showing a wave guide and the electric fields when a series of graded drift tubes are placed therein." The right top picture is "a diagram representing the voltage existing across the gaps of the drift tubes." The left picture of the right bottom corner is "a diagrammatic longitudinal sectional view of drift tube ends with the electric field distribution existing across the gap." The right picture of the right bottom corner is "a diagrammatic longitudinal

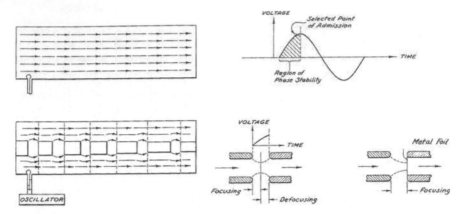

**FIGURE 1.13**: Sketches illustrating the basic principles of the linear accelerator of the Alvarez type. (From L. W. Alvarez, US Patent 2,545,595, 1947 [1].)

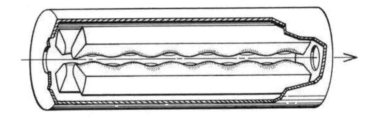

**FIGURE 1.14**: The structure of the RFQ, the radio-frequency quadrupole linear accelerator. The picture shows an early example at Los Alamos National Laboratory, New Mexico, USA. (From K. R. Crandall, et. al., in R. L. Witkover, ed., *Proc. 1979 Linac Conf.*, BNL-51134, 1979 [17]. Courtesy Brookhaven National Laboratory.)

section view of drift tube ends with a focusing foil attached to one tube, and the resulting electric field distribution." For details regarding phase stability (the right top picture), see Section 10.2. For details regarding transverse focusing and defocusing, see Section 10.4.

An interesting **combination** of the need for **bunching, accelerating** and focusing (which is discussed later in detail) is the radio frequency quadrupole (**RFQ**) accelerator. Developed in the late 1960s in the former Soviet Union, RFQs have been widely adopted as injectors for proton and ion accelerators. Fig. 1.14 shows the structure of a RFQ accelerator. The four vanes break the rotational symmetry and produce an electrostatic quadrupole field that oscillates with time. The traveling particles feel the quadrupole field that changes polarity with time and are focused in both transverse planes. The longitudinal electric field that accelerates the charged particles is produced through modulation of the vanes. Similar to the drift tubes, the particles are accelerated throughout the structure when the distance between the peak and the neighboring valley satisfies

$$L_i = \frac{1}{2} v_i T_{\mathrm{rf}}.$$

In general, linacs can provide beams of **high current**, and of higher energies than static accelerators, yet because of the single use of each electric field, they are still rather expensive per MeV. Linacs are frequently used as **pre-accelerators** for accelerators of higher energies. They also have the distinctive advantage that they **avoid synchrotron radiation**, which is often a limiting factor in circular accelerators for light particles. This aspect is very important for electron and positron high energy accelerators such as the Stanford Linear Collider (SLC) at SLAC National Accelerator Laboratory, California, USA. It is the main reason for the interest in next generation Linear Colliders, such as plans being considered for an International Linear Collider (ILC), where a pair of two linacs shoot electrons and positrons at each other at very high energy.

Recently, linear accelerators have been widely used in producing a **free electron laser (FEL)**, whose high peak brightness and short pulse duration has opened up unprecedented opportunities for scientific investigations. Fig. 1.15 shows the setup of the first FEL experiment. Electrons go through a magnetic device called the **undulator**, which consists of alternating magnetic poles. As a result, the trajectory of such an electron is very similar to a sine function, causing the emitted photon field to add coherently. Together with the large number of periods, the peak intensity of the X-ray can be orders of magnitude higher than that from a circular accelerator (see the following subsection). The advantage of a linac is that it can produce an electron beam with smaller emittance and shorter pulse duration.

### 1.3.3 Circular Accelerators

Arguably the simplest circular accelerator is the betatron, which, besides its practical use as a compact accelerator for lower energies, also represents an excellent textbook style application of principles of electrodynamics. In the case of the betatron, the orbit follows a circular shape, which is achieved by a magnetic field. If the motion is perpendicular to the magnetic field, then we have in SI units

$$\frac{mv^2}{\rho} = qvB, \quad \text{and so} \quad \rho = \frac{mv}{qB} = \frac{p}{qB},$$

and so the radius of motion depends only on the momentum and charge of the particle as well as the magnetic field. Note that the equation is correct even in the relativistic case, if $m$ is understood to mean the relativistic mass

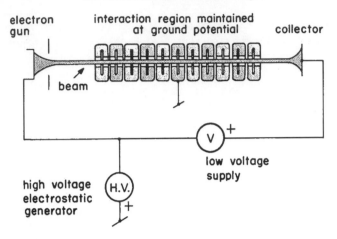

**FIGURE 1.15**:   Sketch of the first Free Electron Laser (FEL). (Reprinted with permission from J. M. J. Madey, *J. Appl. Phys.*, 42:1906, 1971 [47]. Copyright 1971, AIP Publishing LLC.)

$m = \gamma \cdot m_0$. Commonly the ratio of momentum and charge $p/q$ is denoted by $\chi_m$ and called **magnetic rigidity**; we apparently have

$$\chi_m = \frac{p}{q} = B\rho.$$

Because $\chi_m = B\rho$, the magnetic rigidity has the unit Tesla meter (Tm), and is frequently simply referred to as **B rho**.

In the case of the betatron, both **bending** and **acceleration** come from the same source, namely a **magnetic field** the strength of which **increases with time** in such a way that its magnitude matches the increasing energy of the particles to keep them at nearly **constant radius**, and the circular **induced electric field** provides the **acceleration** for the particles. Fig. 1.16 shows a sketch of the betatron [37, 38].

It is worthwhile to note that the basic idea of utilizing an electric field produced by a changing magnetic field also occurs in an application from daily life: certain modern **cooking surfaces**. In this case, the electrons that are accelerated are not within the vacuum of a beam pipe, but merely in the metal that constitutes the bottom of the pot used for cooking; and of course since their mean free path is short, they do not attain high energies before colliding with either other electrons or the lattice atoms, thus transferring their whole kinetic energy to heat.

A quantitative understanding begins with Faraday's law of induction, now one of Maxwell's equations:

$$\vec{\nabla} \times \vec{E} = -\frac{\partial \vec{B}}{\partial t},$$

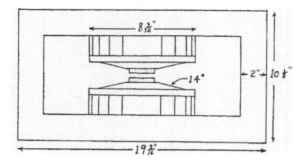

**FIGURE 1.16:** Illustration of the magnet of a betatron, from the original first paper on the subject. (Reprinted Fig. 2 with permission from [37] as follows: D. W. Kerst, *Phys. Rev.*, 60, 47, 1941. Copyright (1941) by the American Physical Society.)

and its integral form over a surface $A$ with the bounding $C$ is

$$\oint_C \vec{E} \cdot d\vec{l} = -\int_A \frac{\partial \vec{B}}{\partial t} \cdot \vec{n} dS.$$

Using the flux of the magnetic field through the surface $\Phi = \int_A \vec{B} \cdot \vec{n} dS$,

$$\oint_C \vec{E} \cdot d\vec{l} = -\frac{d\Phi}{dt}.$$

Here we restrict our interest to circular orbits with a radius $r$, and the surface $A$ is the inside of the circle. Building the magnet rotationally symmetric entails a rotational symmetry of the fields, which simplifies the situation to

$$E_l = -\frac{1}{2\pi r}\frac{d\Phi}{dt} = -\frac{1}{2\pi r}\pi r^2 \frac{d\bar{B}}{dt} = -\frac{r}{2}\frac{d\bar{B}}{dt},$$

where $\bar{B}$ is the average magnetic field enclosed by the orbit. Thus, by denoting the strength of $E_l$ simply by $E$,

$$E = \frac{r}{2}\frac{d|\bar{B}|}{dt},$$

and below we denote $|\bar{B}|$ by $\bar{B}$ for simplicity. Thus we obtain for the momentum $p = mv$

$$\frac{d}{dt}(mv) = qE = q\frac{r}{2}\frac{d\bar{B}}{dt} \Rightarrow mv = qr\frac{\bar{B}}{2}.$$

On the other hand, it is necessary that the centrifugal force on the orbit with radius $r$ is compensated by the Lorentz force at that radius, which requires

$$\frac{mv^2}{r} = qvB(r) \Rightarrow mv = qrB(r).$$

Thus, altogether we obtain the following relationship between the field $B(r)$ at the orbit $r$ and the average field:

$$B\left(r\right) = \frac{\bar{B}}{2}.$$

This equation of central importance is often called the **betatron condition**. It requires a magnetic field that is **stronger in the center** than where the particles move, which can be achieved by suitably **shaping the poles** of the magnet.

In principle the temporal behavior of $\bar{B}$ is irrelevant, and in practice one usually tries to ramp it quickly, because the **pulsed beam** is only available at the end of ramping. This is usually achieved by making the magnet part of an LC circuit (a resonant circuit, consisting of an inductor $L$ and a capacitor $C$), which also conveniently allows the device to recover the energy stored in the magnetic field for the next ramping. For the practical use, it is important to try to limit **Eddy currents** in the iron of the magnets, and in order to maintain the condition $B\left(r\right) = \bar{B}/2$, it is important to control **saturation effects** that may occur at any edges of the magnet.

The transverse confinement of the beam in the betatron is achieved through the inhomogeneity of the outer field, through effects that will be studied in subsequent chapters. The practical use of betatrons is nowadays mostly for electrons, where energies of about 300 MeV have been achieved; for protons, the values are about 50 MeV.

Also in the **microtron**, which was invented by V. Veksler [69], a magnet is used to bend the particles to let them pass through the same source of electric field repeatedly. Different from the betatron, the emphasis here lies on the production of a **continuous beam**. Since this requires that the whole acceleration process must be independent of the specific time of injection, this entails that the **magnetic field is constant in time**. Thus an external voltage source is needed; as discussed above, if it is to be used repeatedly, it has to be a time dependent source, and in practice it is chosen to be an RF (radio frequency) cavity. Altogether, the motion follows a sequence of **tangential circles** of increasing radius that touch at the location of the RF cavity, as shown in Fig. 1.17.

In order to **synchronize** the particle's motion and the momentary direction of the magnetic field, the revolution frequency of the RF cavity $\omega_0$ has to be a multiple of the particle's revolution frequency $\omega$, which can be obtained simply from

$$\frac{\gamma m_0 v^2}{r} = qvB \Rightarrow \omega = \frac{v}{r} = \frac{q}{\gamma m_0} B. \tag{1.9}$$

This means it has to be either the motion is such that $\gamma = 1$, which corresponds to non-relativistic motion and hence severely limits the energy, or just enough acceleration is provided in each turn that the revolution frequency decreases to the next multiple of the RF frequency. So the revolution frequencies would

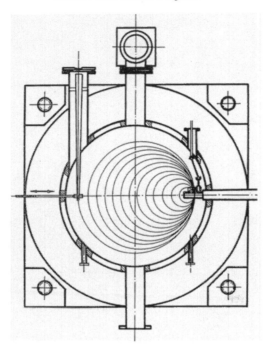

**FIGURE 1.17**: Illustration of the first microtron. (From S. P. Kapitza, *The Microtron*, Harwood Academic, London, 1978 [35]. Fig. 1.9, p. 14. With permission: © Taylor & Francis.)

follow the pattern

$$\omega = \omega_0, \frac{\omega_0}{2}, \frac{\omega_0}{3}, \frac{\omega_0}{4}, \frac{\omega_0}{5}, \ldots. \tag{1.10}$$

This entails that the factor $\gamma$ follows the sequence $\gamma = \gamma_0, 2\gamma_0, 3\gamma_0, 4\gamma_0, \ldots$, which requires $\Delta\gamma = 1$ per turn. Since $E = mc^2 = \gamma m_0 c^2$, this means $\Delta E = m_0 c^2$, and thus the necessary energy gain per turn must equal the rest mass energy of the particle under consideration. For electrons, this means $\Delta E = 511$ keV and is thus possible; for protons, $\Delta E = 938$ MeV and this is not easily possible within the confines of a conventional magnet.

A very important further development of the concept of a microtron is based on the fact that if the orbits of the particles are far enough separated so that one can apply different magnetic fields for each orbit and can even change the shape of the orbit away from circular, then by careful choice of the orbit lengths, it is possible to maintain the **synchronicity condition** (1.10) while maintaining the freedom to have any amount of acceleration that is convenient. This is the basic idea behind **CEBAF**, the Continuous Electron Beam Accelerating Facility, at Thomas Jefferson National Accelerator Facility (TJNAF, Jefferson Lab, JLab), Newport News, Virginia, USA.

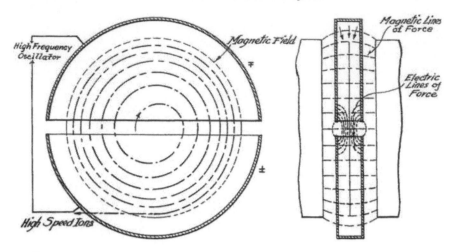

**FIGURE 1.18:**   The principle of the cyclotron, with top view on the left and side view on the right. (From E. O. Lawrence, US Patent 1,948,384, 1932 [40].)

The basic idea of the **cyclotron** is similar to that of the microtron, except that the RF cavity is used more efficiently by providing acceleration twice or even more times per turn, and the orbits roughly follow **concentric circles**. The concept of the cyclotron is shown schematically in Fig. 1.18 [40].

According to eq. (1.9), the revolution frequency is

$$\omega = \frac{q}{\gamma m_0} B, \tag{1.11}$$

and the momentary radius of the orbit is

$$r = \frac{p}{qB}. \tag{1.12}$$

This entails very similar restrictions regarding relativistic effects as in the case of the microtron; as before, any deviation from constancy of the magnetic field prevents continuous injection of the beam and hence leads to a non-continuous outgoing beam. But because the orbits are nearly concentric, it is possible to at least partly compensate the relativistic effects by **increasing B radially** in such a way that the revolution frequency in eq. (1.11) stays constant. This kind of cyclotrons is called the **isochronous cyclotron**. If it is necessary to accelerate different particles in the same machine, then that entails that the actual field profile has to be adjustable, which is usually achieved by having one or several **trim coils**. The superconducting K1200 cyclotron at the National Superconducting Cyclotron Laboratory (NSCL) at Michigan State University, Michigan, USA, allows for such corrections of the profile of the magnetic field.

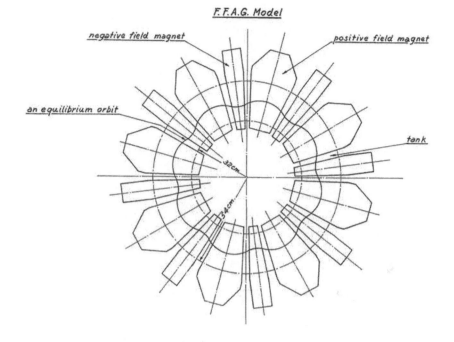

**FIGURE 1.19:** The first model of an FFAG, the Fixed-Field Alternating Gradient accelerator. (From L. W. Jones and K. M. Terwilliger, in E. Regenstreif, ed., *Proc. CERN Symp. High Energy Accelerators and Pion Physics*, CERN 56-25, 1956 [34]. Courtesy CERN.)

If continuity of the beam is not of prime importance, it is possible to make the necessary relativistic corrections due to eq. (1.11) via a decrease of the RF frequency during the acceleration process, which is done in the case of the **synchrocyclotron**. This decrease obviously has to happen very quickly over the few hundred turns of the particles while staying within the accelerating structure, and thus the pulse frequency can still be rather high.

A variant of the cyclotrons that were studied intensively in the 1950s was the **fixed-field alternating gradient (FFAG)** accelerator. Fig. 1.19 shows the drawing of the first of such an accelerator built. It combines the feature of the fixed magnetic field as in a cyclotron and the idea of **alternating gradient focusing** that became widely known in the early 1950s. Although FFAG did not flourish as a high energy accelerator, it has generated renewed interest in the past decade as a candidate to rapidly accelerate decaying particles such as muons and ion beams with large emittance and momentum spread.

For any accelerator, the **ultimate energy limitation** comes from the strength of the magnetic field that is available as the **unavoidable restric-**

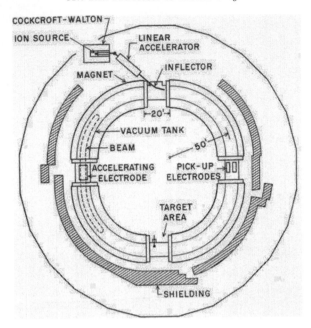

**FIGURE 1.20**:   Sketch of the Bevatron, designed to achieve "**B**illions of **eV** Synchro**tron**," at Lawrence Berkeley National Laboratory, California, USA. (From E. J. Lofgren, in E. Regenstreif, ed., *Proc. CERN Symp. High Energy Accelerators and Pion Physics*, CERN 56-25, 1956 [46]. Courtesy CERN.)

**tion**

$$\frac{p}{q} = B\rho = \chi_m. \tag{1.13}$$

The range of available magnetic fields is rather limited; typical numbers are in the range of 1–2 T for normal conducting dipole magnets, and several times more for superconducting dipole magnets. The superconducting dipole magnets at the Large Hadron Collider (LHC) at the European Organization for Nuclear Research (CERN), near Geneva, in Switzerland and France, operate reliably at 8 T. (See Table 1.1.) Looking beyond the rather stringent requirements for particle accelerators regarding field quality over extended regions and temporal stability, as of 2013 the highest magnetic fields that can be achieved are about 100 T. In fact, the National High Magnetic Field Laboratory (NHMFL), having branches at Florida State University, University of Florida and Los Alamos National Laboratory (LANL), USA, reached 100.75 T at the Los Alamos branch in 2012. The Dresden High Magnetic Field Laboratory (Hochfeld-Magnetlabor Dresden, HLD) at the Helmholtz-Zentrum Dresden-Rossendorf, Germany, reached 91.4 T in 2011, a record at the time, and 94.2 T in 2012.

So for practical purposes, the only way to achieve high energies is to increase

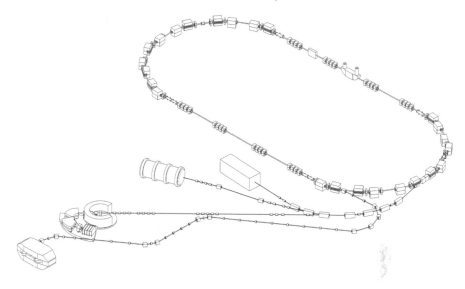

**FIGURE 1.21:** Layout of the Cooler Synchrotron (COSY) ring at the Institute of Nuclear Physics (IKP) at Forschungszentrum Jülich, Germany. (Courtesy Forschungszentrum Jülich GmbH.)

the deflection radius $\rho$. This represents a significant **practical limitation to continuous beam accelerators**, in which $B$ must be time independent and the size of the orbits increases in the acceleration process, since any region in which the beam may come has to be covered by magnetic fields. So for **really high energies**, the only realistic option is to have the particles follow the **same orbit** all the time by **ramping** the magnetic field during acceleration, and thus confine the region that has to be covered by the magnetic field.

Of course this ongoing adjustment of the magnetic field during the acceleration process according to eq. (1.13) to maintain constancy of $\rho$ prevents continuous injection and hence continuous beams. Furthermore, since electric field strengths are comparatively more limited, the fields of the cavities have to be re-utilized many thousands of times, resulting in a rather stretched-out acceleration process, and thus a rather **low repetition rate** of beam pulses.

All these thoughts lead to the concept of the **synchrotron**, in which the magnetic field strength is synchronized with the momentary energy or momentum of the particle so as to maintain a constant location of the reference orbit. The first generation of synchrotrons uses inhomogeneous dipole magnet to bend and confine the beam transversely, which is essentially the same as in a betatron. The only difference is that the acceleration is achieved through RF cavities. Fig. 1.20 shows an example of such a machine. The main limit of this kind of synchrotron is that the transverse focusing force from the gradient magnet is very weak, resulting in large beam pipes and magnets. For this

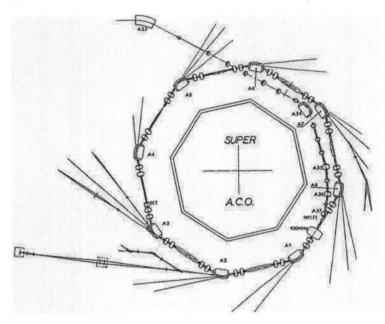

**FIGURE 1.22**: Layout of the Super-ACO light source storage ring at Laboratoire pour l'Utilisation du Rayonnement Electromagnétique, Orsay, France. (From M. P. Level, et. al., in *Proc. PAC 1987*, OSTI ID: 5125784, CONF-870302-Vol.1, 470, 1987 [44].)

**TABLE 1.1:**   Examples of hadron synchrotrons

| Name | Size ($\rho$) | Energy ($E$) | Particles |
|---|---|---|---|
| RHIC | 500 m | 250 GeV | polarized $p$; ions |
| Tevatron | 1 km | 1 TeV | $p, \bar{p}$ |
| LHC | 5 km | 7 TeV | $p$; ions (2.8 TeV/$n$ for Pb) |

reason, they are called **weak focusing synchrotrons**. Alternating gradient focusing offered orders of magnitude stronger focusing, much smaller beam size and much smaller magnets. Since the mid-1950s, **alternating gradient synchrotron**, also called **strong focusing synchrotron**, has replaced the weak focusing synchrotrons. Nowadays, almost all high energy accelerators are strong focusing synchrotrons. Fig. 1.21 shows one example, which is the layout of COSY, the COoler SYnchrotron at Forschungszentrum Jülich [3].

Table 1.1 shows characteristic features of some of hadron synchrotrons. Shown are the Relativistic Heavy Ion Collider (RHIC), at Brookhaven National Laboratory (BNL), Upton, New York, USA, the Tevatron (1987–2011) hosted at Fermi National Accelerator Laboratory (Fermilab, FNAL), Illinois, USA, and the LHC with their approximate dimensions, the maximum energies (per nucleon for ions) for which they are designed, and particles to be

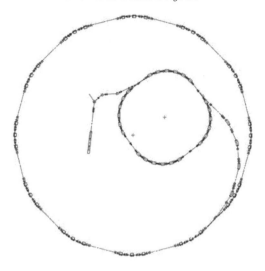

**FIGURE 1.23**:   Layout of the Advanced Light Source (ALS) at Lawrence Berkeley National Laboratory, California, USA. (From Document Control Center of Lawrence Berkeley National Laboratory, Print number: 22Q2593, 1989. Courtesy Lawrence Berkeley National Laboratory.)

accelerated.

The **storage ring** is not an accelerator in the traditional sense, since it holds the energy of the stored beam constant; however, it does not necessarily mean that RF cavities are not needed. In fact, due to synchrotron radiation, all the electron and the high energy proton storage rings use RF cavities to maintain the energy of the beam. Naturally, a synchrotron often can play the role of a storage ring as well.

The time that the particles stay in the storage rings ranges from minutes to days. In the case of the Tevatron, where the circumference is 6.28 km, the time for one operation while having collisions for high energy experiments is about 8 hours, which is 28800 sec. So the particles circulate through the ring

$$n = \frac{3 \times 10^8 \text{m}/\sec \cdot 28800\,\sec}{6.28 \times 10^3 \text{m}} \approx 10^9 \text{ turns.}$$

As a comparison, the operation duration without collision is more than 100 hours at the Tevatron. The LHC has similar numbers. Thus, even more so than in the case of the synchrotron, one of the main design problems and physically perhaps the greatest challenge is to try to ensure that particles actually stay contained over this large number of turns. Because the motion is nonlinear, this immediately leads to questions of nonlinear dynamics with all their complicated and interesting aspects.

One of the applications of storage rings is the **collider**, where counter rotating beams are brought to collision at various points around the ring. At very high energies, colliders have a significant **energy advantage** over fixed target machines because a very large fraction of the beams' energies can be converted to reaction energy. As a detailed study of the relativistic dynamics shows, this is not at all the case for fixed target cases; in fact, conservation of energy and momentum severely limits the energy that can be set free. Large scaled circular colliders are those listed in Table 1.1, and the tunnel with circumference 27 km, hosting the LHC currently, was earlier used for the Large Electron-Positron Collider (LEP) ($e^+, e^-$; $\sim 100$ GeV, 1989–2000). Besides the energy advantage, storage rings also have the disadvantage of the slow ramping times typical for synchrotrons; however, once the beam is stored, it is essentially **continuous** again.

But also for situations that require the beam to hit a **fixed target**, storage rings often offer an advantage over the use of synchrotrons by themselves, because it is often possible to extract the beam much more slowly than in the case of the synchrotron, resulting in a more easily manageable duty cycle and reducing the problem of overflowing the electronics in the detectors. In this method of **ultra-slow extraction**, the nonlinear dynamics of the device is adjusted very carefully and gently, as over time a larger and larger part of the originally stored emittance becomes unstable. If it is possible to control the location around the ring where the spilling occurs, then the spilled particles can be directed toward the fixed target as needed. One storage ring where this approach is utilized is COSY, the cooler synchrotron and storage ring, at Forschungszentrum Jülich, Germany, shown in Fig. 1.21.

Another application of the storage ring that has become one of the most productive tools for scientific research is the **synchrotron light source**. Although not as majestic as the giant high energy colliders, there are many synchrotron light sources throughout the world, and each facility hosts many users from almost all disciplines of the sciences. In the light source, the probe for the experiments is the light (from far infrared to hard X-ray) radiated by the electrons when the orbits are bent in the ring, which is generated through the process of **synchrotron radiation**. Figs. 1.22 [44] and 1.23 [42] show a couple of synchrotron light sources. As the electron mass is so small, in principle, any bending magnet can be used to produce light due to synchrotron radiation. But in addition, in the straight sections of a light source ring, often **wigglers** and **undulators**, which consist of alternating short bending magnets, are placed to produce more intense and coherent light. In such a way, each light source ring can hold tens of light beamlines, much more than the number of interacting locations that a high energy physics or nuclear physics collider can have for the collider experiments.

# Chapter 2

## Linear Beam Optics

In the discussion of the basic physical principles of the various types of accelerators, we casually neglected the fact that it is necessary to take care of more than one particle. In fact, all the above accelerators have to be able to simultaneously deal with an ensemble of particles with similar phase space coordinates, which is what the sources deliver, and hence with a **beam**. As outlined above, a detailed understanding of the motion of the beam requires the study of the motion of the **reference particle** as well as the motion of the **relative coordinates**.

In the case of **accelerators**, our demands on the relative motion are mostly that the beam does not become unreasonably large, and hence that the motion is somehow **bounded** within a suitable volume of phase space. While this appears to be a modest wish for long single pass accelerators, and more so for repetitive systems, this problem actually turns out to be rather nontrivial.

For other types of systems, more specific requirements have to be made for the beam. For example, to maximize the number of collisions at an interaction region of a collider, it is important to "squeeze" together the spatial coordinates of the beam, which under conservation of phase space volume then requires the momentum coordinates becoming large. Devices like particle spectrographs or electron microscopes have different and often even more involved requirements.

In all of these cases, it is important to study the relative motion carefully. As a first step, the motion is linearized, and for higher precision, the nonlinear effects of the motion have to be studied. Because the volume in phase space occupied by a beam is small, these nonlinear effects are often treated in a **perturbative way**, in which the first order corresponds to linear motion, and nonlinear motion appears as higher order (see Table 2.1).

**TABLE 2.1:** Classification of effect

| | |
| --- | --- |
| zeroth order | motion of reference particle |
| first order | linear motion |
| second+higher orders | nonlinear motion |

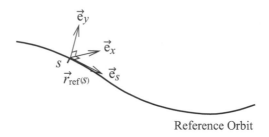

Reference Orbit

**FIGURE 2.1:**   Reference orbit, arc length $s$ along it and local coordinates.

## 2.1   Coordinates and Maps

Usually when studying dynamics, the time $t$ plays the role of the independent variable, and we study the motion of positions $\vec{x}$ and velocities $\vec{v}$ or momenta $\vec{p}$ as coordinates. Using the Lagrange mechanism, it is easy to transfer to new coordinates, in particular the coordinates that describe the **relative dynamics** around the reference orbit. Furthermore, instead of using $t$, we usually use the **arc length** $s$ along the **reference orbit** as an independent variable. Fig. 2.1 illustrates the concept.

For the understanding of the motion in relative coordinates, let us assume we have studied and understood the motion of the reference orbit. In case there is no field at all, this reference orbit will merely follow a **straight line**. Furthermore, there are many devices used in accelerators that have fields, but along one given straight line, all the fields vanish, and the device is lined up in such a way that the reference particle follows this line. Another important device uses magnetic fields, and along the reference orbit one tries to hold the magnetic fields constant, in which case the reference orbit is **circular**, at least within the element. In all other cases, it is usually necessary to describe the reference orbit by **numerically integrating** the equations of motion.

We assume the position and momenta of the reference particle $\vec{r}_{\text{ref}}(s)$, $\vec{p}_{\text{ref}}(s)$ are known. Here the momentum $\vec{p}$ is the dynamical momentum as in eq. (1.4). As a technical detail, let us also assume that for all points $s$, we have $\vec{p}_{\text{ref}}(s) \nparallel \vec{e}_{z\text{Lab}}$, i.e., the motion is never pointing vertically straight (which for most real accelerators is no limitation whatsoever). Let furthermore $r_{\text{tube}}$ be smaller than the minimum radius of curvature that the reference orbit experiences in the section of the device that we want to study. We now consider a "**flexible tube**" of radius $r_{\text{tube}}$ centered around the reference orbit, and restrict the particles that we want to describe to only those within the tube, as Fig. 2.2 illustrates the situation. Again, for practical devices this represents hardly a limitation; for example, in the LHC (see Table 1.1), the "tube" would be more than 2 km wide, much larger than the region required

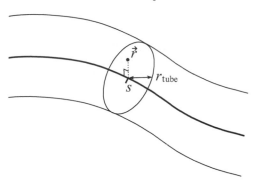

**FIGURE 2.2:** Motion of particles inside the tube with radius $r_{\text{tube}}$ around the reference orbit.

by the beam particles.

For any particle within the tube, there is now a **closest point** on the reference orbit; because only particles within the tube are allowed, this point is indeed **unique**. Let $s$ be the arc length at this point, and $\vec{r}_{\text{ref}}(s)$ the position of the reference particle on the reference orbit. Then the relative coordinates of the point $\vec{r}$ are obviously $\vec{r} - \vec{r}_{\text{ref}}(s)$.

Let now $\vec{e}_s$ be a unit vector in the direction of $\vec{p}_{\text{ref}}$. Consider now the plane perpendicular to $\vec{e}_s$. Of all the unit vectors in this plane, let $\vec{e}_y$ be the one with the largest vertically upward component; because in our setup $\vec{p}_{\text{ref}}$ and hence $\vec{e}_s$ are not allowed to go vertically straight, this vector is well defined. Finally choose a third vector $\vec{e}_x$ as $\vec{e}_x = \vec{e}_y \times \vec{e}_s$. Because $\vec{e}_y$ has a maximum vertically upward component, $\vec{e}_x$ has a vanishing vertical component and hence lies in the horizontal plane.

Denote now by "$x$" the component of $\vec{r} - \vec{r}_{\text{ref}}(s)$ in the direction of $\vec{e}_x$, and by "$y$" the component of $\vec{r} - \vec{r}_{\text{ref}}(s)$ in the direction of $\vec{e}_y$. Similarly, define $p_x$ and $p_y$ to be the momentum components of $\vec{p} - \vec{p}_{\text{ref}}$ in the directions $\vec{e}_x$ and $\vec{e}_y$.

Using $\{x, p_x, y, p_y\}$, the motion in the transversal plane, defined by $\vec{e}_x$ and $\vec{e}_y$, can be described, and it is called the **transversal dynamics**. However, considering how a beam is formed as we have seen in Chapter 1, we have to consider that the energy of a particle $E$ in the beam can be different from that of the reference particle $E_{\text{ref}}$, even if it is only slightly so. The energy difference of the particles as well as the geometry of the orbits also results in the difference of the travel time $t$ of the particles, called the **time-of-flight**. Thus, the energy and the time-of-flight have to be considered when studying the motion of a beam, and it is called the **longitudinal dynamics**.

As we will see later, the transversal motion and the longitudinal motion are in general coupled, except for special cases. Altogether, we will describe the motion of the beam in six coordinates $\{x, p_x, y, p_y, E, t\}$. The actual choice of coordinate quantities requires a careful consideration, as it eventually determines how widely the resulting derivations are compatible with general concepts in physics and mathematics. Below, the quantities with the subscript 0 are meant to indicate the reference particle.

Before launching the motion, we denote the energy deviation of a particular particle of interest by $\delta$ by defining

$$\delta = \frac{K - K_0}{K_0},$$

where $K$ is the initial kinetic energy of the particle under consideration while $K_0$ is that of the reference particle. Finally, we introduce a space-like variable $l$

$$l = \kappa\,(t - t_0)$$

being the deviation of the time-of-flight $t$ from that of the reference particle, multiplied by a constant $\kappa$ that has the dimension of velocity. Specifically,

$$\kappa = -v_0 \frac{\gamma_0}{1 + \gamma_0}, \tag{2.1}$$

using the absolute value of the velocity of the reference particle $v_0$ and the associated $\gamma_0$, which can be expressed as

$$\gamma_0 = \frac{1}{\sqrt{1 - v_0^2/c^2}} = \frac{\sqrt{(\vec{p}_0 c)^2 + (mc^2)^2}}{mc^2} = \frac{E_0}{mc^2},$$

by referring to eq. (1.6). The specific form of $\kappa$, especially the fractional factor involving $\gamma_0$, is important for generating what turns out to be a canonical pair of coordinates $(l, \delta)$; the details go beyond the scope of this book, and we refer to [5] for details.

Then we form the **vector $\vec{Z}$ of particle optical coordinates** as

$$\vec{Z} = \begin{pmatrix} x \\ a & = p_x/p_0 \\ y \\ b & = p_y/p_0 \\ l & = \kappa\,(t - t_0) \\ \delta & = (K - K_0)/K_0 \end{pmatrix}, \tag{2.2}$$

where $p_0$ is some previously chosen scaling momentum; a natural choice is to select the momentum of the reference particle at the beginning. Likewise, $K_0$ is a previously chosen scaling energy, for example the kinetic energy of the reference particle, and similarly, $\kappa$ is a scaling quantity introduced in eq. (2.1).

Note that due to the definition of $\vec{Z}$, the reference particle itself corresponds to $\vec{Z} = \vec{0}$, and hence the vector $\vec{Z}$ does indeed describe the relative motion. In a seemingly simple way, most of the problems of beam physics now revolve around the question as to how $\vec{Z}$ evolves as a function of $s$.

In light of this, the entire action of a beam physics device can now be expressed by how it manipulates the coordinates in $\vec{Z}$. In fact, usually a set of initial conditions $\vec{Z}_0$ at position $s_0$ uniquely determines the future evolution and hence $\vec{Z}$ at any later position $s$. While a common notion, mathematically this **determinism** of classical mechanics rests on some subtle assumptions about the details of the fields that are allowed in the motion; but further details are beyond this book.

Assuming that indeed $\vec{Z}_0$ at $s_0$ uniquely determines the future evolution, we can define a function relating the initial conditions at $s_0$ to the conditions at $s$ via

$$\vec{Z}(s) = \mathcal{M}(s, s_0)\left(\vec{Z}(s_0)\right).$$

The function $\mathcal{M}(s_0, s)$, which formally summarizes the entire action of the system, is of great importance for the description and analysis of beam physics systems. It is often called the **transfer function**, the **transfer map**, or simply the **map** of the system. Note that the transfer maps satisfy the relationship

$$\mathcal{M}(s_2, s_1) \circ \mathcal{M}(s_1, s_0) = \mathcal{M}(s_2, s_0), \tag{2.3}$$

which merely says that transfer maps of systems can be built up from the transfer maps of the pieces.

Since $\mathcal{M}$ describes the motion in relative coordinates, we always have

$$\mathcal{M}(\vec{0}) = \vec{0}.$$

Furthermore, since by the very definition of a beam, the coordinates of $\vec{Z}$ are "small," $\mathcal{M}$ is usually only **weakly nonlinear**. Because of this, its determination and analysis is very amenable to **perturbative techniques**. The first step in this process is to consider only the linearization $\hat{M}$ of $\mathcal{M}$, the so-called **linear map**. Let $\mathcal{N} = \mathcal{M} - \hat{M}$ be the remaining purely nonlinear part, so that we have

$$\mathcal{M} = \hat{M} + \mathcal{N}.$$

The linear map $\hat{M}$ is simultaneously the **most important** and the **easiest** to study. The treatment of the nonlinear part $\mathcal{N}$ is much more complicated, and only later in the book will we address a small part of the problems associated with its treatment. More details can be found, for example, in [5].

In the following section, we will make a short excursion to a field that at first glance appears disconnected from beam physics, namely the field of glass optics. However, a closer look shows that glass optics, which has existed long before the name beam physics was introduced, certainly belongs to this field: the ensembles of light particles or rays typically associated with questions of glass optics form a beam not only in the conventional meaning of the word, but also under the more formal definition.

## 2.2  Glass Optics

As one may recall from a basic course in optics, a distinction is made between so-called "**Gaussian optics**," which indeed turns out to just mean linear motion, and "**aberrations**" that describe nonlinear effects. Optics has developed its very own jargons and techniques, some of which are connected to complicated geometric ideas, and in our opinion it is historically unfortunate that optics has not been treated with the methods of the **transfer map**. We shall remedy this situation here by simultaneously providing a short introduction on Gaussian optics in an appealing and unified way, and also develop our skills in dealing with linear maps.

For simplicity, let us restrict ourselves to systems that are rotationally symmetric, like most glass optical systems; it will be quite clear as we go what has to be done to treat non-rotationally symmetric systems. In this rotationally symmetric case, two variables are enough to study the motion; we here choose them as the position $x$ and the slope $a$ of a ray. The transfer map of an optical system then expresses how $(x, a)$ behave as they transfer a system, and we have

$$\begin{pmatrix} x_2 \\ a_2 \end{pmatrix} = \mathcal{M} \begin{pmatrix} x_1 \\ a_1 \end{pmatrix}.$$

In fact, if we restrict ourselves to linear motion, then this can be expressed in terms of a **transfer matrix**

$$\hat{M} = \begin{pmatrix} (x|x) & (x|a) \\ (a|x) & (a|a) \end{pmatrix}.$$

Note that the notation for the matrix elements is such that the quantity **before** the vertical line "|" describes the **row**, and that **after** the vertical line describes the column. We remind again that knowing matrices of pieces allows the computation of matrices of more complicated systems, which is here achieved by mere matrix multiplication. Indeed, if $\hat{M}_1$ through $\hat{M}_n$ are the matrices for the subsystems, then because of the associativity of matrix multiplication, we obtain for the ray after the last subsystem:

$$\begin{pmatrix} x_{n+1} \\ a_{n+1} \end{pmatrix} = \hat{M}_n \left( \cdots \left( \hat{M}_1 \begin{pmatrix} x_1 \\ a_1 \end{pmatrix} \right) \cdots \right) = \left( \hat{M}_n \cdots \hat{M}_1 \right) \begin{pmatrix} x_1 \\ a_1 \end{pmatrix}.$$

So we have shown that the matrix of a combined system equals to product of matrices of subsystems. Since especially on computers it is very simple to multiply matrices, this is the method of choice for the basic design of optical systems. In the following, we hence derive the forms of the matrices of common optical elements.

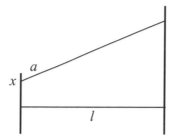

**FIGURE 2.3:** A ray passing through a drift.

### 2.2.1 The Drift

The simplest part of glass optical elements is a region which does not contain any material, the **drift**. The final position and slope $x_2$ and $a_2$ after a drift of length $l$ can be connected very simply to the initial values $x_1$ and $a_1$, as shown in Fig. 2.3

$$x_2 = x_1 + a_1 \cdot l, \qquad a_2 = a_1.$$

This obviously can be written in a matrix form as

$$\begin{pmatrix} x_2 \\ a_2 \end{pmatrix} = \begin{pmatrix} 1 & l \\ 0 & 1 \end{pmatrix} \begin{pmatrix} x_1 \\ a_1 \end{pmatrix}.$$

For the later discussion it is important to note that the matrix $\begin{pmatrix} 1 & l \\ 0 & 1 \end{pmatrix}$ depends only on the characteristic properties of the element, which here is the length $l$. On the other hand, the vector $(x_1, a_1)$ depends only on the parameters of the ray. Altogether, a drift performs a linear transformation in $x$, $a$ space. Note that the determinant of the drift matrix is unity.

As a small exercise, let us now consider a combination of two drifts of lengths $l_1$ and $l_2$. For the value of the coordinates $(x_3, a_3)$ after the combination of the two drifts, we have

$$\begin{pmatrix} x_3 \\ a_3 \end{pmatrix} = \begin{pmatrix} 1 & l_2 \\ 0 & 1 \end{pmatrix} \begin{pmatrix} x_2 \\ a_2 \end{pmatrix} = \begin{pmatrix} 1 & l_2 \\ 0 & 1 \end{pmatrix} \begin{pmatrix} 1 & l_1 \\ 0 & 1 \end{pmatrix} \begin{pmatrix} x_1 \\ a_1 \end{pmatrix} = \begin{pmatrix} 1 & l_1 + l_2 \\ 0 & 1 \end{pmatrix} \begin{pmatrix} x_1 \\ a_1 \end{pmatrix}.$$

Here the necessary **composition** of maps just reduces to a common **multiplication** of transfer matrices. And the result is not surprising, the effect of two subsequent drifts is just the same as that of a drift of the combined length.

### 2.2.2 The Thin Lens

Besides empty space, glass optical devices contain lenses that change the direction of the light ray. Here we are primarily interested in the thin lens, a

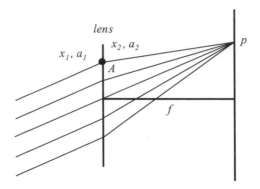

**FIGURE 2.4:** A bundle of parallel rays passing through a focusing lens.

somewhat idealized device without any length, which is characterized by the following facts that are also illustrated in Fig. 2.4.

1. Positions are not changed, but directions are changed.

2. Any bundle of parallel light is unified in one point a distance $f$ after the lens.

3. A ray lighting the center of the lens goes straight through.

The quantity $f$ that describes the lens is called the **focal length**. Let us now consider a ray passing through the lens. From Fig. 2.4 we find

$$x_2 = x_1, \qquad p = f \cdot a_1, \qquad x_1 + a_2 \cdot f = p,$$

from which we infer

$$x_2 = x_1, \quad a_2 = -\frac{x_1}{f} + a_1.$$

This relationship can be written in a matrix form as

$$\begin{pmatrix} x_2 \\ a_2 \end{pmatrix} = \begin{pmatrix} 1 & 0 \\ -1/f & 1 \end{pmatrix} \begin{pmatrix} x_1 \\ a_1 \end{pmatrix}. \tag{2.4}$$

As in the case of the drift, the matrix $\begin{pmatrix} 1 & 0 \\ -1/f & 1 \end{pmatrix}$ depends only on the focal length $f$, the characteristic property of the lens, whereas the vector $(x_1, a_1)$ depends on the ray. Note that the determinant of the matrix $\begin{pmatrix} 1 & 0 \\ -1/f & 1 \end{pmatrix}$ is unity.

The simple thin lens we have discussed here, the so-called focusing Gaussian lens, represents quite an approximation for several reasons. First, any real lens performs a refraction at two different surfaces, so positions do change as one

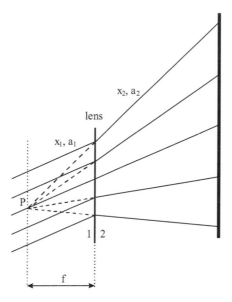

**FIGURE 2.5**:   A bundle of parallel rays passing through a defocusing lens.

goes through the lens. Furthermore, for most lenses it is not really true that parallel rays all meet at a point a distance $f$ behind the lens. This is connected to the fact that lenses are usually ground with spherical surfaces because anything else is technically difficult. Furthermore, the glass has dispersion, so different colors are affected differently. We note however that Snell's law still allows us to determine the true transfer map of a thick, spherical lens in a rather straightforward way. It is important to note, however, that this transfer map will no longer be linear.

Quite interesting is the combination of two glass lenses, which can apparently be described by multiplying their matrices. Note that, always, the matrix of the **first element** is on the **right**. We obtain

$$\begin{pmatrix} 1 & 0 \\ -1/f_2 & 1 \end{pmatrix} \begin{pmatrix} 1 & 0 \\ -1/f_1 & 1 \end{pmatrix} = \begin{pmatrix} 1 & 0 \\ -1/f_1 - 1/f_2 & 1 \end{pmatrix}.$$

So the combination of two lenses provides the same effect as one lens with focus length $f$, where $1/f = 1/f_1 + 1/f_2$. This is of course a famous law of optics, the derivation of which is all but trivial in the matrix context. Indeed the efficiency of the matrix approach becomes clear when observing how to prove this law using the standard geometric method of optics textbooks.

In a similar way as the focusing thin lens we can also treat the defocusing thin lens. In this case, the basic properties can be found as illustrated in Fig. 2.5.

1. Positions are not changed, but directions are changed.

2. Any bundle of parallel light exits the lens in such a way that it appears to come from a point a distance $f$ in front of the lens.

3. A ray lighting the center of the lens goes straight through.

In a similar way as before, we can use basic geometry to determine the action of the lens. From Fig. 2.5, we find

$$x_2 = x_1, \qquad p = -f \cdot a_1, \qquad p = x_2 - f \cdot a_2.$$

Similar to before, we obtain $a_2 = x_1/f + a_1$, which is in a matrix form

$$\begin{pmatrix} x_2 \\ a_2 \end{pmatrix} = \begin{pmatrix} 1 & 0 \\ 1/f & 1 \end{pmatrix} \begin{pmatrix} x_1 \\ a_1 \end{pmatrix}.$$

This is essentially the same matrix as before, except that now the sign of the matrix element $(a|x)$ has changed. Indeed, using the standard convention to count defocusing lenses with a negative focal length, the matrix has even exactly the same form as before.

## 2.2.3   The Thin Mirror

Besides lenses, mirrors are probably the second most important optical device, and there are also focusing and defocusing mirrors. Different from the lens, the reference orbit flips direction when hitting the mirror. A thin focusing mirror is defined by what it does to an ensemble of parallel light via three conditions, illustrated in Fig. 2.6.

1. Positions are not changed, but directions are changed.

2. Any bundle of parallel light that is reflected by the mirror will meet in a point a distance $f$ in front of the mirror.

3. A ray hitting the center of the mirror is reflected such that its outgoing angle equals its incoming angle.

A similar argument as in the case of the focusing lens shows that the transfer matrix of the focusing mirror is

$$\hat{M} = \begin{pmatrix} 1 & 0 \\ -1/f & 1 \end{pmatrix}.$$

There is also a defocusing mirror, defined by three conditions:

1. Positions are not changed, but directions are changed.

2. Any bundle of parallel light that is reflected by the mirror seems to emerge from a point a distance $f$ behind the mirror.

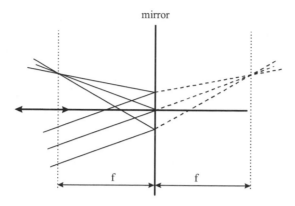

**FIGURE 2.6**:  A bundle of parallel rays is reflected by the focusing mirror.

3. A ray hitting the center of the mirror is reflected such that its outgoing angle equals its incoming angle.

A similar argument shows that also in this case, we have the transfer matrix

$$\hat{M} = \begin{pmatrix} 1 & 0 \\ -1/f & 1 \end{pmatrix},$$

where the convention to count the focal length $f$ of a defocusing element negative is used.

So apparently mathematically, lenses and mirrors behave the same, aside from the fact that they reverse the reference orbit. The choice of which to use in practice depends on a variety of practical factors. For situations requiring only small apertures like in most camera lenses, glass lenses are easily made, and have an advantage because of the straight beam path. For situations requiring large apertures, like in big telescopes, mirrors are the primary choice because it is much easier to manufacture and support large mirrors than large lenses. It is also easier to produce non-spherical shapes for mirrors than for lenses. Finally, mirrors have the additional advantage that they treat light of different colors equally; they do not show the dispersion commonly observed in glass lenses.

## 2.2.4   Liouville's Theorem for Glass Optics

As a direct consequence of the matrix notation for glass optics introduced above, for any combination of lenses, drifts and mirrors, we can prove a special case of **Liouville's theorem**: *The volume of phase space occupied by the beam is conserved.*

Indeed, let us assume that we have an optical system consisting of $n$ ele-

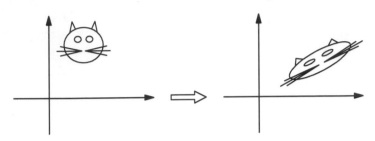

**FIGURE 2.7**:  Liouville's theorem. The volume of phase space occupied by the beam is conserved.

ments with matrices $\hat{M}_i$. Then we have

$$\begin{pmatrix} x_{n+1} \\ a_{n+1} \end{pmatrix} = \hat{M}_n \left( \hat{M}_{n-1} \left( \cdots \hat{M}_1 \begin{pmatrix} x_1 \\ a_1 \end{pmatrix} \cdots \right) \right) = \left( \hat{M}_n \cdot \hat{M}_{n-1} \cdots \hat{M}_1 \right) \begin{pmatrix} x_1 \\ a_1 \end{pmatrix}.$$

The determinants of each of the matrices $\hat{M}_i$ are just unity, as they are all either drifts, lenses or mirrors, so the determinant of the product is unity. Under linear transformations, volumes in space transform with the size of the determinant, thus the volume is indeed conserved. Fig. 2.7 illustrates this situation.

An interesting and remarkable consequence of Liouville's theorem is the famous **recurrence theorem of Poincaré**. Let us assume we have some motion in $n$-dimensional phase space and also that we know that the motion is bounded in all phase space variables. Let us further assume that the motion obeys Liouville's theorem, which as we shall see later is the case for all Hamiltonian systems, and let the motion be deterministic. Then Poincaré's recurrence theorem states that *for any given $\varepsilon$, the system after sufficient time comes back to its original state within a tolerance of at most $\varepsilon$.*

Before we sketch the proof of Poincaré's theorem, let us illustrate some of its consequences. Consider for example a box with classical gas particles that are initially all located in one side of the box and kept there by a wall as shown in Fig. 2.8. After the wall is removed, the gas particles will distribute in the box evenly, as we expect from classical statistical mechanics, increasing their entropy. But their phase space is bounded, as the particles cannot leave the box, and each particle's momentum is limited by the total heat energy contained in the box.

So as time progresses, according to Poincaré, they will at one time in the future just recollect on one side of the box, and by re-inserting the wall, they will be caught again on one side, in crass contradiction to the entropy principle.

There are many other examples. If we have a particle beam in an accelerator that we know is stable, it will eventually come back as close as we want in phase space, which is an effect that is actually observed somewhat routinely in

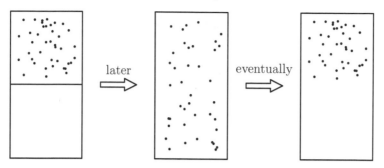

**FIGURE 2.8**:  Poincaré's recurrence theorem.  After sufficient time, the system returns close to its original state.

tracking pictures. Even for our daily life, there are important consequences. If the universe is Hamiltonian and it does not expand indefinitely, then up to minute details, history will keep repeating itself. So we will all be born again, and we will all make the same mistakes all over, but since now we cannot remember anything about our past life, also next time we will not remember our current life.

Now let us sketch the proof of the recurrence theorem. Let an $\varepsilon$ be given, and consider an $\varepsilon$-ball with volume $V_\varepsilon$ in phase space. Consider its motion by regular time steps $\Delta t$. Since the total available phase space volume is finite, say $V_p$, after at most $V_p/V_\varepsilon$ time steps, the image of the ball must reach a part of phase space it has touched before, i.e., it must overlap a previous image of the ball. Let us assume this happens after $N$ steps and the previous image is that after $n$ steps, with $n < N$. But if the images after $n$ steps $I_n$ and after $N$ steps $I_N$ overlap, so must the images after $(n-1)$ and $(N-1)$ steps, respectively. And continuing backwards, so must the images after 0 and $(N-n)$ steps; hence, after $(N-n)$ steps, we touch the original $\varepsilon$-ball again.

## 2.3   Special Optical Systems

In this section we want to apply the matrix techniques to the study of certain special categories of systems. In particular, we associate certain fundamental properties of systems with properties of the matrix. We begin with the imaging systems.

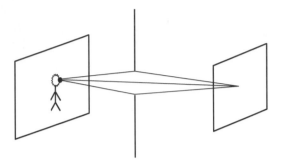

**FIGURE 2.9**:   Sketch of an imaging system.

### 2.3.1   Imaging (Point–to–Point, • •) Systems

Imaging systems or point–to–point systems are perhaps the most important systems in optics, and they deserve some special attention. Suppose we study the action of a slide projector. At one end of the projector, light is sent through the slide. Suppose the slide shows a man wearing a gold earring. The image of this man is to appear on the screen, and the gold earring is to appear at one particular location. This requires that all light passing through the golden spot on the slide and emanating in various directions has to be re-united at one spot on the screen, as shown in Fig. 2.9.

This means that the final position of a ray is independent of its initial angle and it only depends on the initial position. In terms of transfer matrices

$$\hat{M} = \begin{pmatrix} (x|x) & (x|a) \\ (a|x) & (a|a) \end{pmatrix},$$

this means that the element $(x|a)$ has to vanish:

$$(x|a) = 0.$$

Obviously the element $(x|x)$ also has an important interpretation: it is the **magnification** of the system.

$$(x|x) : \quad \text{magnification.}$$

Besides the case of the slide projector, many other devices use imaging. They include the camera, the overhead projector, the eye, the photographic microscope, the electron microscope, as well as particle spectrographs. We will discuss in detail some such devices in Chapter 7.

It is worthwhile to study how imaging systems can be made. First, we observe that a drift is imaging if and only if $l = 0$, while it is a rather boring choice. A single lens is also always imaging as long as there are no drifts before and after, but that is another boring choice. The first interesting imaging system is the DLD (drift-lens-drift) system, consisting of a drift, a lens and another drift. The transfer matrix of the DLD system is given by

$$\hat{M} = \begin{pmatrix} 1 & l_2 \\ 0 & 1 \end{pmatrix} \begin{pmatrix} 1 & 0 \\ -1/f & 1 \end{pmatrix} \begin{pmatrix} 1 & l_1 \\ 0 & 1 \end{pmatrix} = \begin{pmatrix} 1 & l_2 \\ 0 & 1 \end{pmatrix} \begin{pmatrix} 1 & l_1 \\ -1/f & 1 - l_1/f \end{pmatrix}$$

$$= \begin{pmatrix} 1 - l_2/f & l_1 + l_2 - l_1 l_2/f \\ -1/f & 1 - l_1/f \end{pmatrix}.$$

If such a system is supposed to be imaging, we have to satisfy $(x|a) = l_1 + l_2 - l_1 l_2/f = 0$, which is equivalent to

$$\frac{1}{l_1} + \frac{1}{l_2} = \frac{1}{f}.$$

This is another important result of conventional optics, which here is obtained in an almost trivial way. If the DLD system is made to be imaging, the magnification is given by

$$(x|x) = 1 - \frac{l_2}{f} = -\frac{l_2}{l_1}.$$

This principle is used in several different devices. In the slide projector, $l_1$ is very small and $l_2$ is very large, thus it provides a large magnification. Probably the most important imaging system is the eye. Here the situation is just the opposite. $l_1$ is large and $l_2$ is small, allowing for large things to be mapped on the small retina of the eye.

It is interesting to study the combination of two imaging systems:

$$\begin{pmatrix} (x|x)_2 & 0 \\ (a|x)_2 & (a|a)_2 \end{pmatrix} \begin{pmatrix} (x|x)_1 & 0 \\ (a|x)_1 & (a|a)_1 \end{pmatrix} = \begin{pmatrix} (x|x)_2 (x|x)_1 & 0 \\ (a|x)_2 (x|x)_1 + (a|a)_2 (a|x)_1 & (a|a)_2 (a|a)_1 \end{pmatrix}.$$
$$(2.5)$$

As is to be expected, the total system is again imaging, and the magnification is $(x|x)_2 (x|x)_1$, just the product of the individual magnifications.

## 2.3.2  Parallel–to–Point ($\parallel \bullet$) Systems

As we saw above, the human eye observing a nearby object is one of the prime examples of an imaging system. But what happens if the eye looks at things farther and farther away, in particular at the stars, a pastime of the human race and scientists for eternity? The length of the first drift $l_1$ becomes larger and larger, and for all practical purposes the light coming from one star reaches the eye as a parallel bundle. So what the eye is to interpret now is the angle under which the light comes in, and hence the position on the retina should depend only on the initial angle at which the light strikes the eye, but not on the initial position.

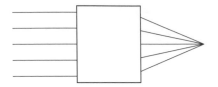

**FIGURE 2.10**: Sketch of a parallel–to–point system.

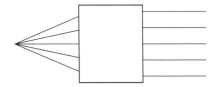

**FIGURE 2.11**: Sketch of a point–to–parallel system.

This is an example of parallel–to–point systems as illustrated in Fig. 2.10. A parallel–to–point system requires that

$$(x|x) = 0.$$

If we look at the eye as a DLD system, this requires

$$(x|x) = 1 - l_2/f = 0 \quad \Longrightarrow \quad l_2 = f,$$

while $l_1$ is arbitrary. Thus the retina has to be exactly at the focal length; almost as important is that the distance to the object is arbitrary since we cannot change our distance to the stars significantly. Another important parallel–to–point system is the photographic camera.

### 2.3.3 Point–to–Parallel (• ∥) Systems

Another important class of systems is the point–to–parallel systems. In these systems, the final slope depends only on the initial position, but not on the initial slope as illustrated in Fig. 2.11. So we have

$$(a|a) = 0.$$

Examples include the flashlight, the microscope, and laser and particle beam transports over long distances such as those considered for the SDI Transport (Strategic Defense Initiative) considered by the US government in the 1980s.

As an example, let us try to achieve a point–to–parallel system with a DLD combination. We obtain

$$(a|a) = 1 - l_1/f = 0 \quad \Longrightarrow \quad l_1 = f,$$

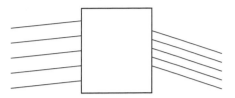

**FIGURE 2.12:** Sketch of a parallel–to–parallel system.

as we may have expected. Note that there is no condition on $l_2$.

From the transfer matrices, it follows rather directly that the combination of a point–to–parallel and a parallel–to–point system forms a point–to–point system.

$$\begin{pmatrix} 0 & (x|a)_2 \\ (a|x)_2 & (a|a)_2 \end{pmatrix}\begin{pmatrix} (x|x)_1 & (x|a)_1 \\ (a|x)_1 & 0 \end{pmatrix} = \begin{pmatrix} (x|a)_2(a|x)_1 & 0 \\ (a|x)_2(x|x)_1+(a|a)_2(a|x)_1 & (a|x)_2(x|a)_1 \end{pmatrix}.$$

Using the relaxed eye as the parallel–to–point system, we can thus build a microscope by placing a suitable point–to–parallel system in front of the eye. It is interesting to see how the lengths in a point–to–parallel system have to be chosen; by requiring $(a|a) = 0$, we obtain $l_1 = f$, while $l_2$ is arbitrary. The first part is as expected; the latter part is fairly important for the operation of a microscope because it allows the eye to move with respect to the microscope.

### 2.3.4 Parallel–to–Parallel ($\parallel \parallel$) Systems

The final important system is the parallel–to–parallel system illustrated in Fig. 2.12. By placing it between the eye and the stars, a magnification of angles can be achieved. This is the principle of the telescope.

The system has to be such that the final slope depends on the initial slope, but not on the initial position, which requires

$$(a|x) = 0.$$

The **magnification** is given by $(a|a)$.

$$(a|a): \quad \text{magnification.}$$

If we try to achieve this with a DLD system, then we have to satisfy $(a|x) = -1/f$ to be 0, which is impossible. This entails that a telescope has to contain at least two lenses.

So let us consider an LDL (lens-drift-lens) system.

$$\hat{M} = \begin{pmatrix} 1 & 0 \\ -1/f_2 & 1 \end{pmatrix}\begin{pmatrix} 1 & l \\ 0 & 1 \end{pmatrix}\begin{pmatrix} 1 & 0 \\ -1/f_1 & 1 \end{pmatrix} = \begin{pmatrix} 1-l/f_1 & l \\ -1/f_2-1/f_1+l/f_1f_2 & 1-l/f_2 \end{pmatrix}.$$

And we have to satisfy

$$(a|x) = -\frac{1}{f_2} - \frac{1}{f_1} + \frac{l}{f_1 f_2} = 0 \quad \Longrightarrow \quad l = f_1 + f_2,$$

which is a well-known condition for Newtonian or Galilean telescopes. The magnification of the telescope is given by

$$(a|a) = 1 - \frac{l}{f_2} = 1 - \frac{1}{f_2}(f_1 + f_2) = -\frac{f_1}{f_2}.$$

Thus, it requires $f_1 \gg f_2$ to obtain large magnification. Since there is a limit on how short $f_2$ can be, it is thus necessary to make $f_1$ large, which entails what the rather large size telescopes usually have.

### 2.3.5   Combination Systems

Often the question arises to what extent it is possible to simultaneously satisfy the requirements for the above systems. To some extent this is possible, but the fact that the determinant of the total system has to be unity due to Liouville's theorem for glass optics imposes some restrictions.

A closer look shows that

1. $\bullet\ \bullet$ and $\|\ \|$ is possible: $(x|a) = (a|x) = 0$.

2. $\|\ \bullet$ and $\bullet\ \|$ is possible: $(x|x) = (a|a) = 0$.

   All other cases are impossible because they would require a zero determinant.

Another important question is what happens when two systems satisfying certain properties are combined into one system; for example, we already saw in eq. (2.5) that two point–to–point systems placed behind each other again produce a point–to–point system. A more detailed analysis shows that of the sixteen cases describing combinations of two systems, eight cases lead to another special system

$$\begin{aligned}
\bullet\bullet + \bullet\bullet = \bullet\bullet\,, \qquad & \|\,\| + \|\,\| = \|\,\|\,,\\
\bullet\bullet + \bullet\,\| = \bullet\,\|\,, \qquad & \|\,\| + \|\,\bullet = \|\,\bullet\,,\\
\bullet\,\| + \|\,\bullet = \bullet\bullet\,, \qquad & \|\,\bullet + \bullet\,\| = \|\,\|\,,\\
\bullet\,\| + \|\,\| = \bullet\,\|\,, \qquad & \|\,\bullet + \bullet\bullet = \|\,\bullet\,.
\end{aligned}$$

The entries in the table above are easy to memorize because it contains just those combinations for which the second symbol of the first system equals the first symbol of the second system, and the final result is obtained by dropping the two identical symbols. So in compact notation, we have:

$$\text{If } A, B, C \in \{\bullet,\ \|\,\}, \quad \text{then} \quad AB + BC = AC.$$

# Chapter 3

## Fields, Potentials and Equations of Motion

For the study of transfer maps of particle optical systems, first it is necessary to undertake a classification of the possible fields that can occur. All fields are governed by **Maxwell's equations**, which in SI units have the form

$$\text{div}\,\vec{B} = 0, \quad \text{curl}\,\vec{H} = \vec{j} + \frac{\partial \vec{D}}{\partial t},$$

$$\text{div}\,\vec{D} = \rho, \quad \text{curl}\,\vec{E} = -\frac{\partial \vec{B}}{\partial t}. \tag{3.1}$$

In the case of particle optics, we are mostly interested in cases in which there are **no sources** of the fields in the region where the beam is located, so in this region we have $\rho = 0$ and $\vec{j} = \vec{0}$. Of course any beam that is present would represent a $\rho$ and a $\vec{j}$, but these effects are usually considered separately.

In the following, we want to restrict ourselves to **time independent** situations, and neglect the treatment of elements with quickly varying fields including cavities. This limitation in very good approximation also includes slowly time varying fields like the magnetic fields that are increased during the ramping of a synchrotron.

So, Maxwell's equations simplify to

$$\text{div}\,\vec{B} = 0, \quad \text{curl}\,\vec{H} = \vec{0},$$

$$\text{div}\,\vec{D} = 0, \quad \text{curl}\,\vec{E} = \vec{0}, \tag{3.2}$$

where

$$\vec{B} = \mu_0 \vec{H}, \quad \vec{D} = \varepsilon_0 \vec{E}.$$

Because of the vanishing curl, we infer that $\vec{E}$ and $\vec{B}$ have **scalar potentials** $V_E$ and $V_B$ such that

$$\vec{E} = -\vec{\nabla} V_E, \quad \vec{B} = -\vec{\nabla} V_B.$$

Note that here even the magnetic field is described by a scalar potential, and not by the vector potential $\vec{A}$ that always exists. From the first and third equations of (3.2), we infer that both scalar potentials $V_E$ and $V_B$ satisfy Laplace's equation, and we thus have

$$\Delta V_E = 0, \quad \Delta V_B = 0.$$

In order to study the solutions of Laplace's equations for the electric and magnetic scalar potentials, we will proceed for two special cases, each of which will be treated in a coordinate system most suitable for the problem.

---

## 3.1 Fields with Straight Reference Orbit

The first major case of systems is those that have a straight reference orbit. In this case, there is no need to distinguish between particle optical coordinates and Cartesian coordinates, and in particular there is no need to transform Laplace's equation to a new set of coordinates. Many elements with a straight reference orbit possess a certain **rotational symmetry** around the axis of the reference orbit, and it is most advantageous to describe the potential in **cylindrical coordinates** with a longitudinal $z$-axis that coincides with the reference orbit.

### 3.1.1 Expansion in Cylindrical Coordinates

We first begin by expanding the $r$ and $\phi$ components of the potential in Taylor and Fourier series, respectively. However, the dependence on the cylindrical "$z$" coordinate, which here coincides with the particle optical coordinate $s$, is not expanded. So we have

$$V = V(r, \phi, s) = \sum_{k=0}^{\infty} \sum_{l=0}^{\infty} M_{k,l}(s) \cos(l\phi + \theta_{k,l}) r^k. \tag{3.3}$$

In cylindrical coordinates, the Laplacian has the form

$$\Delta V(r, \phi, s) = \frac{1}{r} \frac{\partial}{\partial r} \left( r \frac{\partial V}{\partial r} \right) + \frac{1}{r^2} \frac{\partial^2 V}{\partial \phi^2} + \frac{\partial^2 V}{\partial s^2},$$

thus Laplace's equation is

$$\Delta V = \frac{1}{r} \frac{\partial}{\partial r} \left( r \frac{\partial V}{\partial r} \right) + \frac{1}{r^2} \frac{\partial^2 V}{\partial \phi^2} + \frac{\partial^2 V}{\partial s^2} = 0.$$

We insert the Fourier-Taylor expansion of the potential (3.3) into each term of the Laplacian.

$$r \frac{\partial V}{\partial r} = r \sum_{k=1}^{\infty} \sum_{l=0}^{\infty} M_{k,l}(s) \cos(l\phi + \theta_{k,l}) k r^{k-1}$$

$$= \sum_{k=1}^{\infty} \sum_{l=0}^{\infty} M_{k,l}(s) \cos(l\phi + \theta_{k,l}) k r^k,$$

then the first term is

$$\frac{1}{r}\frac{\partial}{\partial r}\left(r\frac{\partial V}{\partial r}\right) = \frac{1}{r}\sum_{k=1}^{\infty}\sum_{l=0}^{\infty} M_{k,l}(s)\cos\left(l\phi + \theta_{k,l}\right)k^2 r^{k-1}$$

$$= \sum_{k=0}^{\infty}\sum_{l=0}^{\infty} M_{k,l}(s)\cos\left(l\phi + \theta_{k,l}\right)k^2 r^{k-2},$$

where we let the sum start at $k = 0$ in the last step, since there is no contribution anyway because of the factor $k^2$. The second term is

$$\frac{1}{r^2}\frac{\partial^2 V}{\partial\phi^2} = -\frac{1}{r^2}\sum_{k=0}^{\infty}\sum_{l=0}^{\infty} M_{k,l}(s)\cos\left(l\phi + \theta_{k,l}\right)l^2 r^k$$

$$= -\sum_{k=0}^{\infty}\sum_{l=0}^{\infty} M_{k,l}(s)\cos\left(l\phi + \theta_{k,l}\right)l^2 r^{k-2}.$$

The third term is

$$\frac{\partial^2 V}{\partial s^2} = \sum_{k=0}^{\infty}\sum_{l=0}^{\infty} M_{k,l}''(s)\cos\left(l\phi + \theta_{k,l}\right)r^k$$

$$= \sum_{k=2}^{\infty}\sum_{l=0}^{\infty} M_{k-2,l}''(s)\cos\left(l\phi + \theta_{k-2,l}\right)r^{k-2}$$

$$= \sum_{k=0}^{\infty}\sum_{l=0}^{\infty} M_{k-2,l}''(s)\cos\left(l\phi + \theta_{k-2,l}\right)r^{k-2},$$

where we let the sum start at $k = 2$ in the second step, and, further, in the last step, we used the convention that the coefficient $M_{k,l}(s)$ vanish for negative indices. Recognizing that all the terms have the common summations and the factor $r^{k-2}$, we obtain the Laplacian for the Fourier-Taylor expansion of the potential (3.3) as

$$\Delta V = \sum_{k,l=0}^{\infty}\left[M_{k,l}(s)\cos\left(l\phi+\theta_{k,l}\right)\left(k^2 - l^2\right) + M_{k-2,l}''(s)\cos\left(l\phi+\theta_{k-2,l}\right)\right]r^{k-2}.$$

To satisfy Laplace's equation, we obtain a set of conditions for $k, l \geq 0$,

$$M_{k,l}(s)\cos\left(l\phi + \theta_{k,l}\right)\left(k^2 - l^2\right) + M_{k-2,l}''(s)\cos\left(l\phi + \theta_{k-2,l}\right) = 0,$$

where the second term vanishes for $k = 0, 1$ because of the negative indices for $M_{k,l}$.

We begin the analysis of Laplace's equation by studying the case $k = 0$, where only the first term matters. Apparently $M_{0,0}$ and $\theta_{0,0}$ can be chosen freely because $k^2 - l^2 = 0$ for $k = l = 0$. For $l \geq 1$, we infer $M_{0,l} = 0$.

By induction over $k$, we now show that $M_{k,l} = 0$ for all cases where $k < l$. Apparently the statement is true for $k = 0$ as we just showed. Now let us assume that the statement is true up to $k - 1$. If $k < l$, also $k - 2 < l$, and thus $M''_{k-2,l}(s) = 0$. Since $k^2 - l^2 \neq 0$ and $\cos(l\phi + \theta_{k,l}) \neq 0$ for some $\phi$ because $l \neq 0$, this requires

$$M_{k,l}(s) = 0 \quad \text{for} \quad k < l.$$

Thus the infinite matrix $M_{k,l}$ is strictly lower triangular.

We now study the situation for different values of $l$. We first notice that for all $l$, the choices of

$$M_{l,l}(s) \text{ and } \theta_{l,l} \quad \text{are free}$$

because $M''_{l-2,l}(s) = 0$ by the previous observation, and $k^2 - l^2 = 0$ for $k = l$. Next we observe that the value $M_{l+1,l}(s)$ must vanish, because $k^2 - l^2 \neq 0$, but $M''_{l-1,l}(s) = 0$ because of the lower triangularity. Recursively we even obtain that

$$M_{l+1,l}(s), M_{l+3,l}(s), \ldots \quad \text{vanish}.$$

On the other hand, for $k = l + 2$, we obtain that $\theta_{l+2,l} = \theta_{l,l}$, and $M_{l+2,l}(s)$ is uniquely specified by $M_{l,l}(s)$. Applying recursion, we see that in general

$$\theta_{l,l} = \theta_{l+2,l} = \theta_{l+4,l} = \ldots,$$

$$M_{l+2n,l}(s) = \frac{M_{l,l}^{(2n)}(s)}{\prod_{\nu=1}^{n}\left(l^2 - (l + 2\nu)^2\right)}. \tag{3.4}$$

Let us now proceed with the physical interpretation of the result. The number $l$ is called the **multipole order**, as it describes how many oscillations the field will experience in one $2\pi$ sweep of $\phi$. The free term $M_{l,l}(s)$ is called the **multipole strength**, and the term $\theta_{l,l}$ is called the **multipole phase**. Apparently, **frequency $l$ and radial power $k$ are coupled**: The lowest order in $r$ that appears is $l$, and if the multipole strength is $s$-dependent, also the powers $l + 2, l + 4, \ldots$ will appear.

For a multipole of order $l$, the potential has a total of $2l$ maxima and minima, and is so often called a **$2l$ pole**. Often Latin names are used for the $2l$ poles, and they are listed in Table 3.1.

In many cases it is very important to study the Cartesian (and hence also particle optical) form of the fields of the elements. We start with the trivial case with $k = 1$. In this case, the potential is $V = M_{1,1}\cos(\phi + \theta_{1,1})r$. For $\theta_{1,1} = 0$, we obtain $V = M_{1,1} \cdot x$, which corresponds to a uniform field in $x$-direction. For $\theta_{1,1} = \pi/2$, another important sub-case, we obtain $V = -M_{1,1} \cdot y$, which corresponds to a uniform field in $y$-direction. In both of these cases, the reference orbit is indeed a straight line only in the limit of weak fields.

**TABLE 3.1:** A list of multipoles

| $l$ | Leading Term in $V$ | Name |
|---|---|---|
| 0 | $M_{0,0}(s)\cos{(\theta_{0,0})}$ | |
| 1 | $M_{1,1}(s)\cos{(\phi+\theta_{1,1})}\,r$ | Dipole |
| 2 | $M_{2,2}(s)\cos{(2\phi+\theta_{2,2})}\,r^2$ | Quadrupole |
| 3 | $M_{3,3}(s)\cos{(3\phi+\theta_{3,3})}\,r^3$ | Sextupole/Hexapole |
| 4 | $M_{4,4}(s)\cos{(4\phi+\theta_{4,4})}\,r^4$ | Octupole |
| 5 | $M_{5,5}(s)\cos{(5\phi+\theta_{5,5})}\,r^5$ | Decapole |
| 6 | $M_{6,6}(s)\cos{(6\phi+\theta_{6,6})}\,r^6$ | Duodecapole |

### 3.1.2 Quadrupole Fields

The case $k = 2$ leads to quadrupoles, and the potential has the form $V = M_{2,2}\cos{(2\phi+\theta_{2,2})}\,r^2$. Particularly important in practice will be the sub-cases $\theta_{2,2} = 0$ and $\theta_{2,2} = \pi/2$. In the first case, we have

$$V = M_{2,2}\cos{(2\phi)}\,r^2 = M_{2,2}\left(\cos^2\phi - \sin^2\phi\right)r^2 = M_{2,2}\left(x^2 - y^2\right),$$

and in the second case we have

$$V = M_{2,2}\cos\left(2\phi + \frac{\pi}{2}\right)r^2 = -M_{2,2}\sin{(2\phi)}\,r^2$$
$$= -M_{2,2}\left(2\sin\phi\cos\phi\right)r^2 = -M_{2,2}\cdot 2xy.$$

All other angles $\theta_{2,2}$ lead to formulas that are more complicated; they can be obtained from the ones here by subjecting the $x$, $y$ coordinates to a suitable rotation. This again leads to terms of purely second order.

Because the potential is quadratic, the resulting fields $\vec{E}$ or $\vec{B}$ are **linear**. Indeed, the quadrupole is the **only $s$-independent element that leads to linear motion** similar to that in glass optics, and thus has great importance.

In the electric case, one usually chooses $\theta_{2,2} = 0$, thus having $V = M_{2,2}(x^2 - y^2)$ and resulting in the fields

$$E_x = -2M_{2,2}\cdot x, \quad E_y = 2M_{2,2}\cdot y.$$

The fields extend throughout the length of the device, and thus provide **strong focusing**. Different from the case of glass optics, it turns out that the motion **cannot be rotationally symmetric** anymore. If there is focusing in the $x$-direction, there is defocusing in the $y$-direction, and vice versa. This effect, completely due to Maxwell's equations, turns out to be perhaps the biggest nuisance in beam physics; i.e., if one uses piecewise $s$-independent particle optical elements, the horizontal and vertical **planes are always different from each other**.

To make an electrostatic device that produces a quadrupole field, it is best to machine the electrodes along the equipotential surfaces, and utilize the fact that if a sufficient amount of boundary information is specified, the field is

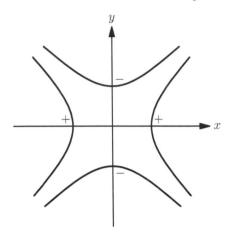

**FIGURE 3.1**:   Ideal electrodes of an electrostatic quadrupole.

uniquely determined, and hence must be as specified by the formula used to determine the equipotential surfaces in the first place. So in practice, the electrodes of an electric quadrupole often look as shown in Fig. 3.1.

In the magnetic case, one chooses $\theta_{2,2} = \pi/2$, thus having $V = -M_{2,2} \cdot 2xy$ and resulting in

$$B_x = 2M_{2,2} \cdot y, \quad B_y = 2M_{2,2} \cdot x,$$

and looking at the Lorentz forces that a particle moving mostly in $s$-direction experiences, we again see that if there is focusing in $x$-direction, there is defocusing in $y$-direction and vice versa.

### 3.1.3   Sextupole and Higher Multipole Fields

To study higher orders in $k$, let us consider the case $k = 3$. For $\theta_{3,3} = 0$, we obtain

$$V = M_{3,3} \cos\left(3\phi\right) r^3 = M_{3,3} \left(\cos^3 \phi - 3 \cos \phi \sin^2 \phi\right) r^3 = M_{3,3} \left(x^3 - 3xy^2\right).$$

In this case, the resulting forces are quadratic, and are thus not suitable for affecting the linear motion; but we shall see later that they are indeed very convenient for the correction of nonlinear motion, and they even have the nice feature of having **no influence** on the linear part of the motion. Another important case for $\theta_{3,3}$ is $\theta_{3,3} = \pi/2$, in which case one can perform a similar argument and again obtain cubic dependencies on the position.

For all the **higher values of** $l$, corresponding to octupoles, decapoles, duodecapoles, etc., the procedure is very similar. We begin with the addition theorem for $\cos(l\phi)$ or $\sin(l\phi)$, and by induction we see that each consists of terms that have a product of precisely $l$ cosines and sines. Since each of these terms is multiplied with $r^l$, each cosine multiplied with one $r$ translates into

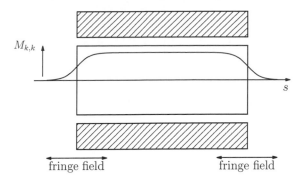

**FIGURE 3.2**: The $s$-dependence of the multipole strength.

an $x$, and each sine multiplied with one $r$ translates into a $y$. The end result is always a polynomial in $x$ and $y$ of exact order $l$.

Because of their nonlinear field dependence, these elements will prove to have no effect on the motion up to order $l-1$, and thus allow us to selectively influence the higher orders of the motion without affecting the lower orders. And if it is the crux of particle optical motion that the horizontal and vertical linear motion cannot be affected simultaneously, it is its blessing that the **nonlinear effects can be corrected order-by-order**.

### 3.1.4  $s$–Dependent Fields

In the case where there is no $s$-dependence, the potential terms that we have derived are the only ones; under the **presence of $s$-dependence**, as shown in eq. (3.4), to the given angular dependence there are higher order terms in $r$, the strengths of which are given by the $s$-derivatives of the multipole strength $M_{l,l}$. The computation of their Cartesian form is very easy once the Cartesian form of the leading term is known, because each additional term just differs by the previous one just by the factor of $r^2 = (x^2 + y^2)$.

In practice, of course, $s$-dependence is unavoidable: the field of any particle optical element has to begin and end somewhere, and it usually does this by rising and falling gently with $s$, entailing $s$-dependence as seen in Fig. 3.2. This actually entails another crux of particle optics: even the quadrupoles, **the "linear" elements, have nonlinear effects** at their edges, requiring higher order correction. The corrective elements in turn have higher order edge effects, possibly requiring even higher order correction, etc. In practical terms, charged particle optical systems are designed in such a way that the effect of the higher order field is smaller than that of the lower order field, which ensures that the iterative process converges.

Without $s$-dependence, the case $l = 0$, corresponding to full rotational symmetry, is not very interesting since there will be no field left. This becomes

**FIGURE 3.3**: Scalar potential (solid), longitudinal (dashed) and radial (dotted) field distribution along the $s$-axis of a single (left) and dual (right) step in potential of a rotationally symmetric lens.

clear if we rewrite eq. (3.4) for the case $l = 0$, which is

$$M_{2n,0}(s) = \frac{M_{0,0}^{(2n)}(s)}{\prod_{\nu=1}^{n} (-1)(2\nu)^2} = \frac{M_{0,0}^{(2n)}(s)}{(n!)^2 (-4)^n}. \tag{3.5}$$

When $M_{0,0}^{(2n)}(s) = 0$, we obtain that $V = M_{0,0}$, which is independent of $s$ and $r$. If we consider $s$-dependence, it actually offers a remarkably useful effect. While there is no $r$-dependence in the leading term, the contributions through the derivatives of $M_{0,0}(s)$ entail terms with an $r$-dependence of the form $r^2$, $r^4, \ldots$. Using eq. (3.5), we obtain the Taylor expansion of the potential, which is

$$V(r,s) = \sum_{n=0}^{\infty} \frac{1}{(n!)^2 (-4)^n} M_{0,0}^{(2n)}(s) r^{2n}$$

$$= M_{0,0}(s) - \frac{1}{4} M_{0,0}^{(2)}(s) r^2 + \frac{1}{(2!)^2 \cdot 4^2} M_{0,0}^{(4)}(s) r^4 - \cdots . \tag{3.6}$$

Of these, the $r^2$ term will indeed produce **linear, rotationally symmetric radial fields** and lead to effects similar to those in the glass lens. In practice these fields are not very strong (proportional to $M_{0,0}^{(2)}(s)$, compared to $M_{0,0}^{(1)}(s)$ for the longitudinal field) and restricted to regions where the potential changes and are used in so-called **weak focusing**. In practice, potential changes often occur as transitions between regions of constant potential. This can be done as a single step as shown on the left of Fig. 3.3, or as a dual step as shown on the right of Fig. 3.3, where the latter has the advantage that no net change in potential occurs.

The resulting fields are given by

$$E_s(r,s) = -M_{0,0}^{(1)}(s) + \frac{1}{4} M_{0,0}^{(3)}(s) r^2 - \frac{1}{(2!)^2 \cdot 4^2} M_{0,0}^{(5)}(s) r^4 + \cdots ,$$

$$E_r(r,s) = \frac{1}{2} M_{0,0}^{(2)}(s) r - \frac{1}{4^2} M_{0,0}^{(4)}(s) r^3 + \cdots .$$

The magnetic field components $B_s(r,s)$, $B_r(r,s)$ take the same form. Furthermore, there are usually quite large nonlinearities, and altogether these

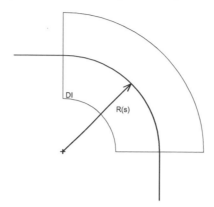

**FIGURE 3.4:** Reference orbit of a bending magnet.

devices are used mostly for low energy, small emittance beams like those found in electron microscopes.

## 3.2 Fields with Planar Reference Orbit

In the case of the straight reference orbit, we saw that Maxwell's equations entail a very clean connection between rotational symmetry and radial potential. As one may expect, in the case of a non-straight reference orbit, this is no longer the case. In this situation, Maxwell's equations have a rather different but not less interesting consequence as long as we restrict ourselves to the case in which the reference orbit stays in one plane.

### 3.2.1 The Laplacian in Curvilinear Coordinates

As it turns out, in this case the arguments to express the Laplacian in the new coordinates are similar to that in cylindrical coordinates. Let us assume that the motion of the reference particle is in a plane, and that all orbits that are on this plane stay in it. Let $R(s)$ be the momentary radius of curvature as shown in Fig. 3.4.

Then we have a situation very similar to cylindrical coordinates $r$, $\phi$, $z$ centered around the momentary origin of $R(s)$. In fact, setting $h(s) = 1/R(s)$, the particle optical coordinates $x$, $y$, $s$ correspond to the cylindrical ones in the following way:

$$z \leftrightarrow y,$$
$$r \leftrightarrow (1 + hx) \cdot R(s),$$
$$\phi \leftrightarrow \frac{s}{R(s)}.$$

As we recall, in cylindrical coordinates the Laplacian had the form

$$\Delta V(r, \phi, z) = \frac{1}{r} \frac{\partial}{\partial r} \left( r \frac{\partial V}{\partial r} \right) + \frac{1}{r} \frac{\partial}{\partial \phi} \left( \frac{1}{r} \frac{\partial V}{\partial \phi} \right) + \frac{\partial^2 V}{\partial z^2}.$$

So we may expect that in particle optical coordinates, we in fact have

$$\Delta V(x, y, s) = \frac{1}{1 + hx} \frac{\partial}{\partial x} \left( (1 + hx) \frac{\partial V}{\partial x} \right) + \frac{1}{1 + hx} \frac{\partial}{\partial s} \left( \frac{1}{1 + hx} \frac{\partial V}{\partial s} \right) + \frac{\partial^2 V}{\partial y^2}.$$

A careful analysis based on the chain rule and determining the proper Jacobian reveals that this is indeed the case. The calculations are rather mechanical and not particularly interesting, but very involved [48], and we skip them for the purposes of this discussion.

### 3.2.2    The Potential in Curvilinear Coordinates

For the potential, we again make an expansion in transversal coordinates, and leave the longitudinal coordinates unexpanded. Since we are working now with $x$ and $y$, both expansions are Taylor, and we have

$$V = V(x, y, s) = \sum_{k=0}^{\infty} \sum_{l=0}^{\infty} a_{k,l}(s) \frac{x^k y^l}{k! l!}. \tag{3.7}$$

This expansion now has to be inserted into the Laplacian in particle optical coordinates. Besides the mere differentiation, we also have to Taylor expand $1/(1 + hx)$ :

$$\frac{1}{1 + hx} = 1 - (hx) + (hx)^2 - (hx)^3 + \cdots.$$

After gathering terms and heavy arithmetic, and again using the convention that coefficients with negative indices are assumed to vanish, we obtain the recursion relation

$$\begin{aligned} a_{k,l+2} = &- a''_{k,l} - kha''_{k-1,l} + kh'a'_{k-1,l} - a_{k+2,l} - (3k+1) ha_{k+1,l} \\ &- 3kha_{k-1,l+2} - k(3k-1)h^2 a_{k,l} - 3k(k-1)h^2 a_{k-2,l+2} \\ &- k(k-1)^2 h^3 a_{k-1,l} - k(k-1)(k-2)h^3 a_{k-3,l+2}. \end{aligned} \tag{3.8}$$

Although admittedly horrible and unpleasant, the formula apparently has the coefficient of highest total order $k + l + 2$ on the left hand side, and thus recursively allows the calculation of coefficients. Indeed, the terms $a_{k,0}(s)$, $a_{k,1}(s)$ can be chosen freely, and all others are uniquely determined through them.

To study the significance of the free terms, let us consider the electric and magnetic case separately. In **the electric field**, in order to ensure that orbits that were in the plane stay there, there must not be any field components in the $y$-direction in the plane corresponding to $y = 0$. Computing the gradient of the potential, we have

$$E_x(x, y = 0) = -\sum_k a_{k,0} \frac{x^{k-1}}{(k-1)!}, \quad E_y(x, y = 0) = -\sum_k a_{k,1} \frac{x^k}{k!} = 0,$$

and looking at $E_y$, we conclude that $a_{k,1} = 0$ for all $k$. So the terms $a_{k,0}$ alone specify the field. Looking at $E_x$, we see that these are just the coefficients that specify the field within the plane, and so **the midplane field determines the entire field**. Furthermore, looking at the details of the recursion relation (3.8), it becomes apparent that all second indices are either $l$ or $l + 2$. This entails that as long as $a_{k,1}$ terms do not appear, also $a_{k,3}, a_{k,5}, \ldots$ terms do not appear. Indeed, the resulting potential is fully symmetric around the plane, and the resulting field lines above and below the plane are mirror images.

In **the magnetic field**, the argument is rather similar. Considering the fields in the plane, we have

$$B_y(x, y = 0) = -\sum_k a_{k,1} \frac{x^k}{k!}, \quad B_x(x, y = 0) = -\sum_k a_{k,0} \frac{x^{k-1}}{(k-1)!} = 0.$$

In order for particles in the midplane to stay there, we must have that $B_x$ vanishes in the midplane, which entails $a_{k,0} = 0$. So in the magnetic case, the coefficients $a_{k,1}$ specify everything. These coefficients, however, again describe the shape of the field in plane, and so again **the midplane field determines the entire field**. In the magnetic case, the potential is fully antisymmetric around the plane, and again the resulting field lines are mirror images of each other.

To summarize the findings,

|  |  |  |
|---|---|---|
| Electric field: | $a_{k,1} = 0$ for all $k$, | $a_{k,0}$ specify everything. |
| Magnetic field: | $a_{k,0} = 0$ for all $k$, | $a_{k,1}$ specify everything. |

To conclude, we note that it is possible to extend the entire discussion also to cases where the motion is not confined to a simple midplane. The derivations connected to this most general case become exceedingly complicated [49, 48] and go beyond what is appropriate for this book.

## 3.3 The Equations of Motion in Curvilinear Coordinates

There are a variety of methods to derive the equations of motion in **curvilinear coordinates** with the arc length $s$ as the independent variable. It is conveniently done in the Lagrangian picture, in which one first expresses Cartesian variables by curvilinear coordinates and rewrites the Lagrangian. Then one proceeds to a Hamiltonian through a Legendre transformation in the common way. In the Hamiltonian picture, it is then possible to perform a change of independent variable from $t$ to $s$ while maintaining the Hamiltonian structure [49].

While very illuminating, the **Lagrangian-Hamiltonian** mechanism is **too involved for our purposes**, and we thus follow a more straightforward, classical way that leads to the same canonical equations of motion. For simplicity, we also restrict ourselves in that the reference orbit is allowed to bend in only one plane.

### 3.3.1 The Coordinate System and the Independent Variable

As a function of the arc length $s$, we first define the momentary **curvature** of the reference orbit as $h(s)$. If the curvature is nonzero, the radius of curvature is then given by $R(s) = 1/h(s)$. We begin by studying the bend angle that the reference orbit experiences as we move from position $s$ to position $\bar{s}$. We have

$$\alpha = \int d\bar{\alpha} = \int_s^{\bar{s}} \frac{d\bar{s}}{R(\bar{s})} = \int_s^{\bar{s}} h(\bar{s})d\bar{s}. \tag{3.9}$$

As described in eq. (1.1), in Cartesian coordinates, the equations of motion have the **Lorentz force** form

$$\frac{d\vec{p}}{dt} = \vec{F} = q\left(\vec{E} + \vec{v} \times \vec{B}\right) = Ze\left(\vec{E} + \vec{v} \times \vec{B}\right),$$

where $\vec{E}$ and $\vec{B}$ are the electric and magnetic fields, $\vec{v}$ is the velocity, and $q = Ze$ is the charge of the particle. Since the left hand side of the equations of motion contain momentum, it is often useful to express the velocity in terms of the momentum. From eq. (1.5),

$$\vec{v} = \frac{d\vec{r}}{dt} = c \cdot \frac{\vec{p}}{\sqrt{\vec{p}^2 + m^2c^2}},$$

which allows to maintain only the momentum $\vec{p}$ in the equations of motion.

For the purpose of our derivation, we rewrite the equation as an integral equation:

$$\vec{p}(\bar{s}) = \vec{p}(s) + \int_{t(s)}^{t(\bar{s})} \vec{F}(t)dt = \vec{p}(s) + \int_s^{\bar{s}} \vec{F}(\bar{s})t'd\bar{s},$$

where we have used $t' = dt/d\bar{s}$, and it is worthwhile to remind ourselves that the force $\vec{F}(\bar{s})$ still depends on both $\vec{x}$ and $\vec{p}$.

As we have progressed from $s$ to $\bar{s}$, the orientation of our locally attached particle optical coordinate system has changed. It was rotated by the angle $\alpha$ of eq. (3.9). By using the rotation matrix

$$\hat{R}(\bar{s}) = \begin{pmatrix} \cos\alpha & 0 & \sin\alpha \\ 0 & 1 & 0 \\ -\sin\alpha & 0 & \cos\alpha \end{pmatrix} = \begin{pmatrix} \cos\int_s^{\bar{s}} h(\bar{s})d\bar{s} & 0 & \sin\int_s^{\bar{s}} h(\bar{s})d\bar{s} \\ 0 & 1 & 0 \\ -\sin\int_s^{\bar{s}} h(\bar{s})d\bar{s} & 0 & \cos\int_s^{\bar{s}} h(\bar{s})d\bar{s} \end{pmatrix},$$

we have the momentum in the new rotated local coordinates $\vec{p}_l(\bar{s}) = (p_x, p_y, p_s)$ as

$$\vec{p}_l(\bar{s}) = \hat{R}(\bar{s}) \cdot \vec{p}(\bar{s})$$
$$= \begin{pmatrix} \cos\int_s^{\bar{s}} h(\bar{s})d\bar{s} & 0 & \sin\int_s^{\bar{s}} h(\bar{s})d\bar{s} \\ 0 & 1 & 0 \\ -\sin\int_s^{\bar{s}} h(\bar{s})d\bar{s} & 0 & \cos\int_s^{\bar{s}} h(\bar{s})d\bar{s} \end{pmatrix} \cdot \left( \vec{p}(s) + \int_s^{\bar{s}} \vec{F}(\bar{s})t'd\bar{s} \right),$$

where we have expressed the last line in the integral form. In order to obtain the rate of change of the momentum $\vec{p}_l$, we differentiate with respect to $\bar{s}$. Noting that

$$\frac{d}{d\bar{s}} \sin \int_s^{\bar{s}} h(\bar{s})d\bar{s} = h(\bar{s}) \cos \int_s^{\bar{s}} h(\bar{s})d\bar{s},$$
$$\frac{d}{d\bar{s}} \cos \int_s^{\bar{s}} h(\bar{s})d\bar{s} = -h(\bar{s}) \sin \int_s^{\bar{s}} h(\bar{s})d\bar{s},$$

evaluate at $\bar{s} = s$ and obtain

$$\vec{p}_l\,'(s) = \left[ \hat{R}(\bar{s})\vec{F}(\bar{s})t' + \hat{R}'(\bar{s}) \left( \vec{p}(\bar{s}) + \int_s^{\bar{s}} \vec{F}(\bar{s})t'd\bar{s} \right) \right]_{\bar{s}=s}$$
$$= \vec{F}(s)t' + \begin{pmatrix} 0 & 0 & h(s) \\ 0 & 0 & 0 \\ -h(s) & 0 & 0 \end{pmatrix} \vec{p}(s). \tag{3.10}$$

Note that the first term depends on the actual forces and the factor $t'$ accounts for the fact that we went to the **arc length** $s$ as an **independent variable**. The second term is a pseudoforce due to the fact that we are located in a rotating frame. Indeed, for $h = 0$, we obtain the conventional result. We also note in passing that if we were to allow out-of-plane motion of the reference orbit, then the matrix $\hat{R}$ would depend on two curvatures. Unfortunately, in this case an additional complication arises from the fact that rotations around different axes do not generally commute [49].

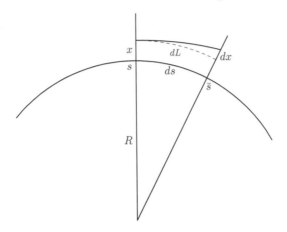

**FIGURE 3.5**:   The curvilinear coordinates in the plane of the reference orbit.

Next we make an observation regarding the rate of change at which distances are covered at different positions $x$. Looking at Fig. 3.5, we observe

$$\frac{dL}{ds} = \frac{R+x}{R} = 1 + hx.$$

Using this and the components of the momentum, we obtain

$$\frac{dx}{ds} = \frac{dL}{ds}\frac{dx}{dL} = (1+hx)\frac{dx}{dL} = (1+hx)\frac{p_x}{p_s},$$

$$\frac{dy}{ds} = \frac{dL}{ds}\frac{dy}{dL} = (1+hx)\frac{dy}{dL} = (1+hx)\frac{p_y}{p_s}. \tag{3.11}$$

For the time-of-flight, we consider the traveling distance divided by the velocity, and we obtain

$$\frac{dt}{ds} = \frac{1}{v}\sqrt{\left(\frac{dx}{ds}\right)^2 + \left(\frac{dy}{ds}\right)^2 + \left(\frac{dL}{ds}\right)^2} = \frac{1}{v}(1+hx)\sqrt{\frac{p_x^2 + p_y^2}{p_s^2} + 1}$$

$$= \frac{1}{v}(1+hx)\frac{p}{p_s}, \tag{3.12}$$

where $p = \sqrt{p_x^2 + p_y^2 + p_s^2}$ has been used.

Altogether, we have so far obtained the equations of motion in local coordinates with $s$ as the independent variable. From there to the particle optical variables, only a small step is left. We remind ourselves that the **particle optical coordinates** are $\{x, a, y, b, l, \delta\}$ as listed in Table 3.2, where $p_0$ and $t_0$ are total momentum and time-of-flight of the reference particle; refer to

**TABLE 3.2:** Optical coordinates

| Coordinate | Phase Space | |
| --- | --- | --- |
| $x$ | Horizontal | Position |
| $a = p_x/p_0$ | | Momentum Slope |
| $y$ | Vertical | Position |
| $b = p_y/p_0$ | | Momentum Slope |
| $l = \kappa \left(t - t_0\right)$ | Longitudinal | Time-of-Flight like Variable |
| $\delta = \left(K - K_0\right)/K_0$ | | Energy Deviation |

eqs. (2.1) and (2.2). Likewise, the subscript 0 will be used below to indicate the respective quantity of the reference particle.

In order to study relativistic effects, it is advantageous to introduce the relativistic measure $\eta$, the ratio of kinetic energy to rest mass energy

$$\eta = \frac{K_0 \left(1 + \delta\right) - ZeV}{mc^2} = \eta_0 \left(1 + \delta\right) - \frac{ZeV}{mc^2}, \tag{3.13}$$

where $m$ is the rest mass. The quantity $V$ is the change in energy that is incurred due to the passage through electric fields; it is given by

$$V = -\int \vec{E}(x, y, s, t) \cdot \vec{v} dt, \tag{3.14}$$

where $\vec{v}$ is the velocity, depending on position and time, of the orbit under consideration. In case the electric fields are time independent, the quantity $V$ is merely the common electrostatic potential, which depends on the position coordinates $(x, y, s)$. For time dependent fields, the quantity $V$ explicitly depends on the specific time dependent orbit taken, which is of importance for the study of dynamics in RF cavities as discussed in Chapter 10.

Since $\gamma mc^2$ represents the total energy, we have

$$\gamma = \frac{1}{\sqrt{1 - v^2/c^2}} = 1 + \eta. \tag{3.15}$$

Using these, we also have

$$\frac{v}{c} = \sqrt{1 - \frac{1}{\gamma^2}} = \sqrt{1 - \frac{1}{\left(1 + \eta\right)^2}} = \frac{\sqrt{2\eta + \eta^2}}{1 + \eta} = \frac{\sqrt{\eta \left(2 + \eta\right)}}{1 + \eta},$$

$$\frac{p}{mc} = \frac{\gamma mv}{mc} = \gamma \frac{v}{c} = \sqrt{\eta \left(2 + \eta\right)}, \quad \text{and} \quad \frac{p}{v} = m \left(1 + \eta\right). \tag{3.16}$$

As a first step, using eq. (3.12), we express the rate of change of the particle optical variable $l = \kappa(t - t_0)$ in terms of particle optical quantities:

$$l' = \frac{dl}{ds} = \kappa \left(t' - t_0'\right) = \kappa \left[\frac{1}{v} \left(1 + hx\right) \frac{p}{p_s} - \frac{1}{v_0}\right]$$

$$= \left[\left(1 + hx\right) \frac{v_0}{p_0} \frac{p}{v} \frac{p_0}{p_s} - 1\right] \frac{\kappa}{v_0} = \left[\left(1 + hx\right) \frac{1 + \eta}{1 + \eta_0} \frac{p_0}{p_s} - 1\right] \frac{\kappa}{v_0}, \tag{3.17}$$

where the relation $t_0' = 1/v_0$ is used because of $p_{s0} = p_0$. The term $p_0/p_s$ appearing above can be expressed by the optical coordinates $a$ and $b$ as

$$\frac{p_0}{p_s} = \frac{p_0}{\sqrt{p^2 - p_x^2 - p_y^2}} = \left(\frac{p^2}{p_0^2} - a^2 - b^2\right)^{-1/2}$$

$$= \left(\frac{\eta(2+\eta)}{\eta_0(2+\eta_0)} - a^2 - b^2\right)^{-1/2}. \tag{3.18}$$

Next, by applying the Lorentz force law to eq. (3.10), we obtain

$$\frac{d}{ds}\left(\frac{p_x}{p_0}, \frac{p_y}{p_0}, \frac{p_s}{p_0}\right) = \vec{F}(s)\,t'\frac{1}{p_0} + \begin{pmatrix} 0 & 0 & h \\ 0 & 0 & 0 \\ -h & 0 & 0 \end{pmatrix}\frac{\vec{p}}{p_0}$$

$$= Ze\left(\vec{E} + \vec{v} \times \vec{B}\right)\frac{t'}{p_0} + h\left(\frac{p_s}{p_0}, 0, -\frac{p_x}{p_0}\right). \tag{3.19}$$

We here introduce the **magnetic rigidity** $\chi_m$ and the **electric rigidity** $\chi_e$,

$$\chi_m = \frac{p}{Ze}, \qquad \chi_e = \frac{pv}{Ze}.$$

As we will see below, the magnetic and the electric rigidities describe directly to what extent the magnetic and the electric fields influence the geometric motion of the particles. The first term of eq. (3.19) can be expressed by using $\chi_{m0}$ and $\chi_{e0}$ as

$$Ze\left(\vec{E} + \vec{v} \times \vec{B}\right)\frac{t'}{p_0} = \frac{Ze}{p_0 v_0}\left(\vec{E} + \vec{v} \times \vec{B}\right)v_0 t' = \frac{\vec{E}}{\chi_{e0}}v_0 t' + \vec{v} \times \frac{\vec{B}}{\chi_{m0}}t'.$$

From eq. (3.12), the factor $v_0 t'$ in the electric term can be written as

$$v_0 t' = \frac{v_0}{v}(1 + hx)\frac{p}{p_s} = (1 + hx)\frac{v_0}{p_0}\frac{p}{v}\frac{p_0}{p_s} = (1 + hx)\frac{1 + \eta}{1 + \eta_0}\frac{p_0}{p_s}.$$

Similarly, the factor $\vec{v}t'$ in the magnetic term can be written as

$$\vec{v}t' = \frac{\vec{v}}{v}(1 + hx)\frac{p}{p_s} = \frac{\vec{p}}{p}(1 + hx)\frac{p}{p_s} = (1 + hx)\frac{\vec{p}}{p_0}\frac{p_0}{p_s},$$

where $\vec{v}/v = \vec{p}/p$ is used because of $\vec{v} \parallel \vec{p}$. Thus,

$$Ze\left(\vec{E} + \vec{v} \times \vec{B}\right)\frac{t'}{p_0} = (1 + hx)\frac{1 + \eta}{1 + \eta_0}\frac{\vec{E}}{\chi_{e0}}\frac{p_0}{p_s} + (1 + hx)\frac{\vec{p}}{p_0} \times \frac{\vec{B}}{\chi_{m0}}\frac{p_0}{p_s}.$$

Continuing from eq. (3.19), we obtain

$$\frac{d}{ds}\left(\frac{p_x}{p_0}, \frac{p_y}{p_0}, \frac{p_s}{p_0}\right)$$

$$= (1 + hx)\frac{1 + \eta}{1 + \eta_0}\frac{\vec{E}}{\chi_{e0}}\frac{p_0}{p_s} + (1 + hx)\frac{\vec{p}}{p_0} \times \frac{\vec{B}}{\chi_{m0}}\frac{p_0}{p_s} + h\left(\frac{p_s}{p_0}, 0, -\frac{p_x}{p_0}\right). \tag{3.20}$$

Finally, we consider the change of the last variable in the particle optical coordinates, $\delta$. Since by definition it describes the deviation of initial kinetic energy of the particle of interest, we have

$$\delta' = 0. \tag{3.21}$$

Note, however, that since $\delta$ describes the deviation from the kinetic energy of the reference particle before the system, in case there is net acceleration or deceleration along the orbits, it may be desirable to periodically absorb the accumulated amounts in the orbit dependent path integral for $V$ in eq. (3.14) into the variable $\delta$. In the case the motion was merely through a static electric field, this will entail that $\delta$ will depend on positional variables. In the case of full time dependence as in the motion in RF cavities discussed in Chapter 10, after the renormalization, $\delta$ will depend on all particle optical coordinates.

### 3.3.2 The Equations of Motion

By observing that $\vec{p}/p_0 = (a, b, p_s/p_0)$, from eqs. (3.11), (3.17), (3.20) and (3.21), we obtain the equations of motion in particle optical coordinates:

$$
\begin{aligned}
x' &= a\,(1 + hx)\,\frac{p_0}{p_s}, \\
a' &= (1 + hx)\left[\frac{1+\eta}{1+\eta_0}\frac{E_x}{\chi_{e0}}\frac{p_0}{p_s} + b\frac{B_s}{\chi_{m0}}\frac{p_0}{p_s} - \frac{B_y}{\chi_{m0}}\right] + h\frac{p_s}{p_0}, \\
y' &= b\,(1 + hx)\,\frac{p_0}{p_s}, \\
b' &= (1 + hx)\left[\frac{1+\eta}{1+\eta_0}\frac{E_y}{\chi_{e0}}\frac{p_0}{p_s} + \frac{B_x}{\chi_{m0}} - a\frac{B_s}{\chi_{m0}}\frac{p_0}{p_s}\right], \\
l' &= \left[(1 + hx)\frac{1+\eta}{1+\eta_0}\frac{p_0}{p_s} - 1\right]\frac{\kappa}{v_0}, \\
\delta' &= 0,
\end{aligned}
\tag{3.22}
$$

where we remind ourselves of the following abbreviations from eqs. (3.13) and (3.18),

$$
\eta = \eta_0\,(1 + \delta) - \frac{ZeV}{mc^2}, \qquad \frac{p_0}{p_s} = \left(\frac{\eta\,(2+\eta)}{\eta_0\,(2+\eta_0)} - a^2 - b^2\right)^{-1/2},
$$

and eq. (2.1),

$$
\kappa = -v_0\frac{\gamma_0}{1+\gamma_0}.
$$

Note that the factor $\kappa/v_0$ in the equation for $l'$ can be expressed in terms of $\eta_0$ instead of $\gamma_0$ using eq. (3.15):

$$
\frac{\kappa}{v_0} = -\frac{1+\eta_0}{2+\eta_0}.
$$

We observe that the horizontal $(x\text{-}a)$ motion is affected mostly by $E_x$ and $B_y$, and the vertical $(y\text{-}b)$ motion is affected mostly by $E_y$ and $B_x$, and it is a direct consequence of the Lorentz force law. When a longitudinal component of the magnetic field $B_s$ is present, it acts to mix the horizontal and the vertical motions through the $B_s$ dependent terms in $a'$ and $b'$, which is even a linear effect. This phenomenon is readily observed in the spiral motion of a charged particle moving through a solenoid if any transversal component of the momentum exists, which is described by $a$ and $b$ in our case.

A careful analysis of the equations of motion reveals that indeed if all the particle optical coordinates are small, so are their derivatives defined through the equations of motion; indeed, the system is weakly nonlinear.

# Chapter 4

## The Linearization of the Equations of Motion

In order to develop a matrix theory of particle optics similar to the Gaussian theory in glass optics, we have to linearize the equations of motion. This procedure is rather similar to other linearizations in physics; in particular, it is very similar to the study of so-called **small oscillations** in mechanics. Since the solutions of linear systems depend linearly on the initial conditions, indeed the resulting transfer maps will be linear as needed. It is worth noting that, although a $6 \times 6$ matrix is required to describe the linear motion, only blocks of $2 \times 2$ and $3 \times 3$ are needed for decoupled linear motion.

We begin the actual process of linearization with the linearization of the fields, which corresponds to quadratic potentials in eq. (3.7). We begin our discussion with the case in which the potentials on the reference orbit vanish, which describes the situation of electric and magnetic multipoles as well as in deflectors. The case of electric and magnetic lenses do require the presence of potentials on axis, and they will be discussed in detail below.

In the electric case, let us assume that there is no potential on axis, i.e., $a_{0,0} = 0$, and that in the midplane, we have

$$E_x = -E_{x0}\left(1 + n_e x\right).$$

Because of the recursion relation for fields, eq. (3.8), we obtain an out-of-plane expansion of

$$E_y = E_{x0}(h + n_e)y,$$

as well as an electrostatic potential

$$V(x, y) = E_{x0}x + \frac{1}{2}E_{x0}(n_e x^2 - (h + n_e)\, y^2),$$

which is chosen in such a way as to vanish on the reference orbit.

In the magnetic case, let the midplane field be given by

$$B_y = B_{y0}(1 + n_b x).$$

Due to the recursion relation, we must then have

$$B_x = B_{y0}n_b y.$$

Before we even discuss linearization, let us consider the zeroth order of the motion: if the system is supposed to be origin preserving, then we must have from the equation of motion for $a'$ in eqs. (3.22) that

$$\frac{E_{x0}}{\chi_{e0}} + \frac{B_{y0}}{\chi_{m0}} = h, \tag{4.1}$$

which in a natural and expected way couples the constant parts of the fields with the curvature of the reference orbit.

Now we begin our process of linearization of the equations of motion (3.22). It is easy to see that

$$x' = a, \quad y' = b.$$

We also obtain

$$\frac{\eta}{\eta_0} = 1 + \delta - \frac{Ze}{\eta_0 mc^2} E_{x0} x,$$

and after more complicated expansions

$$\frac{1+\eta}{1+\eta_0} = 1 + \frac{\eta_0}{1+\eta_0}\delta - \frac{Ze}{(1+\eta_0)mc^2}E_{x0}x,$$

$$\frac{2+\eta}{2+\eta_0} = 1 + \frac{\eta_0}{2+\eta_0}\delta - \frac{Ze}{(2+\eta_0)mc^2}E_{x0}x.$$

Similarly, by using $\sqrt{1+u} =_1 1 + u/2$, $1/(1+u) =_1 1 - u$ for small $u$, we obtain

$$\frac{p_s}{p_0} = \sqrt{\frac{\eta(2+\eta)}{\eta_0(2+\eta_0)} - a^2 - b^2}$$

$$=_1 1 + \frac{1}{2}\left(1 + \frac{\eta_0}{2+\eta_0}\right)\delta - \frac{1}{2}\left(\frac{1}{\eta_0} + \frac{1}{2+\eta_0}\right)\frac{Ze}{mc^2}E_{x0}x$$

$$=_1 1 + \frac{1+\eta_0}{2+\eta_0}\delta - \frac{1+\eta_0}{\eta_0(2+\eta_0)}\frac{Ze}{mc^2}E_{x0}x.$$

Note that the symbol "$=_1$" means we are keeping terms up to first order. After lengthy similar arguments, we also conclude

$$l' =_1 \left[hx - \frac{1}{(1+\eta_0)(2+\eta_0)}\delta + \frac{1}{\eta_0(1+\eta_0)(2+\eta_0)}\frac{Ze}{mc^2}E_{x0}x\right]\frac{\kappa}{v_0}$$

$$=_1 -\left[h\frac{1+\eta_0}{2+\eta_0} + \frac{1}{\eta_0(2+\eta_0)^2}\frac{Ze}{mc^2}E_{x0}\right]x + \frac{1}{(2+\eta_0)^2}\delta,$$

as well as

$$a' =_1 -\left\{h^2 + \frac{E_{x0}}{\chi_{e0}}n_e + \frac{B_{y0}}{\chi_{m0}}n_b + \left[h + \frac{E_{x0}}{\chi_{e0}}\frac{1}{(1+\eta_0)^2}\right]\frac{1+\eta_0}{\eta_0(2+\eta_0)}\frac{Ze}{mc^2}E_{x0}\right\}x$$

$$+ \left[h + \frac{E_{x0}}{\chi_{e0}}\frac{1}{(1+\eta_0)^2}\right]\frac{1+\eta_0}{2+\eta_0}\delta,$$

where the relation (4.1) is used, and

$$b' =_1 \left[ \frac{E_{x0}}{\chi_{e0}}(h + n_e) + \frac{B_{y0}}{\chi_{m0}}n_b \right] y.$$

To summarize, we have the linearized equations of motion as a set:

$$x' =_1 a,$$

$$a' =_1 - \left\{ h^2 + \frac{E_{x0}}{\chi_{e0}}n_e + \frac{B_{y0}}{\chi_{m0}}n_b + \left[ h + \frac{E_{x0}}{\chi_{e0}}\frac{1}{(1+\eta_0)^2} \right] \frac{1+\eta_0}{\eta_0(2+\eta_0)}\frac{Ze}{mc^2}E_{x0} \right\} x$$

$$+ \left[ h + \frac{E_{x0}}{\chi_{e0}}\frac{1}{(1+\eta_0)^2} \right] \frac{1+\eta_0}{2+\eta_0}\delta,$$

$$y' =_1 b,$$

$$b' =_1 \left[ \frac{E_{x0}}{\chi_{e0}}(h + n_e) + \frac{B_{y0}}{\chi_{m0}}n_b \right] y,$$

$$l' =_1 - \left[ h\frac{1+\eta_0}{2+\eta_0} + \frac{1}{\eta_0(2+\eta_0)^2}\frac{Ze}{mc^2}E_{x0} \right] x + \frac{1}{(2+\eta_0)^2}\delta,$$

$$\delta' = 0. \tag{4.2}$$

Now that the equations of motion have been linearized, they have to be studied for a variety of different cases. We begin with the simplest case.

## 4.1   The Drift

In the case of the drift, all fields are 0, and $h = 0$, so the linearized equations of motion have the form

$$x' = a, \qquad a' = 0,$$
$$y' = b, \qquad b' = 0,$$
$$l' = \frac{1}{(2+\eta_0)^2}\delta, \qquad \delta' = 0,$$

where of course only the last equation is of any real interest. These equations are trivial to integrate, and we obtain

$$x_f = x_i + a_i L, \qquad a_f = a_i,$$
$$y_f = y_i + b_i L, \qquad b_f = b_i,$$
$$l_f = \frac{L}{(2+\eta_0)^2}\delta_i + l_i, \qquad \delta_f = \delta_i,$$

where $L$ is the drift length, and they can be written in matrix form as

$$\begin{pmatrix} x_f \\ a_f \\ y_f \\ b_f \\ l_f \\ \delta_f \end{pmatrix} = \begin{pmatrix} 1 & L & 0 & 0 & 0 & 0 \\ 0 & 1 & 0 & 0 & 0 & 0 \\ 0 & 0 & 1 & L & 0 & 0 \\ 0 & 0 & 0 & 1 & 0 & 0 \\ 0 & 0 & 0 & 0 & 1 & D \\ 0 & 0 & 0 & 0 & 0 & 1 \end{pmatrix} \begin{pmatrix} x_i \\ a_i \\ y_i \\ b_i \\ l_i \\ \delta_i \end{pmatrix},$$

where

$$D = \frac{L}{(2 + \eta_0)^2}.$$

First, we observe that, as in the case of glass optics, the determinant is unity. We also observe that the matrix can be grouped to three blocks, corresponding to the $x$-$a$ (horizontal), the $y$-$b$ (vertical) and the $l$-$\delta$ (longitudinal) motions. So, in the linear approximation, those three submotions in the drift are decoupled, without having any mixing. This allows us to study the entire motion in each direction conveniently independently. Later, we often study the motion of a system in decoupled submotions.

It is worthwhile to note that, when we take account of nonlinearity, even the drift motion is no longer simply linear, which may sound striking. This can be seen in the equations of motion (3.22), and the nonlinear effect comes from the factor $p_0/p_s$, which contains the second order contributions of $a$ and $b$. This effect is called the **kinematic correction**, and it also is responsible for nonlinear mixing of submotions in different directions even for the drift.

## 4.2   The Quadrupole without Fringe Fields

More interesting is the case of the quadrupole. Since the reference orbit goes straight, we have $h = 0$.

### 4.2.1   The Electric Quadrupole

For the electric quadrupole, from Section 3.1.2, we have

$$V = M_{2,2} \cos(2\phi) r^2 = M_{2,2} \left( x^2 - y^2 \right),$$

and

$$E_x = -2M_{2,2} \cdot x, \quad E_y = 2M_{2,2} \cdot y,$$

while $\vec{B} = 0$. If $M_{2,2} > 0$ for the positive charge beam, the field acts to focus the beam in the horizontal ($x$) direction, and defocus in the vertical ($y$) direction. The field description above corresponds to the case in the general form that the constant part $E_{x0}$ of $E_x$ is 0 and the factor of the linear term in $x$ is $E_{x0}n_e = -\partial E_x/\partial x = 2M_{2,2}$. The equations of motion have the form

$$x' = a, \qquad a' = -\frac{2M_{2,2}}{\chi_{e0}}x = -\omega^2 x,$$

$$y' = b, \qquad b' = \frac{2M_{2,2}}{\chi_{e0}}y = \omega^2 y,$$

$$l' = \frac{1}{(2+\eta_0)^2}\delta, \qquad \delta' = 0,$$

where

$$\omega = \sqrt{\frac{2M_{2,2}}{\chi_{e0}}}.$$

Apparently we have sine-cosine solutions in the horizontal plane, and sinh-cosh solutions in the vertical plane. For the quadrupole with the length $L$, we have

$$x_f = x_i \cos\omega L + a_i \frac{\sin\omega L}{\omega}, \qquad a_f = -\omega x_i \sin\omega L + a_i \cos\omega L,$$

$$y_f = y_i \cosh\omega L + b_i \frac{\sinh\omega L}{\omega}, \qquad b_f = \omega y_i \sinh\omega L + b_i \cosh\omega L,$$

$$l_f = \frac{L}{(2+\eta_0)^2}\delta_i + l_i, \qquad \delta_f = \delta_i.$$

This can be written in matrix form as

$$
\begin{pmatrix} x_f \\ a_f \\ y_f \\ b_f \\ l_f \\ \delta_f \end{pmatrix}
=
\begin{pmatrix}
\cos(\omega L) & \sin(\omega L)/\omega & 0 & 0 & 0 & 0 \\
-\omega\sin(\omega L) & \cos(\omega L) & 0 & 0 & 0 & 0 \\
0 & 0 & \cosh(\omega L) & \sinh(\omega L)/\omega & 0 & 0 \\
0 & 0 & \omega\sinh(\omega L) & \cosh(\omega L) & 0 & 0 \\
0 & 0 & 0 & 0 & 1 & D \\
0 & 0 & 0 & 0 & 0 & 1
\end{pmatrix}
\begin{pmatrix} x_i \\ a_i \\ y_i \\ b_i \\ l_i \\ \delta_i \end{pmatrix},
$$

where

$$\omega = \sqrt{\frac{2M_{2,2}}{\chi_{e0}}} \quad \text{and} \quad D = \frac{L}{(2+\eta_0)^2}.$$

Similar to the case of the drift, the matrix can be grouped to three blocks, namely the horizontal, the vertical and the longitudinal motions. Again, this holds while we limit ourselves to the linearized motion. We observe that, as in the case of glass optics, the determinant is unity. Furthermore, note that if $M_{2,2} < 0$, $\omega$ is imaginary. In this case, the $x$- and $y$-planes exchange their roles, the quadrupole becomes focusing in the vertical ($y$) direction and defocusing in the horizontal ($x$) direction. To see the focusing action in the similar style to the matrix form of the thin glass focusing lens, eq. (2.4), we consider a thin approximation of the quadrupole. While maintaining the integrated field strength, that is represented by $2M_{2,2} \cdot L$, we make $L \to 0$. Then, we have

$$\begin{pmatrix} \cos(\omega L) & \sin(\omega L)/\omega \\ -\omega \sin(\omega L) & \cos(\omega L) \end{pmatrix} \to \begin{pmatrix} 1 & 0 \\ -(2M_{2,2}/\chi_{e0}) \cdot L & 1 \end{pmatrix}.$$

This corresponds to a thin focusing lens with the focal length $f$ as $1/f = (2M_{2,2}/\chi_{e0}) \cdot L = \omega^2 \cdot L$.

In various cases that will be studied in the following sections, we will observe a harmonic oscillator motion similar to the $x$ plane motion in this section. In such a case, if the system is short it behaves like a thin focusing lens, and the focusing power can be obtained as $1/f = \omega^2 \cdot L$ using the angular frequency $\omega$ of the system, in the same way discussed here.

It is also worthwhile to briefly mention the case of fringe fields. In this case, $M_{2,2}$ changes as a function of $s$. The resulting ordinary differential equation (ODE) is still linear, which entails that the result can be written in matrix form, but in most cases is impossible to solve it analytically.

### 4.2.2   The Magnetic Quadrupole

In the case of the magnetic quadrupole, we have

$$V_b = -2M_{2,2}x \cdot y, \quad B_x = 2M_{2,2}y, \quad B_y = 2M_{2,2}x,$$

while $\vec{E} = 0$. This corresponds to the case in the general form that the constant part $B_{y0}$ of $B_y$ is 0 and the factor of the linear term in $x$ is $B_{y0}n_b = \partial B_y/\partial x = 2M_{2,2}$. This results in the linear equations

$$x' = a, \qquad a' = -\frac{2M_{2,2}}{\chi_{m0}}x = -\omega^2 x,$$

$$y' = b, \qquad b' = \frac{2M_{2,2}}{\chi_{m0}}y = \omega^2 y,$$

$$l' = \frac{1}{(2+\eta_0)^2}\delta, \qquad \delta' = 0.$$

Similar to the electric case, we have introduced

$$\omega = \sqrt{\frac{2M_{2,2}}{\chi_{m0}}},$$

and the resulting transfer matrix is the same as in the case of the electric quadrupole.

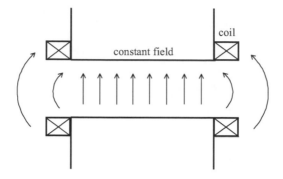

**FIGURE 4.1:** A homogeneous magnetic dipole.

## 4.3 Deflectors

### 4.3.1 The Homogeneous Magnetic Dipole

The next particle optical element we want to study is the magnetic dipole, consisting of a homogeneous, hence constant, magnetic field in the $y$-direction. We consider that the dipole element acts to bend the reference orbit by the bending angle $\phi$ with the bending radius $R_0$, so the curvature is $h = 1/R_0$. We also note that from eq. (4.1), $h = B_{y0}/\chi_{m0}$. In terms of the quantities describing the linearized fields, we have

$$B_{y0} = \text{constant}, \quad n_b = 0,$$

while $\vec{E} = 0$. Keeping in mind magnet design, such a field can be obtained very schematically as shown in Fig. 4.1.

Let us now consider the equations of motion; we obtain

$$x' = a, \qquad a' = -h^2 x + h\frac{1 + \eta_0}{2 + \eta_0}\delta,$$

$$y' = b, \qquad b' = 0,$$

$$l' = -h\frac{1 + \eta_0}{2 + \eta_0}x + \frac{1}{(2 + \eta_0)^2}\delta, \qquad \delta' = 0.$$

First we observe that if we choose $h = 0$, we obtain $a' = 0$, and we have the same situation as in the case of a drift. But even for the case of $h \neq 0$, the motion of the $y$-direction behaves simply like a drift, and we always have

$$y_f = y_i + b_i\, L, \qquad b_f = b_i,$$

where $L$ is the **arc length** of the reference orbit in the dipole and $L = R_0\phi$.

Next we observe that as always, $\delta$ stays constant, and hence in the equation for $a'$ plays the role of a parameter, making the differential equation inhomogeneous. Finally we observe that since $l$ does not couple into the horizontal or vertical motion, we can solve the equation for $l$ after the horizontal motion is analyzed by a mere integration.

In order to solve the horizontal part of the motion, we first solve the **homogeneous** part of the differential equation, which has the form

$$x' = a, \qquad a' = -h^2 x,$$

and we obtain as a solution

$$x_f = x_i \cos \omega L + \frac{1}{\omega} a_i \sin \omega L = x_i \cos \phi + R_0 a_i \sin \phi,$$

$$a_f = -\omega x_i \sin \omega L + a_i \cos \omega L = -\frac{1}{R_0} x_i \sin \phi + a_i \cos \phi,$$

where we have used the angular frequency $\omega$,

$$\omega = h = \frac{1}{R_0}.$$

Altogether, we have a behavior not much different from a focusing quadrupole.

In order to treat the inhomogeneity, we perform a so-called "variation of parameters," that is we make an ansatz of the form

$$x\left(s\right) = \bar{x}_i\left(s\right) \cos \phi + R_0 \bar{a}_i\left(s\right) \sin \phi = \bar{x}_i\left(s\right) \cos \omega s + \frac{1}{\omega} \bar{a}_i\left(s\right) \sin \omega s,$$

$$a\left(s\right) = -\frac{1}{R_0} \bar{x}_i\left(s\right) \sin \phi + \bar{a}_i\left(s\right) \cos \phi = -\omega \bar{x}_i\left(s\right) \sin \omega s + \bar{a}_i\left(s\right) \cos \omega s,$$

where now the original parameters $\bar{x}_i$, $\bar{a}_i$ are viewed as functions of $s$. Inserting into the differential equation, we obtain the following condition:

$$\bar{x}_i'\left(s\right) \cos \omega s + \frac{1}{\omega} \bar{a}_i'\left(s\right) \sin \omega s = 0,$$

$$-\omega \bar{x}_i'\left(s\right) \sin \omega s + \bar{a}_i'\left(s\right) \cos \omega s = \Lambda = h \frac{1 + \eta_0}{2 + \eta_0} \delta_i,$$

using the abbreviation $\Lambda$ for the right hand side in the second equation. Rewriting in matrix form, this reads

$$\begin{pmatrix} \cos(\omega s) & \sin(\omega s)/\omega \\ -\omega \sin(\omega s) & \cos(\omega s) \end{pmatrix} \begin{pmatrix} \bar{x}_i' \\ \bar{a}_i' \end{pmatrix} = \begin{pmatrix} 0 \\ \Lambda \end{pmatrix}.$$

Multiplying with the inverse matrix and integrating, we obtain

$$\bar{x}_i\left(s\right) = \int_0^s \left(-\frac{1}{\omega} \sin \omega s\right) \Lambda ds + x_i = \frac{1}{\omega^2} \Lambda \left(\cos \omega s - 1\right) + x_i,$$

$$\bar{a}_i\left(s\right) = \int_0^s \left(\cos \omega s\right) \Lambda ds + a_i = \frac{1}{\omega} \Lambda \sin \omega s + a_i.$$

So the complete solution of the inhomogeneous part has the form

$$
x(s) = \left[\frac{1}{\omega^2}\Lambda\left(\cos\omega s - 1\right) + x_i\right]\cos\omega s + \frac{1}{\omega}\left[\frac{1}{\omega}\Lambda\sin\omega s + a_i\right]\sin\omega s
$$

$$
= x_i\cos\phi + \frac{1}{\omega}a_i\sin\phi + \frac{1}{\omega^2}\Lambda\left(1 - \cos\phi\right),
$$

$$
a(s) = -\omega\left[\frac{1}{\omega^2}\Lambda\left(\cos\omega s - 1\right) + x_i\right]\sin\omega s + \left[\frac{1}{\omega}\Lambda\sin\omega s + a_i\right]\cos\omega s
$$

$$
= -\omega x_i\sin\phi + a_i\cos\phi + \frac{1}{\omega}\Lambda\sin\phi.
$$

Thus we obtain

$$
x_f = x_i\cos\phi + R_0 a_i\sin\phi + R_0\left(1 - \cos\phi\right)\frac{1 + \eta_0}{2 + \eta_0}\delta_i,
$$

$$
a_f = -\frac{1}{R_0}x_i\sin\phi + a_i\cos\phi + \sin\phi\frac{1 + \eta_0}{2 + \eta_0}\delta_i.
$$

Finally we have to study the case of the time-of-flight part, which as we said before can be obtained by mere integration. We have

$$
l_f = \int_0^L\left[-h\frac{1 + \eta_0}{2 + \eta_0}x + \frac{1}{\left(2 + \eta_0\right)^2}\delta\right]ds + l_i
$$

$$
= \int_0^{R_0\phi}\left[-\frac{1 + \eta_0}{2 + \eta_0}\frac{1}{R_0}x_i\cos\frac{s}{R_0} - \frac{1 + \eta_0}{2 + \eta_0}a_i\sin\frac{s}{R_0}\right.
$$

$$
\left. -\left(1 - \cos\frac{s}{R_0}\right)\left(\frac{1 + \eta_0}{2 + \eta_0}\right)^2\delta_i + \frac{1}{\left(2 + \eta_0\right)^2}\delta_i\right]ds + l_i
$$

$$
= \left[-\frac{1 + \eta_0}{2 + \eta_0}x_i\sin\frac{s}{R_0} + \frac{1 + \eta_0}{2 + \eta_0}R_0 a_i\cos\frac{s}{R_0}\right.
$$

$$
\left. +\left(\frac{1 + \eta_0}{2 + \eta_0}\right)^2 R_0\left(\sin\frac{s}{R_0}\right)\delta_i - \left(\frac{1 + \eta_0}{2 + \eta_0}\right)^2\delta_i s + \frac{1}{\left(2 + \eta_0\right)^2}\delta_i s\right]\Bigg|_0^{R_0\phi} + l_i
$$

$$
= -\frac{1 + \eta_0}{2 + \eta_0}x_i\sin\phi + \frac{1 + \eta_0}{2 + \eta_0}R_0 a_i\left(\cos\phi - 1\right)
$$

$$
+\left(\frac{1 + \eta_0}{2 + \eta_0}\right)^2 R_0\left(\sin\phi\right)\delta_i - \frac{\eta_0}{2 + \eta_0}R_0\phi\delta_i + l_i.
$$

As a result, we see that all the final coordinates indeed depend on all initial coordinates in a linear fashion, and hence the relationship can be written in terms of a transfer matrix. The general shape of this matrix is now

$$
\begin{pmatrix} x_f \\ a_f \\ y_f \\ b_f \\ l_f \\ \delta_f \end{pmatrix} = \begin{pmatrix} \cos\phi & R_0\sin\phi & 0 & 0 & 0 & (x|\delta) \\ -\sin\phi/R_0 & \cos\phi & 0 & 0 & 0 & (a|\delta) \\ 0 & 0 & 1 & R_0\phi & 0 & 0 \\ 0 & 0 & 0 & 1 & 0 & 0 \\ (l|x) & (l|a) & 0 & 0 & 1 & (l|\delta) \\ 0 & 0 & 0 & 0 & 0 & 1 \end{pmatrix} \begin{pmatrix} x_i \\ a_i \\ y_i \\ b_i \\ l_i \\ \delta_i \end{pmatrix}, \quad (4.3)
$$

where $R_0$ and $\phi$ are the bending radius and the bending angle, and the abbreviated matrix elements are

$$
(x|\delta) = -(l|a) = \frac{1+\eta_0}{2+\eta_0} R_0 \left(1-\cos\phi\right), \qquad (a|\delta) = -(l|x) = \frac{1+\eta_0}{2+\eta_0} \sin\phi,
$$

$$
(l|\delta) = -R_0 \left[ \frac{\eta_0}{2+\eta_0}\phi - \left(\frac{1+\eta_0}{2+\eta_0}\right)^2 \sin\phi \right].
$$

We observe that the determinant of the matrix is unity. Note that, while the $y$-$b$ (vertical) motion is decoupled, the $x$-$a$ (horizontal) and the $l$-$\delta$ (longitudinal) motions are coupled. Specifically, the $x$-$a$ motion depends linearly on $\delta$.

The homogeneous dipole magnet we have considered so far has edges that are perpendicular to the reference orbit. So the region where the magnetic field is active corresponds to a sector of a circle, which is the reason such a magnet is often referred to as a **sector magnet**.

## 4.3.2   Edge Focusing

When the reference particle enters and exits a sector dipole magnet, the orbit travels perpendicular to the entrance and the exit edges. When the magnet edge is not perpendicular to the reference orbit, additional focusing and defocusing effects act on the beam, which is called edge focusing. The angle difference from the perpendicular sector magnet case is called the edge angle. Edge focusing is frequently used on purpose to modify the linear properties of the motion, or as a consequence of convenience in manufacturing since it is particularly simple to use a rectangular shape for the magnet, which leads to the so-called parallel-faced dipole. We now study the effects of edge focusing using the matrix form, and compare the result with the sector dipole.

We measure the edge angle $\alpha$ such that the rectangular dipole would have positive edge angles. So, when $\alpha > 0$, a particle that enters or exits the magnet at a positive $x$ location experiences a lesser amount of the bending magnetic field compared to the reference particle. Compared to the sector dipole magnet, this means that the edge line tilts inward for positive $x$ as shown in Fig. 4.2.

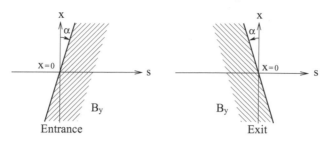

**FIGURE 4.2**:  Entrance and exit edge lines of a dipole magnet with edge angle $\alpha$.

When the edge angle $\alpha$ is sufficiently small, the effect can be approximated as an impulsive effect that changes only the horizontal and the vertical angles of the particle orbit but does not affect position nor the longitudinal motion. The impulsive style of treating such an effect is called "**kick**," and the kick approximation is sometimes used when studying beam optical systems in linear approximations in a similar way as it is in glass optics in the use of the thin lens.

As we will explain in the following, in the kick approximation, the effect of the edge angle $\alpha$ for the homogeneous dipole magnet acts to change only $a$ and $b$ via

$$a_f = a_i + \frac{x_i \tan \alpha}{R_0}, \qquad b_f = b_i - \frac{y_i \tan \alpha}{R_0}, \qquad (4.4)$$

where $R_0$ is the bending radius of the homogeneous magnet, and

$$\frac{1}{R_0} = \frac{B_{y0}}{\chi_{m0}}.$$

The same expression applies to both the entrance edge and the exit edge. Using the abbreviation

$$T = \tan \alpha / R_0,$$

the matrices of the horizontal ($x$) and the vertical ($y$) kicks by the edge angle $\alpha$ are described as

$$\hat{M}_x^{\mathrm{ed}} = \begin{pmatrix} 1 & 0 \\ T & 1 \end{pmatrix}, \qquad \hat{M}_y^{\mathrm{ed}} = \begin{pmatrix} 1 & 0 \\ -T & 1 \end{pmatrix}.$$

Recalling the situation of thin glass lenses, when $\alpha > 0$, the vertical kick acts to focus the beam, and the horizontal kick acts to defocus. Combining the horizontal and the vertical kicks, the effect of the edge is that of a thin quadrupole of the strength $-T$, where always one of the directions experiences focusing, and the other defocusing. When the sign of $\alpha$ is opposite, the effect also becomes opposite.

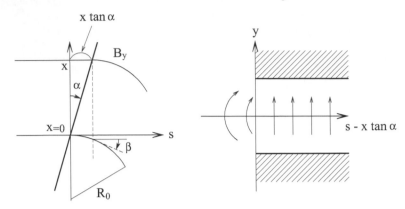

**FIGURE 4.3:**  Mechanism of edge focusing for the horizontal plane (left) and the vertical plane (right).

We use a geometric argument to explain the horizontal kick. For the vertical kick, we use a step function to model $B_y$, and use Maxwell's equation to derive $B_x$ which affects the vertical motion.

We have a homogeneous bending magnet with the entrance edge line tilted by the edge angle $\alpha$ as shown in the left picture of Fig. 4.2. Using the standard step function $H$, also often referred to as the Heaviside function, the vertical field component $B_y$ can be expressed as

$$B_y(x, y, s) = B_{y0} H(s - x \tan \alpha),$$

where $B_{y0}$ is the constant field of the main part of the dipole. The Heaviside step function $H$ is related to the Dirac delta function $\delta$ as

$$H(x) = \int_{-\infty}^{x} \delta(\bar{x}) d\bar{x} = \begin{cases} 1 & \text{for } x > 0 \\ 0 & \text{for } x < 0 \end{cases},$$

$$\delta(x) = \begin{cases} +\infty & \text{for } x = 0 \\ 0 & \text{for } x \neq 0 \end{cases}, \qquad \int_{-\varepsilon}^{\varepsilon} \delta(x) = 1 \quad \text{for any } \varepsilon > 0. \qquad (4.5)$$

The $B_y$ expressed above is of course an idealized situation. In reality there is no magnetic field that can fulfill the above expression while satisfying Maxwell's equations.

Now consider a particle approaching to the entrance of the magnet parallel to the reference orbit, but positioned at $x$. As seen in the left picture of Fig. 4.3, the entering of this particle is delayed by the distance $x \tan \alpha$. In the meantime, the reference particle travels through the magnet for this much of arc length, experiencing a deflection angle amounting to $\beta = x \tan \alpha / R_0$. When observing the situation in the particle optical coordinates that are attached to the reference particle's motion, the particle of interest located at the position $x$ appears to have experienced a change in the direction of motion by $+x \tan \alpha / R_0$. Note that the picture in Fig. 4.3 is drawn exaggerated to emphasize the relevant points. The same explanation applies to the situation at the exit, where the change in the traveling direction appears to be $+x \tan \alpha / R_0$. Thus, the formula for $a_f$ in eq. (4.4) has the $+x_i \tan \alpha / R_0$ term for both the entrance edge and the exit edge.

Next we consider the equation for $b'$ in the set of equations of motion (3.22), and observe that the $B_x$ term is the leading term to affect the vertical motion. As seen in Fig. 4.1, there exists a non-vertical field component around the edges of the magnet off the midplane, i.e., when $y \neq 0$. Also, see the right picture in Fig. 4.3. From Maxwell's equations (3.2), we have the relation

$$\frac{\partial B_x}{\partial y} = \frac{\partial B_y}{\partial x},$$

and using this, we can derive $B_x$ as

$$B_x(x, y, s) = \int \frac{\partial B_y}{\partial x} dy = \int \left[ \frac{\partial}{\partial x} B_{y0} H(s - x \tan \alpha) \right] dy$$
$$= -B_{y0} \tan \alpha \delta(s - x \tan \alpha) \cdot y.$$

Thus, the equation for $b'$ of (3.22) becomes

$$b' = \frac{B_x}{\chi_{m0}} = -\frac{B_{y0}}{\chi_{m0}} \tan \alpha \delta(s - x \tan \alpha) \cdot y,$$

and in the impulsive or kick approximation we obtain

$$b_f = b_i - \frac{B_{y0}}{\chi_{m0}} \tan \alpha \int \delta(s - x \tan \alpha) ds \cdot y_i = b_i - \frac{\tan \alpha}{R_0} y_i.$$

Now, at the exit side, having the opposite sign for the step function, $B_y$ is expressed as

$$B_y(x, y, s) = B_{y0} H(-s - x \tan \alpha),$$

resulting in

$$B_x(x, y, s) = -B_{y0} \tan \alpha \delta(-s - x \tan \alpha) \cdot y.$$

And, we obtain the same result as for the entrance case, namely

$$b_f = b_i - \frac{B_{y0}}{\chi_{m0}} \tan \alpha \int \delta(-s - x \tan \alpha) ds \cdot y_i = b_i - \frac{\tan \alpha}{R_0} y_i.$$

We note that the rise of $B_x$ is caused by the mere tilting of the edge line; thus a longitudinal component $B_s$ also exists, and it can be derived in a similar fashion. But the $B_s$ dependent term in the $b'$ equation of (3.22) also depends on $a$, turning it to be a nonlinear term; thus we do not consider it for this linear kick approximation. Another note is that in the homogeneous sector dipole magnet, $B_x$ does not exist because $\alpha = 0$.

Since the rectangular dipole is rather commonly used, it is worthwhile to calculate the total transfer matrix of a rectangular dipole by combining the edge focusing and the main part of the dipole using eq. (4.3). In this case, the edge angles of the entrance and the exit are the same, namely half of the bending angle and thus $\phi/2$. We study the $x$-$a$-$l$-$\delta$ block and the $y$-$b$ block separately. Below, the abbreviation $T = \tan(\phi/2)/R_0$ is used.

First, we calculate the combination of the main part and one edge for the $x$-$a$-$l$-$\delta$ block:

$$
\hat{M}_x^{di} \hat{M}_x^{ed} = \begin{pmatrix} \cos\phi & R_0 \sin\phi & 0 & (x|\delta)_{di} \\ -\sin\phi/R_0 & \cos\phi & 0 & (a|\delta)_{di} \\ (l|x)_{di} & (l|a)_{di} & 1 & (l|\delta)_{di} \\ 0 & 0 & 0 & 1 \end{pmatrix} \begin{pmatrix} 1 & 0 & 0 & 0 \\ T & 1 & 0 & 0 \\ 0 & 0 & 1 & 0 \\ 0 & 0 & 0 & 1 \end{pmatrix}
$$

$$
= \begin{pmatrix} \cos\phi + TR_0\sin\phi & R_0\sin\phi & 0 & (x|\delta)_{di} \\ -\sin\phi/R_0 + T\cos\phi & \cos\phi & 0 & (a|\delta)_{di} \\ (l|x)_{di} + T(l|a)_{di} & (l|a)_{di} & 1 & (l|\delta)_{di} \\ 0 & 0 & 0 & 1 \end{pmatrix},
$$

where the $(1,1)$ and the $(2,1)$ components of the matrix multiplication are simplified to 1 and $-T$ as follows.

$$
\cos\phi + TR_0\sin\phi = \cos^2\frac{\phi}{2} - \sin^2\frac{\phi}{2} + \tan\frac{\phi}{2} \cdot 2\sin\frac{\phi}{2}\cos\frac{\phi}{2} = 1,
$$

$$
-\sin\phi/R_0 + T\cos\phi = -T\left[2\sin\frac{\phi}{2}\cos\frac{\phi}{2} \middle/ \tan\frac{\phi}{2} - \cos^2\frac{\phi}{2} + \sin^2\frac{\phi}{2}\right] = -T.
$$

So, we obtain the matrix of the $x$ block as

$$
\hat{M}_x = \hat{M}_x^{ed} \hat{M}_x^{di} \hat{M}_x^{ed} = \begin{pmatrix} 1 & 0 & 0 & 0 \\ T & 1 & 0 & 0 \\ 0 & 0 & 1 & 0 \\ 0 & 0 & 0 & 1 \end{pmatrix} \begin{pmatrix} 1 & R_0\sin\phi & 0 & (x|\delta)_{di} \\ -T & \cos\phi & 0 & (a|\delta)_{di} \\ (l|x)_{di} + T(l|a)_{di} & (l|a)_{di} & 1 & (l|\delta)_{di} \\ 0 & 0 & 0 & 1 \end{pmatrix}
$$

$$
= \begin{pmatrix} 1 & R_0\sin\phi & 0 & (x|\delta)_{di} \\ 0 & 1 & 0 & T(x|\delta)_{di} + (a|\delta)_{di} \\ T(l|a)_{di} + (l|x)_{di} & (l|a)_{di} & 1 & (l|\delta)_{di} \\ 0 & 0 & 0 & 1 \end{pmatrix},
$$

where the same simplification happened for the $(2,2)$ component, and

$$
(x|\delta)_{di} = \frac{1+\eta_0}{2+\eta_0} R_0 (1 - \cos\phi),
$$

$$
(l|\delta)_{di} = -R_0 \left[\frac{\eta_0}{2+\eta_0}\phi - \left(\frac{1+\eta_0}{2+\eta_0}\right)^2 \sin\phi\right],
$$

and

$$T(x|\delta)_{\text{di}} + (a|\delta)_{\text{di}} = -\left[T(l|a)_{\text{di}} + (l|x)_{\text{di}}\right]$$

$$= \frac{1}{R_0} \tan\frac{\phi}{2} \frac{1+\eta_0}{2+\eta_0} R_0 \left(1 - \cos\phi\right) + \frac{1+\eta_0}{2+\eta_0} \sin\phi$$

$$= \frac{1+\eta_0}{2+\eta_0} \left[\tan\frac{\phi}{2}\left(1 - \cos\phi\right) + \sin\phi\right] = \frac{1+\eta_0}{2+\eta_0} \cdot 2\tan\frac{\phi}{2}.$$

The matrix of the $y$-$b$ block is

$$\hat{M}_y = \begin{pmatrix} 1 & 0 \\ -T & 1 \end{pmatrix} \begin{pmatrix} 1 & R_0\phi \\ 0 & 1 \end{pmatrix} \begin{pmatrix} 1 & 0 \\ -T & 1 \end{pmatrix} = \begin{pmatrix} 1 - TR_0\phi & R_0\phi \\ -T(2 - TR_0\phi) & 1 - TR_0\phi \end{pmatrix},$$

where

$$1 - TR_0\phi = 1 - \phi\tan\frac{\phi}{2}, \qquad -T(2 - TR_0\phi) = -\frac{1}{R_0}\tan\frac{\phi}{2}\left[2 - \phi\tan\frac{\phi}{2}\right].$$

In summary, we obtain

$$\begin{pmatrix} x_f \\ a_f \\ y_f \\ b_f \\ l_f \\ \delta_f \end{pmatrix} = \begin{pmatrix} 1 & R_0\sin\phi & 0 & 0 & 0 & (x|\delta) \\ 0 & 1 & 0 & 0 & 0 & (a|\delta) \\ 0 & 0 & 1-\phi\tan(\phi/2) & R_0\phi & 0 & 0 \\ 0 & 0 & (b|y) & 1-\phi\tan(\phi/2) & 0 & 0 \\ (l|x) & (l|a) & 0 & 0 & 1 & (l|\delta) \\ 0 & 0 & 0 & 0 & 0 & 1 \end{pmatrix} \begin{pmatrix} x_i \\ a_i \\ y_i \\ b_i \\ l_i \\ \delta_i \end{pmatrix},$$

where the abbreviated matrix elements are

$$(b|y) = -\frac{1}{R_0}\tan\frac{\phi}{2}\left[2 - \phi\tan\frac{\phi}{2}\right],$$

$$(x|\delta) = -(l|a) = \frac{1+\eta_0}{2+\eta_0}R_0\left(1 - \cos\phi\right), \qquad (a|\delta) = -(l|x) = \frac{1+\eta_0}{2+\eta_0}\cdot 2\tan\frac{\phi}{2},$$

$$(l|\delta) = -R_0\left[\frac{\eta_0}{2+\eta_0}\phi - \left(\frac{1+\eta_0}{2+\eta_0}\right)^2\sin\phi\right].$$

Note that the determinant of the matrix is unity, which also can be deduced from that the determinant of all the contributing matrices is unity.

To conclude, let us compare the characteristic effects of the rectangular dipole and the sector dipole in the limit of small deflection angle $\phi$. The $x$-$a$ matrix and the $y$-$b$ matrix of both dipoles can then be approximated as

Sector dipole: $\qquad \hat{M}_{xa} \to \begin{pmatrix} 1 & R_0\phi \\ -\phi/R_0 & 1 \end{pmatrix}, \quad \hat{M}_{yb} \to \begin{pmatrix} 1 & R_0\phi \\ 0 & 1 \end{pmatrix},$

Rectangular dipole: $\quad \hat{M}_{xa} \to \begin{pmatrix} 1 & R_0\phi \\ 0 & 1 \end{pmatrix}, \quad \hat{M}_{yb} \to \begin{pmatrix} 1 & R_0\phi \\ -\phi/R_0 & 1 \end{pmatrix}.$

Thus we have the interesting effect that the characteristic behavior in the horizontal plane and the vertical plane is exchanged.

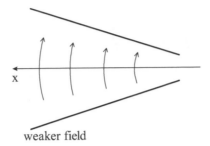

weaker field

**FIGURE 4.4:** An inhomogeneous sector magnet.

### 4.3.3 The Inhomogeneous Sector Magnet

In the case of an inhomogeneous sector, there is a magnetic field that is constant in $s$-direction, but not constant in the $x$-direction; rather it has the shape

$$B_y = B_{y0} \cdot \left(1 - n\frac{x}{R_0}\right).$$

From the recursion relations for the fields, we infer that the corresponding horizontal field is

$$B_x = -B_{y0} \cdot n\frac{y}{R_0}.$$

$n_b$ in eqs. (4.2) is $n_b = -n/R_0$, and $h = 1/R_0$. In general terms, such a field is obtained by changing the distance between what generates the fields (coils or iron) as a function of $x$, similar to what is shown in Fig. 4.4 for the case of $n > 0$.

We have the linearized equations of motion as

$$x' = a, \qquad a' = -\left(h^2 - \frac{B_{y0}}{\chi_{m0}}\frac{n}{R_0}\right)x + h\frac{1+\eta_0}{2+\eta_0}\delta = -h^2(1-n)\,x + h\frac{1+\eta_0}{2+\eta_0}\delta,$$

$$y' = b, \qquad b' = -\frac{B_{y0}}{\chi_{m0}}\frac{n}{R_0}y = -h^2 n y,$$

$$l' = -h\frac{1+\eta_0}{2+\eta_0}x + \frac{1}{(2+\eta_0)^2}\delta, \qquad \delta' = 0.$$

We observe that the horizontal motion is similar to the case of the homogeneous sector dipole, except that the strength of focusing now also depends on $n$, the field inhomogeneity. Different from the homogeneous sector dipole, there is now an effect in the vertical direction, which can be either focusing or defocusing, depending on the sign of $n$.

The solution of these equations of motion proceeds in the same way as before, first solve the homogeneous system, then address the inhomogeneity arising from $\delta$ via variation of parameters, and finally solve for $l$ by a mere integration. In horizontal and vertical directions, the homogeneous solution corresponds to harmonic oscillators with frequencies

$$\omega_x = h\sqrt{1-n}, \qquad \omega_y = h\sqrt{n}.$$

For $0 < n < 1$, the magnet is focusing in both planes. An interesting case occurs for $n = 1/2$, in which case the magnet focuses $x$ and $y$ identically and represents a nice equivalent of the glass lens.

The remainder of the derivation is tedious algebra, and we will only list the result here.

$$
\begin{pmatrix} x_f \\ a_f \\ y_f \\ b_f \\ l_f \\ \delta_f \end{pmatrix} = \begin{pmatrix} (x|x) & (x|a) & 0 & 0 & 0 & (x|\delta) \\ (a|x) & (a|a) & 0 & 0 & 0 & (a|\delta) \\ 0 & 0 & (y|y) & (y|b) & 0 & 0 \\ 0 & 0 & (b|y) & (b|b) & 0 & 0 \\ (l|x) & (l|a) & 0 & 0 & 1 & (l|\delta) \\ 0 & 0 & 0 & 0 & 0 & 1 \end{pmatrix} \begin{pmatrix} x_i \\ a_i \\ y_i \\ b_i \\ l_i \\ \delta_i \end{pmatrix}, \quad (4.6)
$$

where

$$(x|x) = (a|a) = \cos\left(\sqrt{1-n}\,\phi\right),$$

$$(x|a) = \frac{R_0}{\sqrt{1-n}} \sin\left(\sqrt{1-n}\,\phi\right), \qquad (a|x) = -\frac{\sqrt{1-n}}{R_0} \sin\left(\sqrt{1-n}\,\phi\right),$$

$$(y|y) = (b|b) = \cos\left(\sqrt{n}\,\phi\right),$$

$$(y|b) = \frac{R_0}{\sqrt{n}} \sin\left(\sqrt{n}\,\phi\right), \qquad (b|y) = -\frac{\sqrt{n}}{R_0} \sin\left(\sqrt{n}\,\phi\right),$$

$$(x|\delta) = -(l|a) = \frac{1+\eta_0}{2+\eta_0} \frac{R_0}{1-n} \left[1 - \cos\left(\sqrt{1-n}\,\phi\right)\right],$$

$$(a|\delta) = -(l|x) = \frac{1+\eta_0}{2+\eta_0} \frac{1}{\sqrt{1-n}} \sin\left(\sqrt{1-n}\,\phi\right),$$

$$(l|\delta) = -R_0 \left(\frac{1+\eta_0}{2+\eta_0}\right)^2 \left\{ \left[\frac{1}{1-n} - \frac{1}{(1+\eta_0)^2}\right]\phi - \frac{1}{(1-n)^{3/2}} \sin\left(\sqrt{1-n}\,\phi\right) \right\},$$

and the determinant is unity.

### 4.3.4 The Inhomogeneous Electric Deflector

Rather commonly known is the motion of a particle in an electric capacitor. Neglecting fringe fields, it follows a parabola as shown in Fig. 4.5. For particle optical purposes, such an arrangement is not particularly suitable for two reasons. Firstly, the reference orbit has a curvature that depends on $s$, which makes the differential equations non-autonomous. Secondly, the potential along the reference orbit changes with $s$, which complicates the dynamics. Both of these problems do not appear if instead of a straight capacitor, one

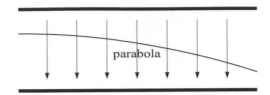

**FIGURE 4.5:** An electric capacitor consisting of two parallel plates. The orbit of a particle is parabolic.

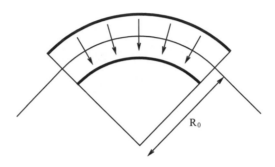

**FIGURE 4.6:** A concentric electric deflector.

chooses a curved one in such a way that the reference orbit is concentric together with the plates as shown in Fig. 4.6.

To obtain the linearized equations of motion for such a device, we first observe that

$$\frac{E_{x0}}{\chi_{e0}} = h$$

while $\vec{B} = 0$, and using eqs. (3.16),

$$\chi_{e0} = \frac{p_0 v_0}{Ze} = \frac{1}{Ze} \cdot mc\sqrt{\eta_0 (2 + \eta_0)} \cdot c \frac{\sqrt{\eta_0 (2 + \eta_0)}}{1 + \eta_0} = \frac{mc^2}{Ze} \frac{\eta_0 (2 + \eta_0)}{1 + \eta_0}.$$

We describe the linearized electric field using the field inhomogenuity $n$, similar to the magnetic case, as

$$E_x = -E_{x0} \cdot \left(1 - n\frac{x}{R_0}\right),$$

hence $n_e$ in eqs. (4.2) is given by $n_e = -n/R_0$, and $h = 1/R_0$. The linearized equations of motion are

$$x' = a, \quad a' = -h^2 \left[2 - n + \frac{1}{(1+\eta_0)^2}\right] x + h \left[1 + \frac{1}{(1+\eta_0)^2}\right] \frac{1+\eta_0}{2+\eta_0} \delta,$$

$$y' = b, \quad b' = h^2 (1 - n) y,$$

$$l' = -h \frac{1+\eta_0}{2+\eta_0} \left[1 + \frac{1}{(1+\eta_0)^2}\right] x + \frac{1}{(2+\eta_0)^2} \delta, \qquad \delta' = 0.$$

Since the electrostatic deflectors are used primarily for low energy electrons and ions (usually below 100 keV) due to the difficulty of achieving high static voltages, the particles are non-relativistic, i.e., $\eta_0 \ll 1$. As a result, the equations of motion can be simplified to a more familiar form

$$x' = a, \qquad a' = -h^2 (3 - n) x + h\delta,$$

$$y' = b, \qquad b' = -h^2 (n - 1) y, \qquad l' = -hx + \frac{1}{4}\delta, \qquad \delta' = 0.$$

We observe that for $1 < n < 3$, both $x$ and $y$ planes are focusing, different from the case of quadrupoles where always one plane defocuses; but the amount of focusing in the $x$ and $y$ planes is different. Indeed, similar to the inhomogeneous dipole magnet, the transfer matrix is

$$\begin{pmatrix} x_f \\ a_f \\ y_f \\ b_f \\ l_f \\ \delta_f \end{pmatrix} = \begin{pmatrix} (x|x) & (x|a) & 0 & 0 & 0 & (x|\delta) \\ (a|x) & (a|a) & 0 & 0 & 0 & (a|\delta) \\ 0 & 0 & (y|y) & (y|b) & 0 & 0 \\ 0 & 0 & (b|y) & (b|b) & 0 & 0 \\ (l|x) & (l|a) & 0 & 0 & 1 & (l|\delta) \\ 0 & 0 & 0 & 0 & 0 & 1 \end{pmatrix} \begin{pmatrix} x_i \\ a_i \\ y_i \\ b_i \\ l_i \\ \delta_i \end{pmatrix},$$

where

$$(x|x) = (a|a) = \cos\left(\sqrt{3 - n}\phi\right),$$

$$(x|a) = \frac{R_0}{\sqrt{3 - n}} \sin\left(\sqrt{3 - n}\phi\right), \qquad (a|x) = -\frac{\sqrt{3 - n}}{R_0} \sin\left(\sqrt{3 - n}\phi\right),$$

$$(y|y) = (b|b) = \cos\left(\sqrt{n - 1}\phi\right),$$

$$(y|b) = \frac{R_0}{\sqrt{n - 1}} \sin\left(\sqrt{n - 1}\phi\right), \qquad (b|y) = -\frac{\sqrt{n - 1}}{R_0} \sin\left(\sqrt{n - 1}\phi\right),$$

$$(x|\delta) = -(l|a) = \frac{R_0}{3 - n} \left[1 - \cos\left(\sqrt{3 - n}\phi\right)\right],$$

$$(a|\delta) = -(l|x) = \frac{1}{\sqrt{3 - n}} \sin\left(\sqrt{3 - n}\phi\right),$$

$$(l|\delta) = -R_0 \left[\left(\frac{1}{3 - n} - \frac{1}{4}\right)\phi - \frac{1}{(3 - n)^{3/2}} \sin\left(\sqrt{3 - n}\phi\right)\right],$$

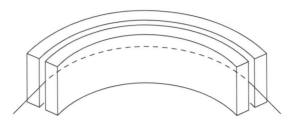

**FIGURE 4.7:** An electric deflector with cylindrical plates.

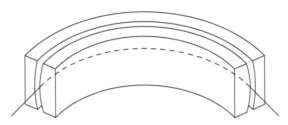

**FIGURE 4.8:** An electric deflector with spherical plates.

and the determinant is unity.

So far, no assumptions have been made about the vertical shapes of the electrodes, and in fact a variety of choices exist. Two common situations are the cylindrical plates and the spherical plates, as shown in Fig. 4.7 and Fig. 4.8, respectively.

In the case of the **cylindrical** field, we have from Gauss' law that $E \propto 1/R$, which implies the expansion

$$E_x = -E_{x_0}\frac{R_0}{R_0 + x} =_1 -E_{x_0}\left(1 - \frac{x}{R_0}\right),$$

and hence corresponds to $n = 1$. The transfer matrix is

$$\begin{pmatrix} x_f \\ a_f \\ y_f \\ b_f \\ l_f \\ \delta_f \end{pmatrix} = \begin{pmatrix} \cos(\sqrt{2}\phi) & (R_0/\sqrt{2})\sin(\sqrt{2}\phi) & 0 & 0 & 0 & (x|\delta) \\ -(\sqrt{2}/R_0)\sin(\sqrt{2}\phi) & \cos(\sqrt{2}\phi) & 0 & 0 & 0 & (a|\delta) \\ 0 & 0 & 1 & R_0\phi & 0 & 0 \\ 0 & 0 & 0 & 1 & 0 & 0 \\ (l|x) & (l|a) & 0 & 0 & 1 & (l|\delta) \\ 0 & 0 & 0 & 0 & 0 & 1 \end{pmatrix}\begin{pmatrix} x_i \\ a_i \\ y_i \\ b_i \\ l_i \\ \delta_i \end{pmatrix},$$

where

$$(x|\delta) = -(l|a) = \frac{R_0}{2}\left[1 - \cos\left(\sqrt{2}\phi\right)\right], \qquad (a|\delta) = -(l|x) = \frac{1}{\sqrt{2}}\sin\left(\sqrt{2}\phi\right),$$

$$(l|\delta) = -R_0\left[\frac{1}{4}\phi - \frac{1}{2\sqrt{2}}\sin\left(\sqrt{2}\phi\right)\right].$$

For $\phi = \pi/\sqrt{2} = 127.28°$, the deflector is imaging in the horizontal plane, which has been used as energy spectrometers.

In the **spherical** case, we have $E \propto 1/R^2$ and thus

$$E_x = -E_{x_0} \left( \frac{R_0}{R_0 + x} \right)^2 =_1 -E_{x_0} \left( 1 - 2 \frac{x}{R_0} \right),$$

and $n = 2$. The transfer matrix is

$$\begin{pmatrix} x_f \\ a_f \\ y_f \\ b_f \\ l_f \\ \delta_f \end{pmatrix} = \begin{pmatrix} \cos \phi & R_0 \sin \phi & 0 & 0 & 0 & (x|\delta) \\ -\sin \phi/R_0 & \cos \phi & 0 & 0 & 0 & \sin \phi \\ 0 & 0 & \cos \phi & R_0 \sin \phi & 0 & 0 \\ 0 & 0 & -\sin \phi/R_0 & \cos \phi & 0 & 0 \\ -\sin \phi & (l|a) & 0 & 0 & 1 & (l|\delta) \\ 0 & 0 & 0 & 0 & 0 & 1 \end{pmatrix} \begin{pmatrix} x_i \\ a_i \\ y_i \\ b_i \\ l_i \\ \delta_i \end{pmatrix},$$

where

$$(x|\delta) = -(l|a) = R_0 \left( 1 - \cos \phi \right), \qquad (l|\delta) = -R_0 \left( \frac{3}{4} \phi - \sin \phi \right).$$

When $\phi = \pi = 180°$, this deflector forms a simultaneous image in both plane, also known as a **stigmatic image**. It has been and still is widely used as an energy spectrometer. It is also called a hemispherical analyzer, which is the main workhorse in the field of angle-resolved photoemission spectroscopy (ARPES).

## 4.4   Round Lenses

We now address the important class of so-called round lenses, which owe their name to their rotational symmetry along the beam axis. Electric round lenses are usually made of arrangements of rotationally symmetric metallic plates or tubes concentric with the reference orbit, each of which is held at a certain potential; and magnetic round lenses are usually made of solenoids carrying current and concentric with the reference orbit.

The rotational symmetry apparently entails that any fields in $x$ and $y$ directions are equal, and Maxwell's equations immediately show that this can only happen if there are also fields in the direction of the axis, which requires that the potential changes along the reference axis. Specifically, the potential for the rotationally symmetric case is described in eq. (3.6), and is given by

$$V =_2 V_0 (s) - \frac{1}{4} V_0'' (s) r^2 =_2 V_0 (s) - \frac{1}{4} V_0'' (s) \left( x^2 + y^2 \right)$$

up to second order in $r$. Therefore, the electric field, linear in $x$ and $y$, is given by

$$E_x =_1 \frac{1}{2} V_0''(s)\, x, \quad E_y =_1 \frac{1}{2} V_0''(s)\, y, \quad E_s =_1 -V_0'(s),$$

where $E_x$ and $E_y$ can also be expressed in terms of $E_s$ as

$$E_x =_1 -\frac{1}{2} E_s'(s)\, x, \quad E_y =_1 -\frac{1}{2} E_s'(s)\, y.$$

Fully analogously, we obtain in the magnetic case that

$$B_x =_1 \frac{1}{2} V_0''(s)\, x, \quad B_y =_1 \frac{1}{2} V_0''(s)\, y, \quad B_s =_1 -V_0'(s),$$

where $B_x$ and $B_y$ can also be expressed in terms of $B_s$ as

$$B_x =_1 -\frac{1}{2} B_s'(s)\, x, \quad B_y =_1 -\frac{1}{2} B_s'(s)\, y.$$

We note that the expansion order listed in the subscript of the equal sign here in this section only applies to the variables $r$, $x$, $y$, $a$, $b$ and $\delta$, while the dependence on $s$ is explicitly retained and not expanded. Fig. 3.3 shows typical resulting electric and magnetic scalar potentials as well as the resulting radial and axial fields for such cases.

Before embarking on detailed studies of the dynamics in electrostatic and magnetic round lenses, we illustrate one of their important characteristics. We determine the average radial electric or magnetic field along a straight line a fixed distance $r$ away from the center. We perform the averaging from $-S$ to $S$ where $S$ is chosen large enough that all fields vanish at $\pm S$. We obtain

$$\int_{-S}^{S} E_r\, ds = \int_{-S}^{S} B_r\, ds$$
$$= \int_{-S}^{S} \frac{1}{2} V_0''(s)\, r\, ds = \frac{1}{2} r\, [V_0'(S) - V_0'(-S)] = 0. \tag{4.7}$$

So all radial **field components average out to zero**. This is in stark contrast to for example the electric and magnetic quadrupoles or the combined function bending magnets, where the integrand of the radial field is constant throughout the integration.

Consider now the case of a thin round lens in which a particle does not change position much, similar to the situation in the idealized thin lens. In this case the average in the above field integrals will be responsible for the directional offset the particle experiences, and so this offset is zero. Thus any focusing action the particle may experience must come through **secondary effects** and is the result of a then incomplete cancellation of radial field contributions. This situation is sometimes referred to as **weak focusing**. These effects will be studied in detail below for both the electrostatic and magnetic cases.

Since the case of rotationally symmetric $s$-dependent potential is not covered by the assumption used for the linearized equations of motion (4.2) derived in the beginning of this chapter, we go back to the general equations of motion (3.22) in Chapter 3. In this case, $h = 0$, and hence the set of equations is given by

$$x' = \frac{p_0}{p_s} a, \qquad a' = \frac{1 + \eta}{1 + \eta_0} \frac{E_x}{\chi_{e0}} \frac{p_0}{p_s} - \frac{B_y}{\chi_{m0}} + \frac{B_s}{\chi_{m0}} \frac{p_0}{p_s} b,$$

$$y' = \frac{p_0}{p_s} b, \qquad b' = \frac{1 + \eta}{1 + \eta_0} \frac{E_y}{\chi_{e0}} \frac{p_0}{p_s} + \frac{B_x}{\chi_{m0}} - \frac{B_s}{\chi_{m0}} \frac{p_0}{p_s} a,$$

$$l' = \left( \frac{1 + \eta}{1 + \eta_0} \frac{p_0}{p_s} - 1 \right) \frac{\kappa}{v_0}, \qquad \delta' = 0, \tag{4.8}$$

where

$$\eta = \eta_0 (1 + \delta) - \frac{ZeV}{mc^2}, \qquad \frac{p_0}{p_s} = \left( \frac{\eta (2 + \eta)}{\eta_0 (2 + \eta_0)} - a^2 - b^2 \right)^{-1/2},$$

$$\frac{\kappa}{v_0} = -\frac{1 + \eta_0}{2 + \eta_0}, \qquad \chi_{m0} = \frac{p_0}{Ze}, \qquad \chi_{e0} = \frac{p_0 v_0}{Ze}.$$

In the following subsections, we study the two important regularly employed classes of lenses, the magnetic round lens and the electric round lens. As we will see, different from the cases of the electric and magnetic quadrupoles, their focusing properties arise from different mechanisms, and the magnetic and electric round lenses require a quite different treatment.

## 4.4.1    The Electrostatic Round Lens

Electrostatic round lenses are arrangements of metallic electrodes with rotational symmetry that act as equipotential surfaces and thus determine the on-axis potential. Electrostatic round lenses are frequently used in electron microscopes and also in the shaping of low-energy beams near the source. For a rotationally symmetric electrostatic lens, we have the electric field, linear in $x$ and $y$, given as

$$E_x = \frac{1}{2} V_0'' (s) x, \qquad E_y = \frac{1}{2} V_0'' (s) y, \qquad E_s = -V_0' (s),$$

and the corresponding electrostatic potential

$$V = V_0 (s) - \frac{1}{4} V_0'' (s) \left( x^2 + y^2 \right).$$

Applying these to eqs. (4.8), we have

$$x' = \frac{p_0}{p_s}a, \qquad a' = \frac{1+\eta}{1+\eta_0}\frac{1}{\chi_{e0}}\frac{p_0}{p_s}\frac{1}{2}V_0''(s)\,x,$$

$$y' = \frac{p_0}{p_s}b, \qquad b' = \frac{1+\eta}{1+\eta_0}\frac{1}{\chi_{e0}}\frac{p_0}{p_s}\frac{1}{2}V_0''(s)\,y,$$

$$l' = \left(\frac{1+\eta}{1+\eta_0}\frac{p_0}{p_s} - 1\right)\frac{\kappa}{v_0}, \qquad \delta' = 0. \tag{4.9}$$

The $x$-$a$ part and the $y$-$b$ part are decoupled, and they have the same form, so we only need to study one of them.

For the further study, it is convenient to express the $a'$ equation above in various ways. Using the relation $p/v = m(1+\eta)$ from eq. (3.16) and $\vec{v}/v = \vec{p}/p$ because of $\vec{v} \parallel \vec{p}$, we can write

$$a' = \frac{1+\eta}{1+\eta_0}\frac{1}{\chi_{e0}}\frac{p_0}{p_s}\frac{1}{2}V_0''(s)\,x = \frac{1}{2\chi_{e0}}\frac{v_0}{v_s}V_0''(s)\,x. \tag{4.10}$$

We note that in linearization, $v_s$, the $s$ component of $\vec{v}$, equals the velocity of the reference particle. Furthermore we observe that the focusing effect rests on the assumption that traveling through the potential leads to appreciable changes in velocity, which, for practically achievable voltages, limits the use of the effect to the near non-relativistic regime. So in the following we perform our argument in the non-relativistic limit and have

$$\frac{v_0}{v_s} = \frac{\sqrt{2K_0/m}}{\sqrt{2K(s)/m}} = \frac{\sqrt{K_0}}{\sqrt{K_0 - ZeV_0(s)}}.$$

Here $K(s)$ denotes the momentary kinetic energy of the reference particle, which of course changes as a function of position in the lens due to the change of $V_0(s)$, while $K_0$ is the constant kinetic energy of the reference particle before the round lens.

The above differential equations are all that is needed to determine the transfer matrix for the linearized motion of an electrostatic round lens with a certain potential distribution $V_0(s)$. However, in the following, we try to obtain a better understanding of the situation, and in particular we show the reasons why these lenses are generally focusing.

### 4.4.1.1  Hard Edge Fringe Fields

As a first step towards understanding the behavior of electrostatic round lenses, we discuss the fringe field effects appearing in an abrupt transition into a region of axial field from a field-free region. In practice this situation arises when transitioning through a hole of small aperture in a large charged metal plate. When passing through the plate leading to the idealized hard edge at $s = 0$, $E_s$ and $E_s'$ are expressed in terms of the Heaviside step function $H$ and the Dirac delta function $\delta$ defined in eqs. (4.5) via

$$E_s(s) = E_0 H(s), \qquad E'_s(s) = E_0 \delta(s),$$

which in terms of potentials corresponds to

$$V_0(s) = V - sE_0 H(s), \qquad V'_0(s) = -E_0 H(s), \qquad V''_0(s) = -E_0 \delta(s),$$

where we assume that the transition happens at an initial potential $V$, and we use the impulsive kick approximation in a similar manner as applied to the dipole edge focusing in Section 4.3.2. The position $x$ is unaffected over the infinitely short transition at $s = 0$, and we obtain

$$a_f = a_i + \frac{x_i}{2\chi_{e0}} \int_{0-}^{0+} \frac{\sqrt{K_0}}{\sqrt{K_0 - ZeV_0(s)}} V''_0(s)\, ds$$

$$= a_i - x_i \frac{E_0}{2\chi_{e0}} \frac{\sqrt{K_0}}{\sqrt{K_0 - ZeV}} = a_i + x_i \alpha \frac{\sqrt{K_0}}{\sqrt{K_0 - ZeV}},$$

where we use the abbreviation

$$\alpha = -\frac{E_0}{2\chi_{e0}}.$$

In an analogous way we can treat the transition from a region with field $E_0$ into a field free region, and obtain

$$E_s(\bar{s}) = E_0 H(-\bar{s}), \qquad E'_s(\bar{s}) = -E_0 \delta(\bar{s}),$$

where the exit is at $\bar{s} = 0$, so that we obtain

$$a_f = a_i + x_i \frac{E_0}{2\chi_{e0}} \frac{\sqrt{K_0}}{\sqrt{K_0 - ZeV}} = a_i - x_i \alpha \frac{\sqrt{K_0}}{\sqrt{K_0 - ZeV}}.$$

To summarize, the $2 \times 2$ transfer matrices for $(x, a)$ at the entrance and the exit are given as

$$\hat{M}^{\text{in}} = \begin{pmatrix} 1 & 0 \\ \alpha\sqrt{K_0/(K_0 - ZeV)} & 1 \end{pmatrix}, \quad \hat{M}^{\text{out}} = \begin{pmatrix} 1 & 0 \\ -\alpha\sqrt{K_0/(K_0 - ZeV)} & 1 \end{pmatrix},$$

$$\tag{4.11}$$

where $\alpha = -E_0/(2\chi_{e0})$ is used. We emphasize that $V$ is the momentary value of the potential, and $V$ appearing in $\hat{M}^{\text{in}}$ may differ from that in $\hat{M}^{\text{out}}$.

Overall we have kick effects that, similar to other fringe field cases, leave the position unaffected, but change the directions by an amount proportional to position. For a particle of positive charge, stepping into a positive field from a field free region leads to a focusing effect. Changing the sign of the field or stepping out of a positive field leads to a defocusing effect.

#### 4.4.1.2 The Mechanism of Focusing

We now consider in a very general manner the situation of a weak or short electrostatic lens. We observe that in eq. (4.7), one important effect has been neglected, namely the fact that the particle necessarily gains and loses energy as it travels through the lens. As a matter of fact, any transverse field bends a particle more readily when it is applied where the particle has lower velocity.

Consider a weak lens that consists of a potential that in the region from $-S$ to $+S$ is first constant, then rises, then plateaus, and then falls off to its original constant value, similar to the case shown on the right in Fig. 3.3. Necessarily the regions of positive second derivative appear near the minima of the potential function, while the regions of negative second derivative correspond to maxima of the potential function. So while traveling through the lens, the particle is more sensitive to the focusing fields than to the defocusing fields; and even though the focusing and defocusing portions cancel, the net effect is that the **orbit experiences focusing**. To observe this quantitatively, we transform the integral in eq. (4.10) via integration by parts, and obtain

$$a_f = a_i + \left. \frac{x_i}{2\chi_{e0}} \frac{\sqrt{K_0}}{\sqrt{K_0 - ZeV_0(s)}} V_0'(s) \right|_{-S}^{S} - \frac{x_i}{4\chi_{e0}} \int_{-S}^{S} \frac{\sqrt{K_0} ZeV_0'(s)}{[K_0 - ZeV_0(s)]^{\frac{3}{2}}} V_0'(s)\, ds.$$

The first term vanishes, and using $\chi_{e0} = p_0 v_0/(Ze)$, which non-relativistically reduces to

$$\chi_{e0} = \frac{2K_0}{Ze}, \tag{4.12}$$

the expression takes the form

$$a_f = a_i - 2x_i \int_{-S}^{S} \left[ \frac{K_0}{K_0 - ZeV_0(s)} \right]^{\frac{3}{2}} \left[ \frac{V_0'(s)}{2\chi_{e0}} \right]^2 ds.$$

Since the integrand involves only non-negative terms, the integral itself is non-negative; and if there is any place where the potential on axis $V_0(s)$ actually changes, then the integral is actually positive. Thus the change of $a$ is negative, corresponding to focusing.

As we have seen, other than restricting ourselves to a thin or weak lens, this result has been obtained by mere manipulations of the integral without any assumption of the specific form of $V_0(s)$ except for its constancy at $\pm S$. This is actually a small example of various similar manipulations that have been developed in the past; for example one can show the focusing property even in the extended case [60, 67]. More impressively, conceptually similar but practically more involved arguments can be used to prove Scherzer's theorems [62] about signs of higher order aberrations.

The resulting formula can be used to obtain an approximation of the focal length of a thin or weak electrostatic lens. However, other coordinates than our $x$ and $a$ are used frequently in the electron optics community; in particular, the coordinate $x$ is scaled to $\widetilde{x}$ by a factor depending on the kinetic energy of the particle. This leads to less change in the value of the position coordinate $\widetilde{x}$ as we travel through a lens. This nonlinear transformation leads to the approximation of $\widetilde{x}$=constant being better than that of $x$=constant, which in turn leads to better estimates for focal length. There is a large amount of work on this topic, but we forgo the details and refer to some of the literature [56, 60, 67].

### 4.4.1.3 The Plate Lens

We now address a particular type of lens for which it is possible to derive the transfer matrix analytically. We consider combinations of individual plates placed perpendicular to the optical axis, each of which has a small hole in its center through which the beam travels. Each of these plates represents a hard edge fringe field as discussed above.

The plates are held at different potentials and form equipotential surfaces. Assuming that the plates extend to infinity, the electric field between two successive plates is that of a common plate capacitor, and it points in the direction of the reference axis.

For any electrostatic lens it is desirable that the field far away vanishes. Compared to the case of the hard edge fringe field, this requires the use of at least two plates. Assuming the field between the plates to be $E_0$ and their distance to be $S$, i.e.,

$$E_s(s) = \begin{cases} E_0 & \text{for } 0 \leq s \leq S \\ 0 & \text{for } s < 0, \ s > S \end{cases}, \tag{4.13}$$

the potential at the second plate is $V = -SE_0$, and we have the potential function

$$V_0(s) = \begin{cases} 0 & \text{for } s < 0 \\ -sE_0 & \text{for } 0 \leq s \leq S \\ -SE_0 & \text{for } s > S \end{cases}.$$

The effects at $s = 0$ and $s = S$ are merely those of hard edge fringe fields as discussed before. It is worthwhile to point out that if $V > 0$, which entails that $E_0 < 0$, a particle of positive charge has lower energy at the second plate, and thus is more susceptible to the transverse electric fields there. From the discussion of hard edge fringe fields, we know that the transverse fields in both places are of opposite sign but identical magnitude, thus the second field, which happens to lead to focusing, has a more significant effect.

Using eqs. (4.11), we have the transfer matrices for the kicks at $s = 0$ and $s = S$ as follows. At $s = 0$, we have

$$V_0(0) = 0, \quad K(0) = K_0, \quad p_s(0) = p_0 = \sqrt{2mK_0}, \tag{4.14}$$

and at $s = S$,

$$V_0(S) = -SE_0, \quad K(S) = K_0 + ZeSE_0,$$
$$p_s(S) = p_S = \sqrt{2m\left(K_0 + ZeSE_0\right)}.$$

Thus, we obtain the transfer matrices at $s = 0$ and $s = S$ as

$$\hat{M}_0 = \begin{pmatrix} 1 & 0 \\ \alpha & 1 \end{pmatrix}, \qquad \hat{M}_S = \begin{pmatrix} 1 & 0 \\ -\alpha(p_0/p_S) & 1 \end{pmatrix}, \qquad (4.15)$$

where again the abbreviation

$$\alpha = -\frac{E_0}{2\chi_{e0}}$$

is used.

In order to study the two-plate lens quantitatively, we need to still determine the transfer matrix of the space between the plates. In this region, a charged particle experiences uniform acceleration or deceleration without change in transverse momentum, i.e., $a_f = a_i$. However, the position changes in a similar mechanism to that of a drift, but including the effect of the change of the kinetic energy. For the region between the electrodes, we have from the $x'$ equation of (4.9), and using $p_s(s) = \sqrt{2m\left(K_0 - ZeV_0(s)\right)}$,

$$\Delta x = x_f - x_i = a_i \int_0^S \frac{p_0}{p_s(s)} ds = a_i \int_0^S \frac{\sqrt{K_0}}{\sqrt{K_0 + ZesE_0}} ds$$
$$= a_i \frac{2}{ZeE_0} \sqrt{K_0} \left( \sqrt{K_0 + ZeSE_0} - \sqrt{K_0} \right) = a_i \frac{2}{ZeE_0} \frac{p_0\left(p_S - p_0\right)}{2m}$$
$$= a_i S \cdot \frac{2p_0}{p_S + p_0} = a_i L_S,$$

where the following relation is used to simplify the last step,

$$p_S^2 - p_0^2 = 2m\left(K(S) - K_0\right) = 2mZeSE_0, \qquad (4.16)$$

and the abbreviation

$$L_S = S \cdot \frac{2p_0}{p_S + p_0}$$

is used. Thus the transfer matrix of the gap between the plates is

$$\hat{M}_g = \begin{pmatrix} 1 & S \cdot 2p_0/(p_S + p_0) \\ 0 & 1 \end{pmatrix} = \begin{pmatrix} 1 & L_S \\ 0 & 1 \end{pmatrix}. \qquad (4.17)$$

We note that if there is no change in the kinetic energy and thus $p_S = p_0$, this matrix agrees with the matrix of a drift with length $S$. On the other hand, if the system is accelerating, we have $L_S < S$, while for a decelerating system, $L_S > S$, reflecting the behavior expected from a simple geometric analysis.

Combining the above matrices, we construct the transfer matrix of the electrostatic round lens with two plates as the field described in eq. (4.13).

$$\hat{M} = \hat{M}_S \cdot \hat{M}_g \cdot \hat{M}_0 = \begin{pmatrix} 1 & 0 \\ -\alpha(p_0/p_S) & 1 \end{pmatrix} \begin{pmatrix} 1 & L_S \\ 0 & 1 \end{pmatrix} \begin{pmatrix} 1 & 0 \\ \alpha & 1 \end{pmatrix}$$

$$= \begin{pmatrix} 1 + \alpha L_S & L_S \\ \alpha\left[1 - (p_0/p_S)(1 + \alpha L_S)\right] & 1 - \alpha L_S(p_0/p_S) \end{pmatrix}.$$

This is a rather compact representation of the matrix. However, to more clearly observe the influence of the defining parameters $V$ and $S$, we now also express the matrix in terms of these quantities. We note that $\alpha L_S$ can be expressed in terms of momenta using eqs. (4.16), (4.12) and (4.14), and we obtain

$$\alpha L_S = -\frac{E_0}{2\chi_{e0}} \cdot S \frac{2p_0}{p_S + p_0} = -\frac{1}{2} \frac{Ze}{2K_0} \frac{p_S^2 - p_0^2}{2mZe} \frac{2p_0}{p_S + p_0} = \frac{p_0 - p_S}{2p_0}. \qquad (4.18)$$

Using this, the elements of $\hat{M}$ can be organized as

$$(x|x) = 1 + \alpha L_S = \frac{3p_0 - p_S}{2p_0},$$

$$(a|a) = 1 - \alpha L_S \frac{p_0}{p_S} = 1 + \frac{p_S - p_0}{2p_S} = \frac{3p_S - p_0}{2p_S},$$

$$(a|x) = \alpha\left[1 - \frac{p_0}{p_S}(1 + \alpha L_S)\right] = \alpha\left(1 - \frac{p_0}{p_S} \frac{3p_0 - p_S}{2p_0}\right) = -3\alpha \frac{p_0 - p_S}{2p_S}$$

$$= -3\alpha^2 \frac{p_0}{p_S} L_S,$$

resulting in

$$\hat{M} = \begin{pmatrix} (3p_0 - p_S)/(2p_0) & L_S \\ -3\alpha^2 L_S(p_0/p_S) & (3p_S - p_0)/(2p_S) \end{pmatrix}. \qquad (4.19)$$

In this representation, it is obvious that the $(a|x)$ element is always negative, and thus we obtain the important conclusion that the two-plate lens always focuses.

As observed above, the simplest plate lens leading to vanishing electric fields at far distance required two plates. However, the potential after the two-plate lens differs from the potential before. Thus, in order to achieve identical potential before and after the lens, which is often desirable in practice, at least three plates are necessary. So we consider a lens that consists of three flat electrodes as shown in Fig. 4.9, where the axial electric field $E_s(s)$ and the potential $V_0(s)$ are given as

$$E_s(s) = \begin{cases} E_0 & \text{for} \quad -S \leq s \leq 0 \\ -E_0 & \text{for} \quad 0 < s \leq S \ , \\ 0 & \text{for} \quad |s| > S \end{cases}$$

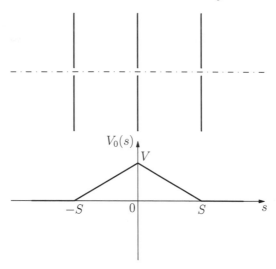

**FIGURE 4.9:** Layout (top) and potential profile (bottom) of the electro-static three-plate round lens.

and

$$V_0(s) = \begin{cases} -sE_0 + V & \text{for } -S \le s \le 0 \\ sE_0 + V & \text{for } 0 < s \le S \,, \\ 0 & \text{for } |s| > S \end{cases}$$

and the potential $V$ at the middle plate is given by

$$V = -SE_0.$$

We observe that we can treat this system as a combination of two of the two-plate electrostatic round lenses already discussed. For this purpose, we first list the momenta at the plates. At $s = \pm S$, we have

$$V_0(\pm S) = 0, \quad K(\pm S) = K_0, \quad p_s(\pm S) = p_0 = \sqrt{2mK_0},$$

and at $s = 0$, we have

$$V_0(0) = V = -SE_0, \quad K(0) = K_0 + ZeSE_0,$$
$$p_s(0) = p_m = \sqrt{2m\left(K_0 + ZeSE_0\right)}.$$

For purpose of clarification we note that if $V > 0$ as shown in Fig. 4.9, we have that $E_0 < 0$ and $p_m < p_0$.

The first half of this lens is simply the two-plate lens discussed above. Thus from eq. (4.19), the transfer matrix for the left half of the system is

$$\hat{M}_L = \begin{pmatrix} (3p_0 - p_m)/(2p_0) & L_S \\ -3\alpha^2 L_S(p_0/p_m) & (3p_m - p_0)/(2p_m) \end{pmatrix},$$

where

$$\alpha = -\frac{E_0}{2\chi_{e0}}, \qquad L_S = S \cdot \frac{2p_0}{p_m + p_0}.$$

The second half consists of the entrance kick with the momentary momentum $p_m$, the gap of length $S$ connecting the two momentum states $p_m$ and $p_0$, and the exit kick with the momentary momentum $p_0$, with the constant electric field $-E_0$ in the gap. The transfer matrices of these components turn out to be $\hat{M}_S$, $\hat{M}_g$ and $\hat{M}_0$ from eqs. (4.15) and (4.17), respectively, where here we use $p_m$ instead of $p_S$. So we obtain the transfer matrix for the right half of the system as

$$\hat{M}_R = \hat{M}_0 \cdot \hat{M}_g \cdot \hat{M}_S = \begin{pmatrix} 1 & 0 \\ \alpha & 1 \end{pmatrix} \begin{pmatrix} 1 & L_S \\ 0 & 1 \end{pmatrix} \begin{pmatrix} 1 & 0 \\ -\alpha(p_0/p_m) & 1 \end{pmatrix}$$

$$= \begin{pmatrix} 1 - \alpha L_S(p_0/p_m) & L_S \\ \alpha\left[1 - (p_0/p_m)(1 + \alpha L_S)\right] & 1 + \alpha L_S \end{pmatrix}$$

$$= \begin{pmatrix} (3p_m - p_0)/(2p_m) & L_S \\ -3\alpha^2 L_S(p_0/p_m) & (3p_0 - p_m)/(2p_0) \end{pmatrix}.$$

We observe much similarity between $\hat{M}_L$ and $\hat{M}_R$; the off-diagonal elements are the same, and the diagonal elements are merely switched. The $(a|x)$ element of both $\hat{M}_L$ and $\hat{M}_R$ is always negative, thus both of the lens halves always focus.

Finally, the calculation of the transfer matrix of the whole lens can be conducted, where the similarity of $\hat{M}_L$ and $\hat{M}_R$ helps the arithmetic in the process. We obtain

$$\hat{M}_T = \hat{M}_R \cdot \hat{M}_L = \begin{pmatrix} (a|a)_L & (x|a)_L \\ (a|x)_L & (x|x)_L \end{pmatrix} \begin{pmatrix} (x|x)_L & (x|a)_L \\ (a|x)_L & (a|a)_L \end{pmatrix}$$

$$= \begin{pmatrix} (x|x)_L(a|a)_L + (x|a)_L(a|x)_L & 2(a|a)_L(x|a)_L \\ (x|x)_L(a|x)_L & (x|x)_L(a|a)_L + (x|a)_L(a|x)_L \end{pmatrix}$$

$$= \begin{pmatrix} 1 - (3/2)(p_0 - p_m)^2/(p_0 p_m) & L_S(3p_m - p_0)/p_m \\ -3\alpha^2 L_S(3p_0 - p_m)/p_m & 1 - (3/2)(p_0 - p_m)^2/(p_0 p_m) \end{pmatrix},$$

where eq. (4.18) is used to simplify the result.

We now consider the focusing property of the three-plate lens. As before, $L_S > 0$, but we now have the factor $3p_0 - p_m$ determining the eventual sign of $(a|x)$. In case the magnitude of the voltage $V$ is small compared to the kinetic energy of the particle, the center momentum $p_m$ is similar to $p_0$, and so $3p_0 - p_m > 0$. On the other hand, in cases of extreme voltage $V$ leading to large acceleration and large center momentum $p_m$, it is conceivable to achieve $p_m > 3p_0$. In this extreme case, the three-plate lens actually defocuses.

## 4.4.2 The Magnetic Round Lens

Magnetic round lenses are arrangements of wires wound equidistant to the reference orbit, either in long arrangements similar to textbook-like solenoids with a large region of nearly constant axial fields, or in short or "thin" arrangements where the field on axis rises, reaches a peak, and then falls off again. Thin magnetic lenses are the main staple used for focusing in electron microscopes, while long magnetic lenses are used in various particle accelerators for guiding the beam, including applications in muon ionization cooling. We have the magnetic field, linear in $x$ and $y$, given as

$$B_x = \frac{1}{2}V_0''(s)\,x, \quad B_y = \frac{1}{2}V_0''(s)\,y, \quad B_s = -V_0'(s),$$

which is derived from the corresponding magnetic scalar potential

$$V = V_0(s) - \frac{1}{4}V_0''(s)\left(x^2 + y^2\right).$$

Different from the electrostatic potential in the case of the electrostatic round lens, the magnetic potential does not enter the equations of motion directly. So we simplify notation by not having it appear in the equations of motion, and rather express the fields in terms of the axial center field $B_s(s)$ as

$$B_x = -\frac{1}{2}B_s'(s)x, \quad B_y = -\frac{1}{2}B_s'(s)y, \quad B_s = B_s(s).$$

Applying this to eqs. (4.8), and linearizing in the similar process for eqs. (4.2), we obtain

$$x' = a, \quad a' = +\frac{B_s}{\chi_{m0}}b + \frac{1}{2}\frac{B_s'}{\chi_{m0}}y,$$

$$y' = b, \quad b' = -\frac{B_s}{\chi_{m0}}a - \frac{1}{2}\frac{B_s'}{\chi_{m0}}x, \quad l' = \frac{1}{(2+\eta_0)^2}\delta, \quad \delta' = 0. \tag{4.20}$$

It is immediately apparent that the motion of the two planes are coupled, which will lead to interesting properties. The longitudinal motion to first order is the same with that of a drift, thus

$$l_f = D\delta_i + l_i = \frac{L}{(2+\eta_0)^2}\delta_i + l_i, \quad \delta_f = \delta_i$$

with length $L$.

To study the transversal motion, we first express the equations of motion (4.20) in vector notation. Using

$$\vec{z} = \begin{pmatrix} x \\ y \end{pmatrix}, \quad \vec{c} = \begin{pmatrix} a \\ b \end{pmatrix}, \quad \hat{J} = \begin{pmatrix} 0 & 1 \\ -1 & 0 \end{pmatrix},$$

we have

$$\vec{z}' = \vec{c}, \quad \vec{c}' = \frac{B_s}{\chi_{m0}}\hat{J}\vec{c} + \frac{B_s'}{2\chi_{m0}}\hat{J}\vec{z}. \tag{4.21}$$

For a particle moving parallel to the $s$-axis without transversal momentum, i.e., $\vec{c}_i = 0$, the second term in the $\vec{c}'$ equation, $B_s'/(2\chi_{m0}) \cdot \hat{J}\vec{z}$, acts to produce a nonzero transversal momentum. Typically, the situation of $B_s' \neq 0$ happens in the fringe field regions near the entrance and the exit, and it is the main reason why particles entering parallel to the $s$-axis will follow a transversely rotating motion inside the magnetic round lens.

In the following, we will employ a dual approach in understanding the motion of particles in the magnetic round lens. The quantitative approach will be based on studying the equations of motion and solving them for certain specific cases. But equally important is the qualitative understanding of where the focusing effects of a magnetic round lens really come from.

### 4.4.2.1 Hard Edge Fringe Fields

We first discuss the fringe field effects appearing in an abrupt transition into a region of axial field from a field-free region. In practice this situation arises at the edge of a solenoid of very small radial aperture. We use the impulsive kick approximation in a similar manner as applied to the dipole edge focusing in Section 4.3.2. When entering into the solenoid of the constant field strength $B_0$ at the idealized hard edge at $s = 0$, $B_s$ and $B_s'$ are expressed in terms of the Heaviside step function $H$ and the Dirac delta function $\delta$ defined in eqs. (4.5) via

$$B_s(s) = B_0 H(s), \qquad B_s'(s) = B_0 \delta(s),$$

where the entrance lies at $s = 0$. Applying them to the equations of motion (4.21), we obtain

$$\vec{c}_f = \vec{c}_i + \int_{0-}^{0+} \left[ \frac{B_0 H(\bar{s})}{\chi_{m0}} \hat{J}\vec{c}_i + \frac{B_0 \delta(\bar{s})}{2\chi_{m0}} \hat{J}\vec{z}_i \right] d\bar{s} = \vec{c}_i + \frac{B_0}{2\chi_{m0}} \hat{J}\vec{z}_i,$$

while $\vec{z}$ is unaffected. Altogether we obtain the transformation relations between the initial conditions $\vec{z}_i$ and $\vec{c}_i$ and the final conditions $\vec{z}_f$ and $\vec{c}_f$ for such an idealized thin edge as

$$\vec{z}_f = \vec{z}_i, \qquad \vec{c}_f = \vec{c}_i + \frac{B_0}{2\chi_{m0}} \hat{J}\vec{z}_i. \tag{4.22}$$

At the exit the same effect happens, but with an opposite sign of $B_s'$, because we now have

$$B_s(\bar{s}) = B_0 H(-\bar{s}), \qquad B_s'(\bar{s}) = -B_0 \delta(\bar{s})$$

where the exit is at $\bar{s} = 0$, so that we obtain

$$\vec{z}_f = \vec{z}_i, \qquad \vec{c}_f = \vec{c}_i - \frac{B_0}{2\chi_{m0}} \hat{J}\vec{z}_i. \tag{4.23}$$

Overall we have kick effects that similar to other fringe field cases leave the position unaffected, but change the directions by an amount proportional to

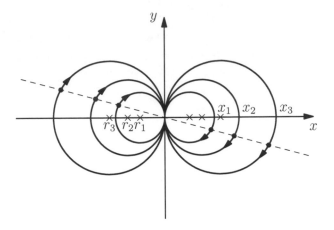

**FIGURE 4.10:** Transversal motion of particles entering at positions $x_i$ initially parallel to the reference axis in the magnetic solenoid. After crossing the fringe fields, the particles execute rotations with radii $r_i$ equaling half of their entrance position.

position. However, very different from the case of the electrostatic round lens, this change in direction in itself is neither focusing nor defocusing but rather happens in azimuthal direction, perpendicular to the radial direction.

### 4.4.2.2 The Mechanism of Focusing

Since the action of fringe fields in the transition into axial magnetic fields leads to azimuthal kicks, one is led to wonder where any focusing effects may arise from. We consider an incoming beam where all trajectories are parallel to the reference axis, i.e., $\vec{c}_i = 0$. Furthermore, because of the rotational symmetry, it is sufficient to limit our attention to a particle starting on the horizontal axis, which means $\vec{z}_f = \vec{z}_i = (x_i, 0)$ in eqs. (4.22). After having moved through the fringe field, the particle has now picked up a velocity component in the negative $y$ direction via $\vec{c}_f = (B_0/2\chi_{m0})\hat{J}\vec{z}_i = (B_0/2\chi_{m0})(0, -x_i)$.

We assume that the particle travels for a while in the constant magnetic field $B_0$. In this field, it performs a rotation due to the nonzero transversal velocity picked up in the fringe field, which is $v_t = (B_0/2\chi_{m0})x_i v_0$. The radius of this rotational motion is given by

$$r = \frac{\gamma m v_t}{Z e B_0} = \frac{\gamma m}{Z e B_0} \cdot \frac{B_0}{2\chi_{m0}} x_i v_0 = \frac{1}{2}x_i,$$

where $\chi_{m0} = p_0/Ze = \gamma m v_0/Ze$ is used. So the rotation radius is exactly half of the radial entrance position of the particle. As it turns out, the factor of $1/2$, which can be traced back to the fact that the radial derivative $\partial B_x/\partial x = -1/2 \cdot B_s'(s)$ is only half of that in the axis direction, will be the key mechanism to focusing.

We illustrate the resulting motion in Fig. 4.10, which shows the resulting transversal orbits for six particles on both sides of the origin that were initially moving parallel to the axis of the solenoid without any transverse motion. Each of these particles performs a circular motion with a radius that equals half of its entrance position. In particular this entails that the particles do not have concentric orbits like in a cyclotron, but rather that after a half period of their oscillations, all these particles will pass through the exact center of the device.

This observation now leads to **focusing**; on their path from their initial position $x_i$ to the position $x = 0$ after a half period, each particle loses distance to the axis and is thus focused.

We further observe that all particles rotate with the same angular frequency given by

$$\omega_c = \frac{v_t}{r} = \frac{Ze}{\gamma m} B_0.$$

This entails that all initially parallel particles reach the origin at the same time, and also that the relative loss of distance to the axis is the same. So the speed with which they move towards the origin is proportional to their initial position, and it is a hallmark of thin lens focusing.

The picture suggests another interesting feature; initially axis-parallel particles that are on a common line upon entering remain on a common line, shown dashed in Fig. 4.10, in their further motion through the region of constant magnetic field. Elementary geometry shows that this is actually the case: by connecting any particle in Fig. 4.10 (dot) and the corresponding center of its orbit (cross), we can see immediately that the angle between the dashed line and the $x$-axis is always half of that between the radius and the $x$-axis. We further observe that when the particles reach the origin and have performed a half revolution around their orbits, the dashed line will coincide with the vertical axis, and thus will have performed a quarter revolution; in fact elementary geometry shows that the dashed line performs a rotation with one half of the rotational frequency of particles.

### 4.4.2.3 The Rotating Coordinate System

These observations of the idealized case now motivate the treatment of the general case, in which fields do not jump abruptly but rather gently depend on $s$. In this case the appearing momentary rotation frequencies of both the particles as well as a possible suitable rotating coordinate system are not constant but will change with the position $s$. Thus, we attempt the Ansatz of introducing new variables $\vec{Z}$ that describe the motion in a rotating coordinate system via

$$\vec{z}(s) = \hat{R}(\theta(s)) \cdot \vec{Z}(s), \tag{4.24}$$

and we will arrive at the expected result at eqs. (4.28). Here, $\hat{R}(\theta)$ is a rotation matrix, and for the further discussion, the following matrix properties are useful:

$$\hat{J} \cdot \hat{J} = -\hat{I}, \quad \text{where} \quad \hat{I} = \begin{pmatrix} 1 & 0 \\ 0 & 1 \end{pmatrix} \quad \text{and} \quad \hat{J} = \begin{pmatrix} 0 & 1 \\ -1 & 0 \end{pmatrix},$$

$$\hat{R}(\theta) = \begin{pmatrix} \cos\theta & -\sin\theta \\ \sin\theta & \cos\theta \end{pmatrix} = \cos\theta \cdot \hat{I} - \sin\theta \cdot \hat{J}, \quad \text{and} \quad \hat{R}^{-1}(\theta) = \hat{R}(-\theta),$$

$$\hat{J} \cdot \hat{R}(\theta) = \hat{R}(\theta) \cdot \hat{J} = \begin{pmatrix} \sin\theta & \cos\theta \\ -\cos\theta & \sin\theta \end{pmatrix} = \sin\theta \cdot \hat{I} + \cos\theta \cdot \hat{J},$$

$$\frac{d\hat{R}(\theta)}{d\theta} = -\begin{pmatrix} \sin\theta & \cos\theta \\ -\cos\theta & \sin\theta \end{pmatrix} = -\hat{J}\hat{R}(\theta), \quad \text{and} \quad \frac{d\hat{R}(\theta)}{ds} = -\theta'\hat{J}\hat{R}(\theta).$$

$$(4.25)$$

Similar to the definitions of $\vec{z}$ and $\vec{c}$, we denote the new rotating variables $\vec{Z}$ and $\vec{C}$ as

$$\vec{Z} = \begin{pmatrix} X \\ Y \end{pmatrix}, \quad \vec{C} = \begin{pmatrix} A \\ B \end{pmatrix},$$

and define $\vec{C}$ as the first derivative of $\vec{Z}$ with respect to $s$

$$\vec{C}(s) = \vec{Z}'(s) = \frac{d\vec{Z}(s)}{ds}.$$

Note that this does not automatically entail $\vec{c} = \hat{R} \cdot \vec{C}$. Rather, $\vec{c}$ differs from $\hat{R} \cdot \vec{C}$ as we will see now. By computing the derivative of eq. (4.24), we express $\vec{c}(s)$ in terms of $\hat{R}$, $\vec{Z}$ and $\vec{C}$.

$$\vec{c}(s) = \vec{z}'(s) = \frac{d\hat{R}(\theta)}{ds}\vec{Z}(s) + \hat{R}(\theta)\vec{Z}'(s) = -\theta'\hat{J}\hat{R}(\theta)\vec{Z}(s) + \hat{R}(\theta)\vec{C}(s). \quad (4.26)$$

In turn, this allows us to express $\vec{C}(s)$ in terms of $\vec{z}$ and $\vec{c}$.

$$\vec{C}(s) = \hat{R}^{-1}(\theta) \cdot \left[\vec{c}(s) + \theta'\hat{J}\hat{R}(\theta)\vec{Z}(s)\right] = \hat{R}(-\theta)\vec{c}(s) + \theta'\hat{J}\hat{R}(-\theta)\vec{z}(s),$$

where eqs. (4.24) and (4.25) are used. Next, we calculate the derivative of $\vec{C}(s)$, expressed in terms of $\vec{z}$ and $\vec{c}$.

$$\vec{C}'(s) = \frac{d\hat{R}(-\theta)}{ds}\vec{c}(s) + \hat{R}(-\theta)\vec{c}'(s)$$

$$+ \theta''\hat{J}\hat{R}(-\theta)\vec{z}(s) + \theta'\hat{J}\frac{d\hat{R}(-\theta)}{ds}\vec{z}(s) + \theta'\hat{J}\hat{R}(-\theta)\vec{z}'(s)$$

$$= \left(\theta''\hat{J} - \theta'^2\hat{I}\right)\hat{R}(-\theta)\vec{z}(s) + 2\theta'\hat{J}\hat{R}(-\theta)\vec{c}(s) + \hat{R}(-\theta)\vec{c}'(s),$$

where the relations (4.25) and $\vec{c} = \vec{z}'$ are used.

We now insert the equation of motion (4.21) for $\vec{c}\,'$ above, and express $\vec{C}'(s)$ in terms of $\vec{Z}$ and $\vec{C}$ using eqs. (4.24) and (4.26).

$$\vec{C}'(s) = \left( \theta'' \hat{J} - \theta'^2 \hat{I} \right) \hat{R}(-\theta) \vec{z}(s) + 2\theta' \hat{J} \hat{R}(-\theta) \vec{c}(s)$$

$$+ \hat{R}(-\theta) \left[ \frac{B_s}{\chi_{m0}} \hat{J} \vec{c}(s) + \frac{B_s'}{2\chi_{m0}} \hat{J} \vec{z}(s) \right]$$

$$= \left[ \left( \theta'' + \frac{B_s'}{2\chi_{m0}} \right) \hat{J} - \theta'^2 \hat{I} \right] \hat{R}(-\theta) \vec{z}(s) + \left( 2\theta' + \frac{B_s}{\chi_{m0}} \right) \hat{J} \hat{R}(-\theta) \vec{c}(s)$$

$$= \left[ \left( \theta'' + \frac{B_s'}{2\chi_{m0}} \right) \hat{J} - \theta'^2 \hat{I} + 2\theta' \left( \theta' + \frac{B_s}{2\chi_{m0}} \right) \hat{I} \right] \vec{Z}(s)$$

$$+ 2 \left( \theta' + \frac{B_s}{2\chi_{m0}} \right) \hat{J} \vec{C}(s),$$

where the relations (4.25) are used.

The resulting equation of motion for $\vec{C}$ looks rather complicated, but a closer inspection shows the same factor appearing repeatedly. Indeed, if we demand

$$\theta' + \frac{B_s}{2\chi_{m0}} = 0,$$

which is equivalent to

$$\theta(s) = - \int_0^s \frac{B_s(\bar{s})}{2\chi_{m0}} d\bar{s}, \tag{4.27}$$

the equation for $\vec{C}'(s)$ greatly simplifies to

$$\vec{C}'(s) = -\theta'^2 \vec{Z}(s).$$

This means that the motions of the two coordinates of the vectors $\vec{Z}$ and $\vec{C}$ fully decouple, and we hence have the following simple set of first order differential equations to describe the motion of $\vec{Z}(s)$ :

$$\vec{Z}'(s) = \vec{C}(s), \qquad \vec{C}'(s) = -\theta'^2 \vec{Z}(s) = - \left( \frac{B_s(s)}{2\chi_{m0}} \right)^2 \cdot \vec{Z}(s). \tag{4.28}$$

So the motion in the rotating system is like a harmonic oscillator with varying strength, which is given by the square of the angular frequency $\theta'$ of the rotation of the coordinate system, which itself is proportional to the longitudinal field $B_s(s)$. It is a quite remarkable and yet simple result that fully describes the linearized motion in the magnetic round lens.

We note that, as a consequence, a short magnetic solenoid is **focusing**, and the focusing power is proportional to $\theta'^2 \propto 1/\chi_{m0}^2 \propto 1/p_0^2$, whereas that of a magnetic quadrupole is proportional to $1/\chi_{m0} \propto 1/p_0$. Therefore the advantage of using magnetic quadrupoles compared to magnetic round lenses becomes more pronounced for beams of higher momentum.

#### 4.4.2.4 The Solenoid with Hard Edge Fringe Fields

We now study an idealized long solenoid with vanishing field outside and with a constant interior field of

$$B_s(s) = B_0, \qquad B_x = B_y = 0.$$

We begin the discussion with the observation that the form of the equation of motion (4.28) entails that the quantities $\vec{Z}$ and $\vec{C}$ vary continuously even when passing through a hard edge fringe field. In fact, different from the situation for $\vec{z}$ and $\vec{c}$, there are no delta functions appearing which led to the discontinuities in eqs. (4.22) and (4.23). So in both the entrance and the exit of the hard edge fringe field, instead of eqs. (4.22) and (4.23), we simply have

$$\vec{Z}_f = \vec{Z}_i, \qquad \vec{C}_f = \vec{C}_i. \tag{4.29}$$

Assuming that the beginning edge of the solenoid is located at $s = 0$, we have that

$$\theta(s) = -\frac{B_0}{2\chi_{m0}} s, \qquad \theta' = \frac{d\theta}{ds} = -\frac{B_0}{2\chi_{m0}} \quad \text{(constant)}.$$

The angular frequency $\theta'$ is now constant on the inside, and we use the abbreviation

$$\omega = -\frac{B_0}{2\chi_{m0}} \quad \text{(constant)}.$$

Thus we obtain a simple harmonic oscillator solution for $\vec{Z}$. Using the initial conditions $\vec{Z}_0$ and $\vec{C}_0$, we have

$$\vec{Z}(s) = \cos(\omega s)\vec{Z}_0 + \frac{1}{\omega}\sin(\omega s)\vec{C}_0,$$
$$\vec{C}(s) = -\omega \sin(\omega s)\vec{Z}_0 + \cos(\omega s)\vec{C}_0. \tag{4.30}$$

Now this solution has to be expressed in terms of the original coordinates $\vec{z}$ and $\vec{c}$.

We begin by observing that outside the beginning of the solenoid, we simply have

$$\vec{Z}_i = \vec{z}_i, \qquad \vec{C}_i = \vec{c}_i. \tag{4.31}$$

Next, because of eqs. (4.29), even just after entering the solenoid, we have $\vec{Z}_i = \vec{z}_i$, $\vec{C}_i = \vec{c}_i$. In the solenoid itself, the quantities $\vec{Z}$ and $\vec{C}$ change according to eqs. (4.30) until the end of the solenoid is reached, where they have the values $\vec{Z}(L)$ and $\vec{C}(L)$. When exiting the solenoid, $\vec{Z}$ and $\vec{C}$ again remain unchanged because of eqs. (4.29). In the outside region, there is no field left, so in eqs. (4.24) and (4.26) we have $\theta' = 0$, thus the transformation equations simplify to

$$\vec{z}(L) = \hat{R}(\omega L) \cdot \vec{Z}(L), \qquad \vec{c}(L) = \hat{R}(\omega L) \cdot \vec{C}(L). \tag{4.32}$$

We now combine the various matrices and use $\varphi = \omega L$. We obtain the resulting $4 \times 4$ transfer matrix $\hat{M}^{\text{out-int-in}}$ of the solenoid of length $L$ for $(x, a, y, b)$, representing first entering into the solenoid, then passing through its interior, and then exiting out of the solenoid, as a combination of eqs. (4.31), (4.30) and (4.32):

$$\hat{M}^{\text{out-int-in}} = \hat{R}(\varphi) \cdot \hat{M}_{HO}(\varphi)$$

$$= \begin{pmatrix} \cos^2 \varphi & \sin \varphi \cos \varphi / \omega & -\sin \varphi \cos \varphi & -\sin^2 \varphi / \omega \\ -\omega \sin \varphi \cos \varphi & \cos^2 \varphi & \omega \sin^2 \varphi & -\sin \varphi \cos \varphi \\ \sin \varphi \cos \varphi & \sin^2 \varphi / \omega & \cos^2 \varphi & \sin \varphi \cos \varphi / \omega \\ -\omega \sin^2 \varphi & \sin \varphi \cos \varphi & -\omega \sin \varphi \cos \varphi & \cos^2 \varphi \end{pmatrix}, \quad (4.33)$$

where we used the $4 \times 4$ matrices for the harmonic oscillator solution

$$\hat{M}_{HO}(\varphi) = \begin{pmatrix} \cos \varphi & \sin \varphi / \omega & 0 & 0 \\ -\omega \sin \varphi & \cos \varphi & 0 & 0 \\ 0 & 0 & \cos \varphi & \sin \varphi / \omega \\ 0 & 0 & -\omega \sin \varphi & \cos \varphi \end{pmatrix}, \quad (4.34)$$

and for the rotation

$$\hat{R}(\varphi) = \begin{pmatrix} \cos \varphi & 0 & -\sin \varphi & 0 \\ 0 & \cos \varphi & 0 & -\sin \varphi \\ \sin \varphi & 0 & \cos \varphi & 0 \\ 0 & \sin \varphi & 0 & \cos \varphi \end{pmatrix}. \quad (4.35)$$

We observe that these matrices commute, i.e., $\hat{M}_{HO} \cdot \hat{R} = \hat{R} \cdot \hat{M}_{HO}$, which helps simplify the matrix arithmetic here and below.

We note that the matrix $\hat{M}^{\text{out-int-in}}$ has unit determinant since both the harmonic oscillator solution matrix $\hat{M}_{HO}$ and the subsequent rotation $\hat{R}$ do. It is also easy to show that the longitudinal angular momentum is conserved, because

$$\vec{z}_f \times \vec{c}_f = \vec{z}(L) \times \vec{c}(L) = (\hat{R}(\omega L) \cdot \vec{Z}(L)) \times (\hat{R}(\omega L) \cdot \vec{C}(L)) = \vec{Z}(L) \times \vec{C}(L)$$
$$= \left( \cos \varphi \vec{Z}_i + \frac{1}{\omega} \sin \varphi \vec{C}_i \right) \times \left( -\omega \sin \varphi \vec{Z}_i + \cos \varphi \vec{C}_i \right) = \vec{Z}_i \times \vec{C}_i$$
$$= \vec{z}_i \times \vec{c}_i.$$

Now we may wonder what happens if we study the motion not only from the field free regions before to the field free region after the solenoid. For this purpose, we first remind ourselves of the $4 \times 4$ transfer matrices of the entrance and the exit edges, which according to eqs. (4.22) and (4.23) are

$$\hat{M}^{\text{in}} = \begin{pmatrix} 1 & 0 & 0 & 0 \\ 0 & 1 & -\omega & 0 \\ 0 & 0 & 1 & 0 \\ \omega & 0 & 0 & 1 \end{pmatrix}, \qquad \hat{M}^{\text{out}} = \begin{pmatrix} 1 & 0 & 0 & 0 \\ 0 & 1 & \omega & 0 \\ 0 & 0 & 1 & 0 \\ -\omega & 0 & 0 & 1 \end{pmatrix}.$$

We observe $(\hat{M}^{\text{out}})^{-1} = \hat{M}^{\text{in}}$, and both of the matrices have determinant 1.

We discussed the behavior of the edges in Section 4.4.2.2, and we can now determine the change in the longitudinal angular momentum across the edges. At the entrance edge, using eqs. (4.22),

$$\vec{z}_f \times \vec{c}_f = \vec{z}_i \times (\vec{c}_i - \omega \hat{J}\vec{z}_i) = \vec{z}_i \times \vec{c}_i - \omega \vec{z}_i \times (\hat{J}\vec{z}_i)$$

$$= \vec{z}_i \times \vec{c}_i - \omega \begin{pmatrix} x_i \\ y_i \end{pmatrix} \times \begin{pmatrix} y_i \\ -x_i \end{pmatrix} = \vec{z}_i \times \vec{c}_i + \omega \left(x_i^2 + y_i^2\right) \vec{e}_s = \vec{z}_i \times \vec{c}_i + \omega r_i^2 \vec{e}_s,$$

where $\vec{e}_s$ is the longitudinal unit vector. Thus the amount of angular momentum generated is $\omega r_i^2$ at the entrance and $-\omega r_i^2$ at the exit.

The transfer matrices for various situations below can be obtained by combining $\hat{M}^{\text{out-int-in}}$, $\hat{M}^{\text{in}}$, $\hat{M}^{\text{out}}$, and their inverse matrices. The transfer matrix of the interior part is obtained by removing the entrance and the exit matrices from $\hat{M}^{\text{out-int-in}}$ as $(\hat{M}^{\text{out}})^{-1} \cdot \hat{M}^{\text{out-int-in}} \cdot (\hat{M}^{\text{in}})^{-1}$, and we obtain

$$\hat{M}^{\text{int}} = \begin{pmatrix} 1 & \sin\varphi\cos\varphi/\omega & 0 & -\sin^2\varphi/\omega \\ 0 & \cos(2\varphi) & 0 & -\sin(2\varphi) \\ 0 & \sin^2\varphi/\omega & 1 & \sin\varphi\cos\varphi/\omega \\ 0 & \sin(2\varphi) & 0 & \cos(2\varphi) \end{pmatrix}.$$

The determinant is again 1. Interestingly, the longitudinal angular momentum is not conserved in general when going through the interior, and the amount of the change is $\omega(r_f^2 - r_i^2)$.

When the particles start inside the solenoid and then exit, which typically happens when the particles are "born" inside the solenoid, we obtain $\hat{M}^{\text{out-int}}$ by removing the entrance matrix from $\hat{M}^{\text{out-int-in}}$, i.e., $\hat{M}^{\text{out-int-in}} \cdot (\hat{M}^{\text{in}})^{-1}$, or equivalently adding the exit matrix to $\hat{M}^{\text{int}}$, i.e., $\hat{M}^{\text{out}} \cdot \hat{M}^{\text{int}}$, as

$$\hat{M}^{\text{out-int}} = \begin{pmatrix} 1 & \sin\varphi\cos\varphi/\omega & 0 & -\sin^2\varphi/\omega \\ 0 & \cos^2\varphi & \omega & -\sin\varphi\cos\varphi \\ 0 & \sin^2\varphi/\omega & 1 & \sin\varphi\cos\varphi/\omega \\ -\omega & \sin\varphi\cos\varphi & 0 & \cos^2\varphi \end{pmatrix},$$

which again has determinant 1. As discussed above, while $\hat{M}^{\text{out-int-in}}$ conserves the longitudinal angular momentum, $\hat{M}^{\text{in}}$ does not, thus the longitudinal angular momentum changes in the process.

Finally, when the particles enter into the solenoid and remain inside, we have $\hat{M}^{\text{int-in}}$ of the form

$$\hat{M}^{\text{int-in}} = \begin{pmatrix} \cos^2\varphi & \sin\varphi\cos\varphi/\omega & -\sin\varphi\cos\varphi & -\sin^2\varphi/\omega \\ -\omega\sin(2\varphi) & \cos(2\varphi) & -\omega\cos(2\varphi) & -\sin(2\varphi) \\ \sin\varphi\cos\varphi & \sin^2\varphi/\omega & \cos^2\varphi & \sin\varphi\cos\varphi/\omega \\ \omega\cos(2\varphi) & \sin(2\varphi) & -\omega\sin(2\varphi) & \cos(2\varphi) \end{pmatrix}$$

as a result of removing the exit matrix from $\hat{M}^{\text{out-int-in}}$, i.e., $(\hat{M}^{\text{out}})^{-1}$ · $\hat{M}^{\text{out-int-in}}$. Again the determinant is 1, and the longitudinal angular momentum changes.

We conclude our discussion with a more detailed comparison of the treatment of the solenoid with the results of the electrostatic round lenses. We note the similarity between the solenoid including entrance and exit fringe fields and the two-plate electrostatic lens. Indeed, the magnetic scalar potential, expressed in terms of the angle $\theta$ in eq. (4.27), has risen steadily inside the solenoid and is reaching a plateau outside of the solenoid, just as the electrostatic potential rose steadily between the plates of the two-plate lens.

It is thus illuminating to study the case of two solenoids of equal length and opposite strength, which will lead to a vanishing magnetic potential change at the end of the second solenoid, and conceptually corresponds to the electrostatic three-plate lens. In that case, the fact that the potential returns to its original constant value entailed that the particle's energies are the same as before. Here by virtue of eq. (4.27), we obtain that the net rotation of the system is zero.

To be quantitative, we can obtain the transfer matrix of this system by combining those of two opposite solenoids with hard edges. We remind ourselves that the solenoid transfer matrix in eq. (4.33) was obtained as the product

$$\hat{M}_1 = \hat{R}(\varphi) \cdot \hat{M}_{HO}(\varphi)$$

of the matrices $\hat{R}(\varphi)$ and $\hat{M}_{HO}(\varphi)$ defined in eqs. (4.35) and (4.34). So the transfer matrix of a solenoid of opposite field is simply given by

$$\hat{M}_2 = \hat{R}(-\varphi) \cdot \hat{M}_{HO}(-\varphi).$$

Now we can use the fact that $\hat{R}(\varphi)$ and $\hat{M}_{HO}(\varphi)$ commute as noted above, and obtain for the combined matrix

$$\hat{M} = \hat{M}_2 \cdot \hat{M}_1 = \hat{R}(-\varphi) \cdot \hat{M}_{HO}(-\varphi) \cdot \hat{R}(\varphi) \cdot \hat{M}_{HO}(\varphi)$$
$$= \hat{R}(-\varphi) \cdot \hat{R}(\varphi) \cdot \hat{M}_{HO}(-\varphi) \cdot \hat{M}_{HO}(\varphi) = \hat{M}_{HO}(2\varphi),$$

where we have used that $\hat{R}(\varphi)^{-1} = \hat{R}(-\varphi)$ and $\hat{M}_{HO}(-\varphi) = \hat{M}_{HO}(\varphi)$. So indeed any rotation is removed, the $x$ and $y$ motion are fully decoupled, and correspond to a simple harmonic oscillator that is equal to that of a solenoid of twice the original length.

---

## 4.5  *Aberration Formulas

In the previous sections we have discussed in detail the linearization of the motion in particle optical coordinates, and the resulting transfer matrices for common particle optical elements. However, in general the motion is not linear, and in many situations it is important to take into account various contributions of the nonlinear effects.

Unfortunately, as straightforward as the determination of linear transfer matrices is in many cases, the determination of nonlinear terms, especially those of higher order, becomes exceedingly more difficult using paper and pencil methods. In the following we will describe a general method that in principle allows the recursive determination of aberrations of higher and higher orders, but which in practice quickly succumbs to a rapid increase in complexity for higher orders.

Let us assume we are given the system described by the following ordinary differential equation (ODE)

$$\frac{d}{ds}\vec{r} = \vec{f}(\vec{r}, s),$$

which satisfies $\vec{f}(\vec{0}, s) = \vec{0}$. We perform a Taylor expansion of the right hand side. Because the system is origin preserving, the first contribution is linear, and altogether we have

$$\frac{d}{ds}\vec{r} = \hat{M}(s) \cdot \vec{r} + \sum_{j=2}^{\infty} \vec{N}_j(\vec{r}, s),$$

where the $\vec{N}_j$ are polynomials of exact order $j$, the coefficients of which may depend on $s$.

The first step in obtaining a perturbative solution of the system is a **linearization** as in the previous sections. We have

$$\frac{d}{ds}\vec{r} = \hat{M}(s) \cdot \vec{r}.$$

For this system, we determine a system of $n$ independent solutions $\vec{l}_k(s)$, $k = 1, \ldots, n$, that satisfy the initial condition

$$\vec{l}_k(0) = (0, 0, \ldots, \underbrace{1}_{k\text{th}}, \ldots 0, 0)^T.$$

We define the matrix

$$\hat{L}(s) = \left(\vec{l}_1(s), \vec{l}_2(s), \ldots, \vec{l}_n(s)\right),$$

and observe that the **general solution** of the linearized problem with initial condition $\vec{r}_i$ is then given by

$$\vec{r}(s) = \hat{L}(s) \cdot \vec{r}_i.$$

In practice, the determination of $\hat{L}$ may be possible in closed form, depending on the structure of $\hat{M}$, or may have to rely on numerical integration. For the special case that $\hat{M}$ is **piecewise constant**, then for every such piece, one can try the ansatz $\vec{l}_k = \vec{v}_k \cdot \exp(\omega_k s)$, which leads to the condition

$$\omega_k \vec{v}_k \exp(\omega_k s) = \hat{M} \cdot \vec{v}_k \exp(\omega_k s),$$

an eigenvector problem. If $\hat{M}$ has $n$ distinct eigenvalues, we are done, and depending on whether $\omega_k$ is real or complex, the solutions can also be expressed in terms of sin, cos or sinh, cosh. In case of multiple eigenvalues, often solutions of the form $s \cdot \sin$, etc., can be found.

The next step consists of an **expansion** of $\vec{r}(s)$ in a Taylor polynomial

$$\vec{r}(s) = \hat{L}(s) \cdot \vec{r}_i + \sum_{j=2}^{\infty} \vec{R}_j (s, \vec{r}_i),$$

where $\vec{R}_j$ denotes a polynomial of exact order $j$ in the initial conditions, the coefficients of which may depend on $s$. We **insert** this expansion into the ODE and obtain

$$\frac{d}{ds}\hat{L}(s) \cdot \vec{r}_i + \sum_{j=2}^{\infty} \frac{d}{ds}\vec{R}_j (s, \vec{r}_i)$$

$$= \hat{M}(s) \cdot \hat{L}(s) \cdot \vec{r}_i + \hat{M}(s) \cdot \sum_{j=2}^{\infty} \vec{R}_j (s, \vec{r}_i) + \sum_{j-2}^{\infty} \vec{Q}_j(s, \hat{L}, \vec{R}_k),$$

where $\vec{Q}_j$'s $(j \geq 2)$ are polynomials of exact order $j$ in $\vec{r}$, which result from inserting $\vec{r}$ into $\vec{N}_j$'s. This insertion leaves no linear or constant parts, which is due to the fact that the ODE is origin preserving. This will prove crucial later in the algorithm for the solution.

We now sort the result by order. The **linear** part has the form

$$\frac{d}{ds}\hat{L}(s) = \hat{M}(s) \cdot \hat{L}(s), \tag{4.36}$$

and the higher order parts, $j \geq 2$, assume the form

$$\frac{d}{ds}\vec{R}_j (s, \vec{r}_i) = \hat{M}(s)\vec{R}_j (s, \vec{r}_i) + \vec{Q}_j(s, \hat{L}, \vec{R}_k), \tag{4.37}$$

where $\vec{Q}_j$ contains only $\vec{R}_k$ with $k < j$. So for $j = 2, 3, \ldots$, we obtain a triangular system of ODEs. It can be solved iteratively in an **order-by-order** manner, and then each of the differential equations for $\vec{R}_j$ contains only lower order terms $\vec{R}_k$ that are already known. In this way, the ODEs **decouple** and become **inhomogeneous**.

Initially at $s = 0$, we have the initial condition $\vec{r}(0) = \vec{r}_i$, and

$$\hat{L}(0) = \hat{I}, \qquad \vec{R}_j (0, \vec{r}) = \vec{0} \quad \text{for all } j = 2, 3, \ldots.$$

In order to solve the inhomogeneous eq. (4.37) of order $j$, we first determine the homogeneous solution, and then perform a so-called variation of parameters. The **homogeneous** solution is exactly the same form as for the linearized part. To obtain the **inhomogeneous** solution, we make the ansatz $\vec{R}_j(s) = \hat{L}(s) \cdot \vec{T}(s)$. Then

$$\frac{d}{ds}\vec{R}_j = \left(\frac{d}{ds}\hat{L}(s)\right) \cdot \vec{T}(s) + \hat{L}(s) \cdot \frac{d}{ds}\vec{T}(s).$$

Using eq. (4.36), the first term in the right hand side is

$$\left(\frac{d}{ds}\hat{L}(s)\right) \cdot \vec{T}(s) = \hat{M}(s) \cdot \hat{L}(s) \cdot \vec{T}(s) = \hat{M}(s) \cdot \vec{R}_j(s).$$

Thus, from eq. (4.37), we obtain

$$\hat{L}(s) \cdot \frac{d}{ds}\vec{T}(s) = \vec{Q}_j(s, \hat{L}, \vec{R}_k),$$

that is

$$\vec{T}(s) = \int_0^s \hat{L}^{-1}(\bar{s})\vec{Q}_j(\bar{s}, \hat{L}, \vec{R}_k)d\bar{s}, \tag{4.38}$$

where the choice of the lower integration boundary as 0 ensures that $\vec{T}(0) = \vec{0}$, which agrees with the initial condition $\vec{R}_j(0) = \vec{0}$. Altogether we have

$$\vec{R}_j(s) = \hat{L}(s) \cdot \int_0^s \hat{L}^{-1}(\bar{s})\vec{Q}_j(\bar{s}, \hat{L}, \vec{R}_k)d\bar{s}.$$

The integral is often referred to as the **aberration integral**, and the integrand $\hat{L}^{-1}\vec{Q}_j$ as the **driving term**. The complete solution then is obtained as

$$\vec{r}(s) = \hat{L}(s)\vec{r}_i + \sum_{j=2}^{\infty} \vec{R}_j(s) = \hat{L}(s)\vec{r}_i + \sum_{j=2}^{\infty} \hat{L}(s) \cdot \int_0^s \hat{L}^{-1}(\bar{s})\vec{Q}_j(\bar{s}, \hat{L}, \vec{R}_k)d\bar{s}.$$

So, once the linear solution is known, everything else just boils down to quadratures. If within a piece in which it is constant, $\hat{M}(s)$ is diagonalizable, the linear solutions can be written as combinations of sin, cos, sinh, cosh and $s$. In other important cases where $\hat{M}(s)$ is singular, often a complete set of linear solutions that are polynomials in $s$ can be obtained.

In both of these cases, the insertion into the polynomials $\vec{R}_j(s)$ leads to terms that are polynomials in sin, cos, sinh, cosh and $s$. By expressing such functions in terms of exponentials times powers of $s$, one can show that the result of any integration can again be expressed as a polynomial of sin, cos, sinh, cosh and $s$.

For practical cases, it is worthwhile to discuss the **complexity** of the procedure. With each new order, the expansion of the ODE becomes more complicated; then all previous orders have to be inserted, multiplied with the linear inverses, and integrated, resulting in substantially more terms than for the previous order. Altogether, the effort **increases extremely dramatically** with the order being considered, and for typical systems, it is practical only to orders around five.

**Computer codes** that use the above procedure usually contain a library of procedures that compute the aberrations for each particle optical element of interest. The aberrations of combined systems is then determined from those of the pieces with the help of a composition procedure. The **Differential Algebraic (DA)** approach, as described in Chapter 5, allows the computation of aberrations to any order in an elegant way without the need of explicit formulas for aberrations.

To illustrate the method of computation of aberrations with a simple **example**, let us consider the differential equations

$$x' = a, \qquad a' = -x + kx^2,$$

which corresponds to the horizontal motion in a quadrupole with a superimposed sextupole. We first perform the **linearization** to obtain

$$\begin{pmatrix} x \\ a \end{pmatrix}' = \begin{pmatrix} 0 & 1 \\ -1 & 0 \end{pmatrix} \begin{pmatrix} x \\ a \end{pmatrix} = \hat{M}(s) \begin{pmatrix} x \\ a \end{pmatrix}.$$

The linear solution then is

$$\begin{pmatrix} x_f \\ a_f \end{pmatrix} = \begin{pmatrix} (x|x) & (x|a) \\ (a|x) & (a|a) \end{pmatrix} \begin{pmatrix} x_i \\ a_i \end{pmatrix} = \begin{pmatrix} \cos s & \sin s \\ -\sin s & \cos s \end{pmatrix} \begin{pmatrix} x_i \\ a_i \end{pmatrix} = \hat{L}(s) \begin{pmatrix} x_i \\ a_i \end{pmatrix}. \quad (4.39)$$

The next step is the **expansion** of the ODE, which is already done in the given differential equations. We then **insert** the solution expanded up to the second order in the initial conditions $x_i$ and $a_i$

$$x(s) = (x|x)x_i + (x|a)a_i + (x|xx)x_i^2 + (x|xa)x_i a_i + (x|aa)a_i^2,$$
$$a(s) = (a|x)x_i + (a|a)a_i + (a|xx)x_i^2 + (a|xa)x_i a_i + (a|aa)a_i^2,$$

into the ODE, and obtain

$$(x|x)'x_i + (x|a)'a_i + (x|xx)'x_i^2 + (x|xa)'x_i a_i + (x|aa)'a_i^2$$
$$= (a|x)x_i + (a|a)a_i + (a|xx)x_i^2 + (a|xa)x_i a_i + (a|aa)a_i^2,$$

and

$$(a|x)'x_i + (a|a)'a_i + (a|xx)'x_i^2 + (a|xa)'x_i a_i + (a|aa)'a_i^2$$
$$= -\left[(x|x)x_i + (x|a)a_i + (x|xx)x_i^2 + (x|xa)x_i a_i + (x|aa)a_i^2\right]$$
$$+ k\left[(x|x)^2 x_i^2 + 2(x|x)(x|a)x_i a_i + (x|a)^2 a_i^2 + \cdots\right],$$

where we can ignore the higher order terms, since we are interested only in order two.

*An Introduction to Beam Physics*

The second order equations then read

$$(x|xx)'x_i^2+(x|xa)'x_ia_i+(x|aa)'a_i^2 = \quad (a|xx)x_i^2+(a|xa)x_ia_i+(a|aa)a_i^2,$$
$$(a|xx)'x_i^2+(a|xa)'x_ia_i+(a|aa)'a_i^2 = -[(x|xx)x_i^2+(x|xa)x_ia_i+(x|aa)a_i^2]$$
$$+ k[(x|x)^2x_i^2+2(x|x)(x|a)x_ia_i+(x|a)^2a_i^2],$$

where the last line proportional to $k$ is the inhomogeneous part $\vec{Q}_2$, and using eq. (4.39),

$$\vec{Q}_2 = \begin{pmatrix} Q_{2x} \\ Q_{2a} \end{pmatrix} = \begin{pmatrix} 0 \\ k\left[\cos^2 s \cdot x_i^2 + 2\cos s \sin s \cdot x_ia_i + \sin^2 s \cdot a_i^2\right] \end{pmatrix}.$$

We make the ansatz $\vec{R}_2(s) = \hat{L}(s)\vec{T}(s)$ :

$$\begin{pmatrix} (x|xx)x_i^2 + (x|xa)x_ia_i + (x|aa)a_i^2 \\ (a|xx)x_i^2 + (a|xa)x_ia_i + (a|aa)a_i^2 \end{pmatrix} = \begin{pmatrix} \cos s & \sin s \\ -\sin s & \cos s \end{pmatrix} \vec{T}(s).$$

From eq. (4.38),

$$\vec{T}(s) = \begin{pmatrix} T_x \\ T_a \end{pmatrix} = \int_0^s \hat{L}^{-1}\vec{Q}_2 d\bar{s} = \int_0^s \begin{pmatrix} \cos\bar{s} & -\sin\bar{s} \\ \sin\bar{s} & \cos\bar{s} \end{pmatrix}\begin{pmatrix} 0 \\ Q_{2a} \end{pmatrix} d\bar{s},$$

so

$$T_x = \int_0^s (-\sin\bar{s} \cdot Q_{2a})\, d\bar{s}$$
$$= k\int_0^s \left(-\cos^2\bar{s}\sin\bar{s} \cdot x_i^2 - 2\cos\bar{s}\sin^2\bar{s} \cdot x_ia_i - \sin^3\bar{s} \cdot a_i^2\right) d\bar{s}$$
$$= k\left[\frac{1}{3}(\cos^3 s - 1)x_i^2 - \frac{2}{3}\sin^3 s \cdot x_ia_i + \left(\cos s - \frac{1}{3}\cos^3 s - \frac{2}{3}\right)a_i^2\right],$$
$$T_a = \int_0^s \cos\bar{s} \cdot Q_{2a} d\bar{s}$$
$$= k\int_0^s \left(\cos^3\bar{s} \cdot x_i^2 + 2\cos^2\bar{s}\sin\bar{s} \cdot x_ia_i + \cos\bar{s}\sin^2\bar{s} \cdot a_i^2\right) d\bar{s}$$
$$= k\left[\left(\sin s - \frac{1}{3}\sin^3 s\right)x_i^2 - \frac{2}{3}(\cos^3 s - 1)x_ia_i + \frac{1}{3}\sin^3 s \cdot a_i^2\right].$$

Then we obtain $\vec{R}_2(s) = \hat{L}(s)\cdot\vec{T}(s)$, which yields the second order elements of the transfer map

$$(x|xx) = k \left( \frac{1}{3} \sin^2 s - \frac{1}{3} \cos s + \frac{1}{3} \right),$$

$$(x|xa) = k \left( -\frac{2}{3} \sin s \cos s + \frac{2}{3} \sin s \right),$$

$$(x|aa) = k \left( \frac{1}{3} \cos^2 s - \frac{2}{3} \cos s + \frac{1}{3} \right),$$

$$(a|xx) = k \left( \frac{2}{3} \sin s \cos s + \frac{1}{3} \sin s \right),$$

$$(a|xa) = k \left( \frac{2}{3} \sin^2 s - \frac{2}{3} \cos^2 s + \frac{2}{3} \cos s \right),$$

$$(a|aa) = k \left( -\frac{2}{3} \sin s \cos s + \frac{2}{3} \sin s \right).$$

In a similar fashion, but with much more effort, one can also compute the third and higher order terms.

# Chapter 5

---

## Computation and Properties of Maps

Up to now, the equations of motion have only been solved perturbatively to the first order. Yet the knowledge of the nonlinear part of the solution is also needed to determine precisely the performance of a device. Traditionally, this is done analytically using perturbation theory (see Section 4.5 for more details). In the past, a tremendous amount of knowledge about aberrations for various kinds of devices has been accumulated. Yet this approach is far from being systematic, making the simulation prone to errors, and it is difficult to obtain an accurate solution for a realistic device where no analytical solution exists. In this chapter, a modern method, the Differential Algebraic (DA) technique, for computing the transfer map to arbitrary order, will be described. But before we embark on this task, we will first classify aberrations that can appear in transfer maps in terms of their symmetries.

---

## 5.1 Aberrations and Symmetries

Recall that the transfer map of an optical system relates final coordinates to initial coordinates via

$$\vec{z}_f = \mathcal{M}(\vec{z}_i),$$

where $\vec{z} = (x, a, y, b, l, \delta)$. In the previous chapters, we were concerned mostly with the linearized part of the map, which describes the major part of the motion and which can be described by transfer matrices. The matrix elements were denoted as $(x, a)$, etc.

In order to study the effects of the motion very precisely, it is necessary to also consider higher order or nonlinear effects. For this purpose we Taylor expand the map (in a rigorous sense the question whether the map can actually be Taylor expanded is rather nontrivial, but we ignore this here), and use names for the coefficients similar to what we had for the linear motion. We write

$$x_f = \sum \left( x | x^{i_x} a^{i_a} y^{i_y} b^{i_b} l^{i_l} \delta^{i_\delta} \right) x^{i_x} a^{i_a} y^{i_y} b^{i_b} l^{i_l} \delta^{i_\delta},$$

$$a_f = \sum \left( a | x^{i_x} a^{i_a} y^{i_y} b^{i_b} l^{i_l} \delta^{i_\delta} \right) x^{i_x} a^{i_a} y^{i_y} b^{i_b} l^{i_l} \delta^{i_\delta},$$

$$y_f = \sum \left( y | x^{i_x} a^{i_a} y^{i_y} b^{i_b} l^{i_l} \delta^{i_\delta} \right) x^{i_x} a^{i_a} y^{i_y} b^{i_b} l^{i_l} \delta^{i_\delta},$$

$$b_f = \sum \left( b | x^{i_x} a^{i_a} y^{i_y} b^{i_b} l^{i_l} \delta^{i_\delta} \right) x^{i_x} a^{i_a} y^{i_y} b^{i_b} l^{i_l} \delta^{i_\delta},$$

$$l_f = \sum \left( l | x^{i_x} a^{i_a} y^{i_y} b^{i_b} l^{i_l} \delta^{i_\delta} \right) x^{i_x} a^{i_a} y^{i_y} b^{i_b} l^{i_l} \delta^{i_\delta},$$

$$\delta_f = \sum \left( \delta | x^{i_x} a^{i_a} y^{i_y} b^{i_b} l^{i_l} \delta^{i_\delta} \right) x^{i_x} a^{i_a} y^{i_y} b^{i_b} l^{i_l} \delta^{i_\delta},$$

where the sums go over all six-tuples $(i_x, i_a, i_y, i_b, i_l, i_\delta)$; for convenience, they are usually sorted by total order. The Taylor coefficients belonging to terms of orders 2 or higher are usually called **aberrations**, as they describe corrections to the linear part of the map that are usually small if the phase space variables are small.

In most cases, the freedom of the aberration coefficients is severely restricted by the presence of a variety of symmetries. First, in many cases the motion of one of the variables does not depend on the values of some other variables. For example, if the motion is time independent, we have

$$\left( Z_j | x^{i_x} a^{i_a} y^{i_y} b^{i_b} l^{i_l} \delta^{i_\delta} \right) = 0 \quad \text{if} \quad i_l \neq 0,$$

where $j = 1, \ldots, 6$ and $Z_j$ is defined in eq. (2.2). Furthermore, in this case we know that the kinetic plus potential energy of the particle is conserved, and we have that

$$\left( \delta | x^{i_x} a^{i_a} y^{i_y} b^{i_b} l^{i_l} \delta^{i_\delta} \right) = 0 \quad \text{except} \quad (\delta | \delta) = 1.$$

### 5.1.1   Horizontal Midplane Symmetry

This is perhaps the most important symmetry in beam physics, as it affects almost all devices: bending elements, quadrupoles, sextupoles, higher order multipoles, cyclotrons and all the combinations of them. It requires that the motion of charged particles is always symmetric around the midplane (the $x$-$z$ plane), which is illustrated in Fig. 5.1.

In a system with midplane symmetry, two particles that are symmetric about the midplane at the beginning **stay symmetric** throughout the system. Suppose that a particle is launched at $(x_i, y_i, d_i, a_i, b_i, t_i)$. After the map $\mathcal{M}$

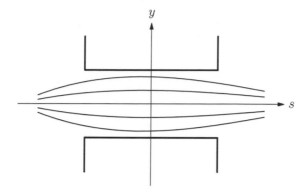

**FIGURE 5.1**: Trajectories of particles in a system with horizontal mid-plane symmetry.

is applied, its coordinates are

$$x_f = m_x(x_i, a_i, y_i, b_i, l_i, \delta_i),$$
$$a_f = m_a(x_i, a_i, y_i, b_i, l_i, \delta_i),$$
$$y_f = m_y(x_i, a_i, y_i, b_i, l_i, \delta_i),$$
$$b_f = m_b(x_i, a_i, y_i, b_i, l_i, \delta_i),$$
$$l_f = m_l(x_i, a_i, y_i, b_i, l_i, \delta_i),$$
$$\delta_f = m_\delta(x_i, a_i, y_i, b_i, l_i, \delta_i).$$

Under the presence of midplane symmetry, a particle that starts at

$$(x_i, -y_i, d_i, a_i, -b_i, t_i)$$

must end at

$$(x_f, -y_f, d_f, a_f, -b_f, t_f),$$

which entails that

$$x_f = m_x(x_i, a_i, -y_i, -b_i, l_i, \delta_i),$$
$$a_f = m_a(x_i, a_i, -y_i, -b_i, l_i, \delta_i),$$
$$-y_f = m_y(x_i, a_i, -y_i, -b_i, l_i, \delta_i),$$
$$-b_f = m_b(x_i, a_i, -y_i, -b_i, l_i, \delta_i),$$
$$l_f = m_l(x_i, a_i, -y_i, -b_i, l_i, \delta_i),$$
$$\delta_f = m_\delta(x_i, a_i, -y_i, -b_i, l_i, \delta_i). \tag{5.1}$$

Thus flipping the signs of $y_i$, $b_i$ simultaneously flips the signs of $y_f$, $b_f$, but leaves $x_f$, $a_f$, $l_f$, $\delta_f$ intact. Flipping the sign of $y_i$, $b_i$ simultaneously produces a sign of $(-1)^{i_y+i_b}$ in each monomial. So $i_y + i_b$ must be odd for $y_f$, $b_f$ and $i_y + i_b$ must be even for all others. So we obtain

$$\left(x|x^{i_x}a^{i_a}y^{i_y}b^{i_b}l^{i_l}\delta^{i_\delta}\right) = 0 \quad \text{for} \quad i_y + i_b \quad \text{odd,}$$
$$\left(a|x^{i_x}a^{i_a}y^{i_y}b^{i_b}l^{i_l}\delta^{i_\delta}\right) = 0 \quad \text{for} \quad i_y + i_b \quad \text{odd,}$$
$$\left(y|x^{i_x}a^{i_a}y^{i_y}b^{i_b}l^{i_l}\delta^{i_\delta}\right) = 0 \quad \text{for} \quad i_y + i_b \quad \text{even,}$$
$$\left(b|x^{i_x}a^{i_a}y^{i_y}b^{i_b}l^{i_l}\delta^{i_\delta}\right) = 0 \quad \text{for} \quad i_y + i_b \quad \text{even,}$$
$$\left(l|x^{i_x}a^{i_a}y^{i_y}b^{i_b}l^{i_l}\delta^{i_\delta}\right) = 0 \quad \text{for} \quad i_y + i_b \quad \text{odd,}$$
$$\left(\delta|x^{i_x}a^{i_a}y^{i_y}b^{i_b}l^{i_l}\delta^{i_\delta}\right) = 0 \quad \text{for} \quad i_y + i_b \quad \text{odd.}$$

For the first order map, the transfer matrix, this leads to the form

$$\hat{M} = \begin{pmatrix} (x|x) & (x|a) & 0 & 0 & (x|l) & (x|\delta) \\ (a|x) & (a|a) & 0 & 0 & (a|l) & (a|\delta) \\ 0 & 0 & (y|y) & (y|b) & 0 & 0 \\ 0 & 0 & (b|y) & (b|b) & 0 & 0 \\ (l|x) & (l|a) & 0 & 0 & (l|l) & (l|\delta) \\ (\delta|x) & (\delta|a) & 0 & 0 & (\delta|l) & (\delta|\delta) \end{pmatrix},$$

which is a form seen earlier for electrostatic and magnetic bending elements. Altogether, the symmetry entails that to any given order, roughly half of all aberrations vanish.

## 5.1.2   Double Midplane Symmetry

Several devices have a midplane symmetry not only around the horizontal plane, but also around a vertical plane. This is the case for all electric cylindrically symmetric devices, as well as quadrupoles, octupoles, and in general $4k$ poles. In this case, in addition to the requirements we just had, we obtain a second set in which the roles of $x$, $a$ and $y$, $b$ are interchanged. In this case we obtain

$$(x|\ldots) = 0 \quad \text{for} \quad i_y + i_b \quad \text{odd} \quad \text{or} \quad i_x + i_a \quad \text{even,}$$
$$(a|\ldots) = 0 \quad \text{for} \quad i_y + i_b \quad \text{odd} \quad \text{or} \quad i_x + i_a \quad \text{even,}$$
$$(y|\ldots) = 0 \quad \text{for} \quad i_y + i_b \quad \text{even} \quad \text{or} \quad i_x + i_a \quad \text{odd,}$$
$$(b|\ldots) = 0 \quad \text{for} \quad i_y + i_b \quad \text{even} \quad \text{or} \quad i_x + i_a \quad \text{odd,}$$
$$(l|\ldots) = 0 \quad \text{for} \quad i_y + i_b \quad \text{odd} \quad \text{or} \quad i_x + i_a \quad \text{even,}$$
$$(\delta|\ldots) = 0 \quad \text{for} \quad i_y + i_b \quad \text{odd} \quad \text{or} \quad i_x + i_a \quad \text{even,}$$

and altogether, about three-fourths of all matrix elements vanish. To first order, the matrix must have the special form

$$
\hat{M} = \begin{pmatrix}
(x|x) & (x|a) & 0 & 0 & 0 & 0 \\
(a|x) & (a|a) & 0 & 0 & 0 & 0 \\
0 & 0 & (y|y) & (y|b) & 0 & 0 \\
0 & 0 & (b|y) & (b|b) & 0 & 0 \\
0 & 0 & 0 & 0 & (l|l) & (l|\delta) \\
0 & 0 & 0 & 0 & (\delta|l) & (\delta|\delta)
\end{pmatrix},
$$

which is what we observed in the case of the drift and the electric and magnetic quadrupoles.

## 5.1.3 Rotational Symmetry

One special case of the double midplane symmetry that we just discussed is the full rotational symmetry that round lenses satisfy. In this case there is a symmetry going beyond what double midplane symmetry requires; the map has to be invariant under a rotation in the $x$-$y$ plane. Let the rotation angle be $\phi$. The linear transformation $\mathcal{R}(Z) = \hat{R} \cdot Z$ is described in terms of the matrix

$$
\hat{R} = \begin{pmatrix}
\cos\phi & 0 & \sin\phi & 0 & 0 & 0 \\
0 & \cos\phi & 0 & \sin\phi & 0 & 0 \\
-\sin\phi & 0 & \cos\phi & 0 & 0 & 0 \\
0 & -\sin\phi & 0 & \cos\phi & 0 & 0 \\
0 & 0 & 0 & 0 & 1 & 0 \\
0 & 0 & 0 & 0 & 0 & 1
\end{pmatrix},
$$

and we must have that the transfer map satisfies

$$
M \circ R = R \circ M. \tag{5.2}
$$

In the variables we are currently using, the study of the influence of the rotation on the map is somewhat cumbersome, and for this purpose it is actually better to choose complex coordinates

$$
z = x + iy, \quad w = a + ib,
$$

as well as their complex conjugates

$$
\bar{z} = x - iy, \quad \bar{w} = a - ib.
$$

In these complex variables, the map $\mathcal{R}$ has the simple diagonal form

$$
\mathcal{R} = \begin{pmatrix}
e^{i\phi} & 0 & 0 & 0 & 0 & 0 \\
0 & e^{i\phi} & 0 & 0 & 0 & 0 \\
0 & 0 & e^{-i\phi} & 0 & 0 & 0 \\
0 & 0 & 0 & e^{-i\phi} & 0 & 0 \\
0 & 0 & 0 & 0 & 1 & 0 \\
0 & 0 & 0 & 0 & 0 & 1
\end{pmatrix},
$$

and its effect in eq. (5.2) is easy to study. It turns out that in the map only those terms that have the form

$$z_f = z_i \cdot f_z(z\bar{z}, w\bar{w}), \quad w_f = w_i \cdot f_w(z\bar{z}, w\bar{w})$$

are allowed to remain. In passing we note that this situation is remarkably similar to what happens in the theory of normal forms of repetitive motion [5].

There are two types of rotational symmetry: One is characterized by the system being invariant under a rotation of **any angle**, which we call **continuous** rotational symmetry; the other is characterized by the system being invariant under a **fixed** angle, which we refer to as **discrete** rotational symmetry. The former is widely seen in light optics where almost all glass lenses are rotational invariant and in electron microscopes where solenoids are the primary focusing elements. The latter is preserved in quadrupoles and all higher multipoles.

For the analysis of both cases we proceed with the above complex coordinates. After expressing $x$, $a$, $y$ and $b$ in terms of $z$, $\bar{z}$, $w$ and $\bar{w}$, the transfer map is transformed into

$$\begin{pmatrix} z_f \\ w_f \end{pmatrix} = \begin{pmatrix} x_f + iy_f \\ a_f + ib_f \end{pmatrix} = \begin{pmatrix} F_z \\ F_w \end{pmatrix} (z_i, \bar{z}_i, w_i, \bar{w}_i, t_i, d_i),$$

where

$$\begin{pmatrix} F_z \\ F_w \end{pmatrix} = \sum_{j_1 j_2 j_3 j_4 j_t j_d} \begin{pmatrix} c_z \\ c_w \end{pmatrix}_{j_1 j_2 j_3 j_4 j_t j_d} z^{j_1} \bar{z}^{j_2} w^{j_3} \bar{w}^{j_4} t^{j_t} d^{j_d}.$$

Note that besides $z$ and $w$, also $\bar{z}$ and $\bar{w}$ will appear, contrary to the familiar Taylor expansion of analytic functions. This is due to the fact that while the original map may be Taylor expandable and hence analytic as a real function, it is not necessary that the resulting complex function is analytic in the sense of complex analysis.

Given the fact that a rotation by $\phi$ transforms $z$ to $e^{i\phi}z$ and $w$ to $e^{i\phi}w$, rotational symmetry requires that a rotation in initial coordinates results in the same transformation in the final coordinates, i.e., $z_f \to e^{i\phi}z_f$, $w_f \to e^{i\phi}w_f$. Inserting this yields

$$(j_1 - j_2 + j_3 - j_4 - 1)\,\phi = 2\pi n \quad \text{for} \quad x, a, y, b \text{ terms},$$
$$(j_1 - j_2 + j_3 - j_4)\,\phi = 2\pi n \quad \text{for} \quad t, d \text{ terms},$$

where $n$ is an integer.

For continuous rotational symmetry, which means invariance for all $\phi$, the $j_i$ ($i = 1, 2, 3, 4$) should be independent of $\phi$. Thus we have

$$j_1 - j_2 + j_3 - j_4 = 1 \quad \text{for} \quad z_f \text{ and } w_f,$$
$$j_1 - j_2 + j_3 - j_4 = 0 \quad \text{for} \quad t_f \text{ and } d_f,$$

**TABLE 5.1:**

Number of aberrations

| Order | $j_1$ | $j_2$ | $j_3$ | $j_4$ |
|-------|-------|-------|-------|-------|
| 1 | 1 | 0 | 0 | 0 |
| | 0 | 0 | 1 | 0 |
| 3 | 2 | 1 | 0 | 0 |
| | 2 | 0 | 0 | 1 |
| | 0 | 1 | 2 | 0 |
| | 0 | 0 | 2 | 1 |
| | 1 | 1 | 1 | 0 |
| | 1 | 0 | 1 | 1 |

which eliminate many terms. First, all terms with $j_1 + j_2 + j_3 + j_4$ even vanish, because $j_1 + j_3$ and $j_2 + j_4$ always have the same parity, which means that $j_1 - j_2 + j_3 - j_4$ is also even. This implies that in rotationally symmetric systems, all even order geometric aberrations disappear. As a summary, all remaining $z$ and $w$ terms up to order 3 are shown in Table 5.1, where the order represents the sum of the $j_i$. To illustrate the characteristic of such a map, let us derive the linear matrix from the conditions above. First define

$$(z|z) = (c_z)_{100000}, \qquad (z|w) = (c_z)_{001000},$$
$$(w|z) = (c_w)_{100000}, \qquad (w|w) = (c_w)_{001000}.$$

The first order map is then given by

$$x_f + iy_f = (z|z)(x_i + iy_i) + (z|w)(a_i + ib_i),$$
$$a_f + ib_f = (w|z)(x_i + iy_i) + (w|w)(a_i + ib_i),$$

which entails that the linear matrix is

$$\hat{M} = \begin{pmatrix} \Re(z|z) & -\Im(z|z) & \Re(z|w) & -\Im(z|w) \\ \Im(z|z) & \Re(z|z) & \Im(z|w) & \Re(z|w) \\ \Re(w|z) & -\Im(w|z) & \Re(w|w) & -\Im(w|w) \\ \Im(w|z) & \Re(w|z) & \Im(w|w) & \Re(w|w) \end{pmatrix}. \tag{5.3}$$

As an example, we show the second order map of a solenoid, which has rotational symmetry, but exhibits a coupling between $x$ and $y$, and $a$ and $b$.

Table 5.2 lists the coefficients of the second order map of a solenoid, which shows that indeed all second order geometric aberrations vanish, which is a consequence of the rotational symmetry.

Eq. (5.3) also shows that a rotationally invariant system preserves midplane symmetry to first order when the first order coefficients are real numbers. In fact a simple argument shows that this is true even for higher orders. In

**TABLE 5.2:**  The second order map of a solenoid (Exponents in the initial variables $x, a, y, b, l, \delta$)

| $x_f$ | $a_f$ | $y_f$ | $b_f$ | $l_f$ | exponents |
|---|---|---|---|---|---|
| 0.999662 | -0.408336e-3 | -0.186647e-1 | 0.761884e-5 | 0 | 100000 |
| 0.799815 | 0.999662 | -0.149334e-1 | -0.186647e-1 | 0 | 010000 |
| 0.186647e-1 | -0.761884e-5 | 0.999662 | -0.408336e-3 | 0 | 001000 |
| 0.149334e-1 | 0.186647e-1 | 0.799815 | 0.999662 | 0 | 000100 |
| 0 | 0 | 0 | 0 | 1.000000 | 000010 |
| 0 | 0 | 0 | 0 | 0.163101 | 000001 |
| 0 | 0 | 0 | 0 | -0.112007e-3 | 200000 |
| 0 | 0 | 0 | 0 | -0.876499e-9 | 110000 |
| 0 | 0 | 0 | 0 | -0.219388 | 020000 |
| 0 | 0 | 0 | 0 | -0.102394e-1 | 011000 |
| 0 | 0 | 0 | 0 | -0.112007e-3 | 002000 |
| 0 | 0 | 0 | 0 | 0.102394e-1 | 100100 |
| 0 | 0 | 0 | 0 | -0.876499e-9 | 001100 |
| 0.370284e-3 | 0.223860e-3 | 0.102326e-1 | -0.836226e-5 | 0 | 100001 |
| -0.438474 | 0.370284e-3 | 0.163792e-1 | 0.102326e-1 | 0 | 010001 |
| -0.102326e-1 | 0.836226e-5 | 0.370284e-3 | 0.223860e-3 | 0 | 001001 |
| 0 | 0 | 0 | 0 | -0.219388 | 000200 |
| -0.163792e-1 | -0.102326e-1 | -0.438474 | 0.370284e-3 | 0 | 000101 |
| 0 | 0 | 0 | 0 | -0.134185 | 000002 |

complex coordinates, eqs. (5.1) are transformed to

$$\begin{pmatrix} \bar{z}_f \\ \bar{w}_f \end{pmatrix} = \begin{pmatrix} F_z \\ F_w \end{pmatrix} (\bar{z}_i, z_i, \bar{w}_i, w_i, t_i, d_i)$$

$$\Rightarrow \begin{pmatrix} z_f \\ w_f \end{pmatrix} = \begin{pmatrix} \bar{F}_z \\ \bar{F}_w \end{pmatrix} (z_i, \bar{z}_i, w_i, \bar{w}_i, t_i, d_i)$$

$$= \sum_{j_1 j_2 j_3 j_4 j_t j_d} \begin{pmatrix} \bar{a}_z \\ \bar{a}_w \end{pmatrix}_{j_1 j_2 j_3 j_4 j_t j_d} z^{j_1} \bar{z}^{j_2} w^{j_3} \bar{w}^{j_4} t^{j_t} d^{j_d},$$

which shows that all coefficients have to be real numbers in order to preserve midplane symmetry.

For discrete rotational symmetry, invariance occurs only when $\phi = 2\pi/k$, where $k$ is an integer. Hence the nonzero terms satisfy

$$j_1 - j_2 + j_3 - j_4 - 1 = nk.$$

In general, a $2k$-pole is invariant under rotation of $\phi = 2\pi/k$. For example, for a quadrupole, we have $k = 2$. Hence the nonzero terms satisfy

$$j_1 - j_2 + j_3 - j_4 = 2n + 1.$$

Like round lenses, systems with quadrupole symmetry are also free of even order geometric aberrations. The linear map of a quadrupole can be obtained from

$$j_1 - j_2 + j_3 - j_4 = \pm 1,$$

which is

$$x_f + iy_f = (z|z)(x_i + iy_i) + (z|w)(a_i + ib_i) + (z|\bar{z})(x_i - iy_i) + (z|\bar{w})(a_i - ib_i),$$
$$a_f + ib_f = (w|z)(x_i + iy_i) + (w|w)(a_i + ib_i) + (w|\bar{z})(x_i - iy_i) + (w|\bar{w})(a_i - ib_i).$$

Since a quadrupole has midplane symmetry, all the coefficients are real numbers. Thus its linear matrix is

$$\begin{pmatrix} (z|z) + (z|\bar{z}) & 0 & (z|w) + (z|\bar{w}) & 0 \\ 0 & (z|z) - (z|\bar{z}) & 0 & (z|w) - (z|\bar{w}) \\ (w|z) + (w|\bar{z}) & 0 & (w|w) + (w|\bar{w}) & 0 \\ 0 & (w|z) - (w|\bar{z}) & 0 & (w|w) - (w|\bar{w}) \end{pmatrix}.$$

For other multipoles, we have

$$j_1 - j_2 + j_3 - j_4 = nk + 1 = \begin{cases} \cdots \\ -k + 1 \\ 1 \\ k + 1 \\ \cdots \end{cases} \quad k = 3, 4, \ldots. \quad (5.4)$$

With midplane symmetry, the linear matrix of a $2k$-pole is

$$\hat{M} = \begin{pmatrix} (z|z) & (z|w) & 0 & 0 \\ (w|z) & (w|w) & 0 & 0 \\ 0 & 0 & (z|z) & (z|w) \\ 0 & 0 & (w|z) & (w|w) \end{pmatrix}. \quad (5.5)$$

Since the linear matrix of a $2k$-pole ($k \leq 3$) is just a drift, it satisfies eq. (5.5). Eq. (5.4) shows that the geometric aberrations appear only for orders of at least $k - 1$. The fact that multipoles do not have dispersion determines that the chromatic aberrations do not appear until order $k$. This can be easily seen from the equations of motion (3.22). Therefore, a $2k$-pole is necessarily a drift up to order $k - 2$.

A lens with rotational symmetry is frequently called a **round lens**. The main examples are magnetic solenoids and electrostatic round lenses. Since rotational symmetry is the highest degree of symmetry a lens can have, round lenses have the fewest number of aberrations. As a result, they are widely used in low energy electron optical devices such as electron microscopes. At high energy, round lenses are too weak to be effective.

### 5.1.4 Symplectic Symmetry

Another important symmetry of the motion is due to the fact that the motion is indeed obtained as the solution of Hamiltonian differential equations.

In this case, one can show that the Jacobian $\hat{M}$ of the transfer map $\mathcal{M}$, i.e., the matrix

$$\hat{M} = \begin{pmatrix} \partial\mathcal{M}_1/\partial z_1 & \partial\mathcal{M}_1/\partial z_2 & \partial\mathcal{M}_1/\partial z_3 & \partial\mathcal{M}_1/\partial z_4 & \partial\mathcal{M}_1/\partial z_5 & \partial\mathcal{M}_1/\partial z_6 \\ \partial\mathcal{M}_2/\partial z_1 & \partial\mathcal{M}_2/\partial z_2 & \partial\mathcal{M}_2/\partial z_3 & \partial\mathcal{M}_2/\partial z_4 & \partial\mathcal{M}_2/\partial z_5 & \partial\mathcal{M}_2/\partial z_6 \\ \partial\mathcal{M}_3/\partial z_1 & \partial\mathcal{M}_3/\partial z_2 & \partial\mathcal{M}_3/\partial z_3 & \partial\mathcal{M}_3/\partial z_4 & \partial\mathcal{M}_3/\partial z_5 & \partial\mathcal{M}_3/\partial z_6 \\ \partial\mathcal{M}_4/\partial z_1 & \partial\mathcal{M}_4/\partial z_2 & \partial\mathcal{M}_4/\partial z_3 & \partial\mathcal{M}_4/\partial z_4 & \partial\mathcal{M}_4/\partial z_5 & \partial\mathcal{M}_4/\partial z_6 \\ \partial\mathcal{M}_5/\partial z_1 & \partial\mathcal{M}_5/\partial z_2 & \partial\mathcal{M}_5/\partial z_3 & \partial\mathcal{M}_5/\partial z_4 & \partial\mathcal{M}_5/\partial z_5 & \partial\mathcal{M}_5/\partial z_6 \\ \partial\mathcal{M}_6/\partial z_1 & \partial\mathcal{M}_6/\partial z_2 & \partial\mathcal{M}_6/\partial z_3 & \partial\mathcal{M}_6/\partial z_4 & \partial\mathcal{M}_6/\partial z_5 & \partial\mathcal{M}_6/\partial z_6 \end{pmatrix},$$

has to satisfy the condition

$$\hat{M}^T \hat{J} \hat{M} = \hat{J}, \tag{5.6}$$

where $\hat{J}$ is the totally antisymmetric matrix

$$\hat{J} = \begin{pmatrix} 0 & 1 & 0 & 0 & 0 & 0 \\ -1 & 0 & 0 & 0 & 0 & 0 \\ 0 & 0 & 0 & 1 & 0 & 0 \\ 0 & 0 & -1 & 0 & 0 & 0 \\ 0 & 0 & 0 & 0 & 0 & 1 \\ 0 & 0 & 0 & 0 & -1 & 0 \end{pmatrix}.$$

The proof of this so-called condition of symplecticity certainly goes beyond this volume and can be found, for example, in [5]. But we can readily see that the symplectic condition, which mixes in a very defined way the terms $\partial\mathcal{M}_i/\partial z_j$ that are themselves power series, entails a large variety of nonlinear restrictions between the aberrations.

From eq. (5.6), we have

$$-\hat{J}\hat{M}^T \hat{J} \hat{M} = \hat{I} \Rightarrow -\hat{M}\hat{J}\hat{M}^T \hat{J} \hat{M} = \hat{M} \Rightarrow -\hat{M}\hat{J}\hat{M}^T \hat{J} = \hat{I},$$

hence

$$\hat{M}\hat{J}\hat{M}^T = \hat{J}.$$

Furthermore, the inverse of a symplectic matrix always exists and can be obtained easily with the help of eq. (5.6), as

$$\hat{J}\hat{M}^T \hat{J} \hat{M} = -\hat{I} \Rightarrow \left(-\hat{J}\hat{M}^T \hat{J}\right) \hat{M} = \hat{I} \Rightarrow \hat{M}^{-1} = -\hat{J}\hat{M}^T \hat{J}.$$

$$\hat{M}^{-1} = -\hat{J}\hat{M}^T \hat{J}. \tag{5.7}$$

And from the following simple arithmetic

$$\left(\hat{M}^{-1}\right)^T \hat{J}\hat{M}^{-1} = \left(\hat{J}\hat{M}^T \hat{J}\right)^T \hat{J} \left(\hat{J}\hat{M}^T \hat{J}\right) = \hat{J}\hat{M}\hat{J}\hat{J}\hat{J}\hat{M}^T \hat{J}$$

$$= -\hat{J}\hat{M}\hat{J}\hat{M}^T \hat{J} = -\hat{J}\hat{J}\hat{J} = \hat{J},$$

it is shown that $\hat{M}^{-1}$ is symplectic as well:

$$\left(\hat{M}^{-1}\right)^T \hat{J} \hat{M}^{-1} = \hat{J}.$$

Now let us obtain the determinant of $\hat{M}$ following Kauderer (page 10 in [36]). Through a series of permutations, we can rewrite $\hat{J}$, which becomes

$$\hat{J} = \begin{pmatrix} \hat{0} & \hat{I} \\ -\hat{I} & \hat{0} \end{pmatrix},$$

where $\hat{I}$ is the $n \times n$ identity matrix. Furthermore, $\hat{M}$ can be written as

$$\hat{M} = \begin{pmatrix} \hat{A} & \hat{B} \\ \hat{C} & \hat{D} \end{pmatrix},$$

where $\hat{A}$, $\hat{B}$, $\hat{C}$ and $\hat{D}$ are $n \times n$ matrices. The symplectic condition becomes

$$\begin{pmatrix} \hat{A}^T & \hat{C}^T \\ \hat{B}^T & \hat{D}^T \end{pmatrix} \begin{pmatrix} \hat{0} & \hat{I} \\ -\hat{I} & \hat{0} \end{pmatrix} \begin{pmatrix} \hat{A} & \hat{B} \\ \hat{C} & \hat{D} \end{pmatrix} = \begin{pmatrix} \hat{0} & \hat{I} \\ -\hat{I} & \hat{0} \end{pmatrix},$$

which leads to the relations

$$-\hat{C}^T \hat{A} + \hat{A}^T \hat{C} = \hat{0}, \qquad -\hat{C}^T \hat{B} + \hat{A}^T \hat{D} = \hat{I},$$
$$-\hat{D}^T \hat{A} + \hat{B}^T \hat{C} = -\hat{I}, \qquad -\hat{D}^T \hat{B} + \hat{B}^T \hat{D} = \hat{0}. \tag{5.8}$$

Furthermore we need one more mathematical theorem, which is

$$\det \begin{pmatrix} \hat{A} & \hat{0} \\ \hat{C} & \hat{D} \end{pmatrix} = \det \hat{A} \cdot \det \hat{D} = \det(\hat{A}\hat{D}).$$

Hence we have

$$\det \begin{pmatrix} \hat{I} & -\hat{A}^{-1}\hat{B} \\ \hat{0} & \hat{I} \end{pmatrix} = 1.$$

Using this, we obtain

$$\det \begin{pmatrix} \hat{A} & \hat{B} \\ \hat{C} & \hat{D} \end{pmatrix} = \det \begin{pmatrix} \hat{A} & \hat{B} \\ \hat{C} & \hat{D} \end{pmatrix} \cdot \det \begin{pmatrix} \hat{I} & -\hat{A}^{-1}\hat{B} \\ \hat{0} & \hat{I} \end{pmatrix}$$

$$= \det \begin{pmatrix} \hat{A} & \hat{0} \\ \hat{C} & \hat{D} - \hat{C}\hat{A}^{-1}\hat{B} \end{pmatrix} = \det \hat{A} \det(\hat{D} - \hat{C}\hat{A}^{-1}\hat{B}).$$

Furthermore, using the relations $\det \hat{A} = \det \hat{A}^T$, $\hat{C}\hat{A}^{-1} = (\hat{A}^T)^{-1}\hat{C}^T$ and $\hat{A}^T \hat{D} - \hat{C}^T \hat{B} = \hat{I}$ that have resulted from eq. (5.8), we obtain

$$\det \begin{pmatrix} \hat{A} & \hat{B} \\ \hat{C} & \hat{D} \end{pmatrix} = \det \hat{A}^T \det \left(\hat{D} - (\hat{A}^T)^{-1}\hat{C}^T \hat{B}\right) = \det \left(\hat{A}^T \hat{D} - \hat{C}^T \hat{B}\right) = 1.$$

One direct consequence of the symplectic condition and the resulting unity of the Jacobian determinant is that the volume of the phase space is conserved under Hamiltonian motion, which is known as Liouville's theorem.

The detailed study of the relationships between matrix elements is cumbersome and can be found in [79]. Here we will restrict our attention to what happens in the linear case. Considering the constant part of the symplectic condition (5.6), we observe that what contributes via the Jacobian is just the transfer map. Let us assume no acceleration and coupling between the transverse planes, which entails that the transfer matrix is given by

$$
\hat{M} = \begin{pmatrix}
(x|x) & (x|a) & 0 & 0 & 0 & (x|\delta) \\
(x|a) & (a|a) & 0 & 0 & 0 & (a|\delta) \\
0 & 0 & (y|y) & (y|b) & 0 & (y|\delta) \\
0 & 0 & (b|y) & (b|b) & 0 & (b|\delta) \\
(l|x) & (l|a) & (l|y) & (l|b) & 1 & (l|\delta) \\
0 & 0 & 0 & 0 & 0 & 1
\end{pmatrix}
= \begin{pmatrix}
\hat{X} & \hat{0} & \hat{D}_x \\
\hat{0} & \hat{Y} & \hat{D}_y \\
\hat{L}_x & \hat{L}_y & \hat{E}
\end{pmatrix},
$$

where

$$
\hat{X} = \begin{pmatrix} (x|x) & (x|a) \\ (x|a) & (a|a) \end{pmatrix}, \quad
\hat{Y} = \begin{pmatrix} (y|y) & (y|b) \\ (b|y) & (b|b) \end{pmatrix}, \quad
\hat{E} = \begin{pmatrix} 1 & (l|\delta) \\ 0 & 1 \end{pmatrix}
$$

$$
\hat{D}_x = \begin{pmatrix} 0 & (x|\delta) \\ 0 & (a|\delta) \end{pmatrix}, \quad
\hat{D}_y = \begin{pmatrix} 0 & (y|\delta) \\ 0 & (b|\delta) \end{pmatrix},
$$

$$
\hat{L}_x = \begin{pmatrix} (l|x) & (l|a) \\ 0 & 0 \end{pmatrix}, \quad
\hat{L}_y = \begin{pmatrix} (l|y) & (l|b) \\ 0 & 0 \end{pmatrix}.
$$

Plugging into the symplectic condition yields the equation

$$
\begin{pmatrix}
\hat{X}^T & \hat{0} & \hat{L}_x^T \\
\hat{0} & \hat{Y}^T & \hat{L}_y^T \\
\hat{D}_x^T & \hat{D}_y^T & \hat{E}^T
\end{pmatrix}
\begin{pmatrix}
\hat{J} & \hat{0} & \hat{0} \\
\hat{0} & \hat{J} & \hat{0} \\
\hat{0} & \hat{0} & \hat{J}
\end{pmatrix}
\begin{pmatrix}
\hat{X} & \hat{0} & \hat{D}_x \\
\hat{0} & \hat{Y} & \hat{D}_y \\
\hat{L}_x & \hat{L}_y & \hat{E}
\end{pmatrix}
=
\begin{pmatrix}
\hat{J} & \hat{0} & \hat{0} \\
\hat{0} & \hat{J} & \hat{0} \\
\hat{0} & \hat{0} & \hat{J}
\end{pmatrix}.
\tag{5.9}
$$

The left hand side is

$$
\begin{pmatrix}
\hat{X}^T \hat{J} \hat{X} + \hat{L}_x^T \hat{J} \hat{L}_x & \hat{L}_x^T \hat{J} \hat{L}_y & \hat{X}^T \hat{J} \hat{D}_x + \hat{L}_x^T \hat{J} \hat{E} \\
\hat{L}_y^T \hat{J} \hat{L}_x & \hat{Y}^T \hat{J} \hat{Y} + \hat{L}_y^T \hat{J} \hat{L}_y & \hat{Y}^T \hat{J} \hat{D}_y + \hat{L}_y^T \hat{J} \hat{E} \\
\hat{D}_x^T \hat{J} \hat{X} + \hat{E}^T \hat{J} \hat{L}_x & \hat{D}_y^T \hat{J} \hat{Y} + \hat{E}^T \hat{J} \hat{L}_y & \hat{D}_x^T \hat{J} \hat{D}_x + \hat{D}_y^T \hat{J} \hat{D}_y + \hat{E}^T \hat{J} \hat{E}
\end{pmatrix}.
$$

After straightforward algebraic manipulation, we have for each term in the left hand side

$$\hat{X}^T \hat{J} \hat{X} + \hat{L}_x^T \hat{J} \hat{L}_x = \begin{pmatrix} 0 & (x|x)(a|a) - (x|a)(a|x) \\ -[(x|x)(a|a) - (x|a)(a|x)] & 0 \end{pmatrix},$$

$$\hat{Y}^T \hat{J} \hat{Y} + \hat{L}_y^T \hat{J} \hat{L}_y = \begin{pmatrix} 0 & (y|y)(b|b) - (b|y)(y|b) \\ -[(y|y)(b|b) - (b|y)(y|b)] & 0 \end{pmatrix},$$

$$\hat{D}_x^T \hat{J} \hat{D}_x + \hat{D}_y^T \hat{J} \hat{D}_y + \hat{E}^T \hat{J} \hat{E} = \begin{pmatrix} 0 & 1 \\ -1 & 0 \end{pmatrix},$$

$$\hat{L}_x^T \hat{J} \hat{L}_y = \begin{pmatrix} 0 & 0 \\ 0 & 0 \end{pmatrix},$$

$$\hat{X}^T \hat{J} \hat{D}_x + \hat{L}_x^T \hat{J} \hat{E} = \begin{pmatrix} 0 & (l|x) + (x|x)(a|\delta) - (a|x)(x|\delta) \\ 0 & (l|a) + (x|a)(a|\delta) - (a|a)(x|\delta) \end{pmatrix},$$

$$\hat{Y}^T \hat{J} \hat{D}_y + \hat{L}_y^T \hat{J} \hat{E} = \begin{pmatrix} 0 & (l|y) + (y|y)(b|\delta) - (b|y)(y|\delta) \\ 0 & (l|b) + (y|b)(b|\delta) - (b|b)(y|\delta) \end{pmatrix}.$$

So, the conditions in eq. (5.9) are described as

$$(x|x)(a|a) - (x|a)(a|x) = 1,$$
$$(y|y)(b|b) - (b|y)(y|b) = 1,$$
$$(l|x) + (x|x)(a|\delta) - (a|x)(x|\delta) = 0,$$
$$(l|a) + (x|a)(a|\delta) - (a|a)(x|\delta) = 0,$$
$$(l|y) + (y|y)(b|\delta) - (b|y)(y|\delta) = 0,$$
$$(l|b) + (y|b)(b|\delta) - (b|b)(y|\delta) = 0. \tag{5.10}$$

The first two of these are familiar and describe the fact that the volume of phase space is preserved under the linear transformations generated by particle optical elements. The other conditions, however, represent the connection between longitudinal and dispersive effects. For them to be satisfied requires the use of specific scaling factors for the variables $l$ and $\delta$, and they are the reason for the specific choice of the variable $\kappa$ in eq. (2.1) in Section 2.1.

Next let us study the case that coupling between horizontal and vertical planes is present, where the Jacobian matrix $\hat{M}$ is a $4 \times 4$ symplectic matrix. Similarly, we can divide $\hat{M}$ into four blocks of $2 \times 2$ matrices, which is

$$\hat{M} = \begin{pmatrix} \hat{A} & \hat{B} \\ \hat{C} & \hat{D} \end{pmatrix}.$$

Plugging into eq. (5.6), we obtain

$$\begin{pmatrix} \hat{A}^T \hat{J} \hat{A} + \hat{C}^T \hat{J} \hat{C} & \hat{A}^T \hat{J} \hat{B} + \hat{C}^T \hat{J} \hat{D} \\ \hat{B}^T \hat{J} \hat{A} + \hat{D}^T \hat{J} \hat{C} & \hat{B}^T \hat{J} \hat{B} + \hat{D}^T \hat{J} \hat{D} \end{pmatrix} = \begin{pmatrix} \hat{J} & \hat{0} \\ \hat{0} & \hat{J} \end{pmatrix},$$

which can be simplified as

$$\det \hat{A} + \det \hat{C} = 1, \quad \det \hat{B} + \det \hat{D} = 1, \quad \hat{A}^T \hat{J} \hat{B} + \hat{C}^T \hat{J} \hat{D} = \hat{0}.$$

In general $\hat{A}$, $\hat{B}$, $\hat{C}$ and $\hat{D}$ are not symplectic matrices. Meanwhile, from eq. (5.7), we have

$$
\begin{aligned}
\hat{M}^{-1} &= - \begin{pmatrix} \hat{J} & \hat{0} \\ \hat{0} & \hat{J} \end{pmatrix} \begin{pmatrix} \hat{A}^T & \hat{C}^T \\ \hat{B}^T & \hat{D}^T \end{pmatrix} \begin{pmatrix} \hat{J} & \hat{0} \\ \hat{0} & \hat{J} \end{pmatrix} = - \begin{pmatrix} \hat{J} \hat{A}^T \hat{J} & \hat{J} \hat{C}^T \hat{J} \\ \hat{J} \hat{B}^T \hat{J} & \hat{J} \hat{D}^T \hat{J} \end{pmatrix} \\
&= \begin{pmatrix} \bar{A} & \bar{C} \\ \bar{B} & \bar{D} \end{pmatrix},
\end{aligned}
\tag{5.11}
$$

where $\bar{A}$ is defined as

$$\bar{A} = - \hat{J} \hat{A}^T \hat{J},$$

and $\bar{B}$, $\bar{C}$ and $\bar{D}$ are defined the same way. In addition, we have

$$\bar{A} = - \begin{pmatrix} 0 & 1 \\ -1 & 0 \end{pmatrix} \begin{pmatrix} a_{11} & a_{21} \\ a_{12} & a_{22} \end{pmatrix} \begin{pmatrix} 0 & 1 \\ -1 & 0 \end{pmatrix} = \begin{pmatrix} a_{22} & -a_{12} \\ -a_{21} & a_{11} \end{pmatrix} = (\det \hat{A}) \cdot \hat{A}^{-1}.$$

Since $\hat{M}^{-1}$ is symplectic, we have

$$
\begin{pmatrix} \bar{A}^T \hat{J} \bar{A} + \bar{B}^T \hat{J} \bar{B} & \bar{A}^T \hat{J} \bar{C} + \bar{B}^T \hat{J} \bar{D} \\ \bar{C}^T \hat{J} \bar{A} + \bar{D}^T \hat{J} \bar{B} & \bar{C}^T \hat{J} \bar{C} + \bar{D}^T \hat{J} \bar{D} \end{pmatrix} = \begin{pmatrix} \hat{J} & \hat{0} \\ \hat{0} & \hat{J} \end{pmatrix},
$$

and hence

$$\det \hat{A} + \det \hat{B} = 1, \quad \det \hat{C} + \det \hat{D} = 1, \quad \bar{A}^T \hat{J} \bar{C} + \bar{B}^T \hat{J} \bar{D} = \hat{0},$$

where the relation $\det \bar{X} = \det \hat{X}$, $(\hat{X} = \hat{A}, \hat{B}, \hat{C}, \hat{D})$ is used. As a result, we obtain a set of important relations

$$\det \hat{A} + \det \hat{B} = 1, \quad \det \hat{D} = \det \hat{A}, \quad \det \hat{C} = \det \hat{B}.$$

It is worth noting that there are only six independent constraints from the symplectic condition for a $4 \times 4$ matrix.

---

## 5.2  Differential Algebras

In this section we will provide an introduction to the theory of Differential Algebras (DA) which enables the computation of transfer maps to an arbitrary order. For reasons of brevity we only provide a limited overview [4]; a more complete treatment can be found for example in [5]. For the sake of clarity, we first address the simplest case of Differential Algebras, mathematically denoted as the structure $_1D_1$.

### 5.2.1   The Structure $_1D_1$

Consider the vector space $R^2$ of ordered pairs $(a_0, a_1)$, $a_0, a_1 \in R$ in which an addition and a scalar multiplication are defined in the usual way:

$$(a_0, a_1) + (b_0, b_1) = (a_0 + b_0, a_1 + b_1), \tag{5.12}$$
$$t \cdot (a_0, a_1) = (t \cdot a_0, t \cdot a_1),$$

for $a_0, a_1, b_0, b_1 \in R$. Besides the above addition and scalar multiplication, a multiplication between vectors is introduced in the following way:

$$(a_0, a_1) \cdot (b_0, b_1) = (a_0 \cdot b_0, a_0 \cdot b_1 + a_1 \cdot b_0), \tag{5.13}$$

for $a_0, a_1, b_0, b_1 \in R$. With this definition of a vector multiplication the set of ordered pairs becomes an algebra, denoted by $_1D_1$.

In the same way as in the case of complex numbers, one can identify $(a_0, 0)$ as the real number $a_0$. Where in the complex numbers, $(0, 1)$ was a root of $-1$, here it has another interesting property:

$$(0, 1) \cdot (0, 1) = (0, 0),$$

which follows directly from eq. (5.13). So $(0, 1)$ is a root of 0. Such a property suggests thinking of $d = (0, 1)$ as something infinitely small, small enough that its square vanishes. Because of this we call $d = (0, 1)$ the differential unit. The first component of the pair $(a_0, a_1)$ is called the real part, and the second component is called the differential part. Using this notation it becomes clear that elements of $_1D_1$ can be written as $a_0 + a_1 \cdot d$, and multiplication amounts to multiplying the polynomials $(a_0 + a_1 \cdot d)$ and $(b_0 + b_1 \cdot d)$ and keeping only terms linear in $d$.

It is easy to verify that $(1, 0)$ is a neutral element of multiplication, because according to eq. (5.13)

$$(1, 0) \cdot (a_0, a_1) = (a_0, a_1) \cdot (1, 0) = (a_0, a_1).$$

It turns out that $(a_0, a_1)$ has a multiplicative inverse if and only if $a_0$ is nonzero. In case $a_0 \neq 0$ the inverse is

$$(a_0, a_1)^{-1} = \left( \frac{1}{a_0}, -\frac{a_1}{a_0^2} \right). \tag{5.14}$$

Using equations it is easy to check that in fact $(a_0, a_1)^{-1} \cdot (a_0, a_1) = (1, 0)$.

An outstanding result of the methods of differential algebras is that differentiation becomes an algebraic problem, and the differential part of the difference

$$f(x + d) - f(x)$$

equals the conventional derivative. Thus, given any differentiable function $f$, we can compute its derivatives by just evaluating the formula and thus obtain

$$f'(x) = \mathcal{D}\left[ f(x + d) - f(x) \right] = \mathcal{D}\left[ f(x + d) \right], \tag{5.15}$$

where $\mathcal{D}$ denotes the differential part. In the last step use has been made of the fact that $f(x)$ has no differential part. Hence Differential Algebras are useful to compute derivatives directly, without requiring an analytic formula for the derivative and without the inaccuracies of numerical techniques.

The computation of derivatives shall be illustrated in an example using the following function:

$$f(x) = \frac{1}{x + 1/x}. \tag{5.16}$$

The derivative of the function is

$$f'(x) = \frac{1/x^2 - 1}{(x + 1/x)^2}.$$

Suppose we are interested in the value of the function and its derivative at $x = 2$. We obtain

$$f(2) = \frac{2}{5}, \quad f'(2) = -\frac{3}{25}.$$

Now take the definition of the function $f$ in eq. (5.16) and evaluate it at $2 + d = (2, 1)$. One obtains:

$$f[(2,1)] = \frac{1}{(2,1) + 1/(2,1)} = \frac{1}{(2,1) + (1/2, -1/4)} = \frac{1}{(5/2, 3/4)}$$

$$= \left(1/(5/2), -(3/4)/(5/2)^2\right) = \left(\frac{2}{5}, -\frac{3}{25}\right).$$

As we can see, after the evaluation of the function the real part of the result is just the value of the function at $x = 2$, whereas the differential part is the derivative of the function at $x = 2$.

By our choice of the starting vector $(2, 1)$, initially the vector contains the value $I(2)$ of the identity function $I : x \to x$ in the first component and the derivative of $I'(2) = 1$ in the second component.

Now assume that in an intermediate step two vectors of value and derivative $(g(2), g'(2))$ and $(h(2), h'(2))$ have to be added. According to (5.12) one obtains $(g(2) + h(2), g'(2) + h'(2))$. But according to the rule for the differentiation of sums, this is just the value and derivative of the sum function $(g + h)$ at $x = 2$.

The same holds for the multiplication: Suppose that two vectors of value and derivatives $(g(2), g'(2))$ and $(h(2), h'(2))$ have to be multiplied. Then according to (5.13) one obtains $(g(2) \cdot h(2), g(2) \cdot h'(2) + g'(2) \cdot h(2))$. But according to the product rule, this is just the value and derivative of the product function $(g \cdot h)$ at $x = 2$.

The evaluation of the function $f$ at $(2, 1)$ can now be viewed as successively combining two intermediate functions $g$ and $h$, starting with the identity function and finally arriving at $f$. At each intermediate step the derivative of the intermediate function is automatically obtained as the differential part according to the above reasoning.

An interesting side aspect is that with the search for a multiplicative inverse in eq. (5.14) one has derived a rule to differentiate the function $f(x) = 1/x$ without explicitly using calculus rules.

After discussing the algebra ${}_1D_1$ and its virtues for the computation of derivatives, we now address the most general Differential Algebra, the structure ${}_nD_v$, which corresponds to the case of power series of $v$ variables to the $n$th order. It will eventually allow us to arithmetically compute partial derivatives of functions of $v$ variables through order $n$.

## 5.2.2    The Structure ${}_nD_v$

We define $N(n,v)$ to be the number of monomials in $v$ variables through order $n$. We will show that

$$N(n,v) = \frac{(n+v)!}{n!v!} = C(n+v,v),$$

where $C(i,j)$ is the familiar binomial coefficient. First note that the number of monomials with **exact** order $n$ equals $N(n,v-1)$. This is true because each monomial of exact order $n$ can be written as a monomial with one variable less times the last variable to such a power that the total power equals $n$. Thus we have

$$N(n,v) = N(n-1,v) + N(n,v-1) :$$

the number of monomials in $v$ variables through order $n$ equals the number of monomials of one order less plus the ones of exact order $n$. This recursive relation is satisfied by $C(n+v,v)$. Since also obviously $C(1+1,1) = 2 = N(1,1)$, the formula follows by induction.

Now assume that all these $N$ monomials are arranged in a certain manner order-by-order. For each monomial $M$ we call $I_M$ the position of $M$ according to the ordering. Conversely, with $M_I$ we denote the $I$th monomial of the ordering. Finally, for an $I$ with $M_I = x_1^{i_1} \cdots x_v^{i_v}$ we define $F_I = i_1! \cdots i_v!$.

We now define an addition, a scalar multiplication and a vector multiplication on $R^N$ in the following way:

$$(a_1, \ldots, a_N) + (b_1, \ldots, b_N) = (a_1 + b_1, \cdots, a_N + b_N),$$
$$t \cdot (a_1, \ldots, a_N) = (t \cdot a_1, \cdots, t \cdot a_N),$$
$$(a_1, \ldots, a_N) \cdot (b_1, \ldots, b_N) = (c_1, \ldots, c_N), \tag{5.17}$$

where the coefficients $c_i$ are defined as follows:

$$c_i = F_i \sum_{\substack{0 \le \nu, \mu \le N \\ M_\nu \cdot M_\mu = M_i}} \frac{a_\nu \cdot b_\mu}{F_\nu \cdot F_\mu}. \tag{5.18}$$

To help clarify these definitions, let us look at the case of two variables and second order. In this case, we have $n = 2$ and $v = 2$. There are $N =$

$C(2 + 2, 2) = 6$ monomials in two variables, namely

$$1, \ x, \ y, \ xx, \ xy, \ yy. \tag{5.19}$$

As an example, using the ordering in (5.19), we have $I_{xy} = 5$ and $M_3 = y$. Using the ordering in (5.19), we obtain for $c_1$ through $c_6$ in eq. (5.18):

$$c_1 = a_1 \cdot b_1,$$
$$c_2 = a_1 \cdot b_2 + a_2 \cdot b_1,$$
$$c_3 = a_1 \cdot b_3 + a_3 \cdot b_1,$$
$$c_4 = 2 \cdot (a_1 \cdot b_4/2 + a_2 \cdot b_2 + a_4 \cdot b_1/2),$$
$$c_5 = a_1 \cdot b_5 + a_2 \cdot b_3 + a_3 \cdot b_2 + a_5 \cdot b_1,$$
$$c_6 = 2 \cdot (a_1 \cdot b_6/2 + a_3 \cdot b_3 + a_6 \cdot b_1/2).$$

On $_nD_v$ we introduce a third operation $\partial_i$ :

$$\partial_\nu(a_1, \ldots, a_N) = (c_1, \ldots, c_N)$$

with

$$c_i = \begin{cases} 0 & \text{if } M_i \text{ has order } n, \\ a_{I_{(M_i \cdot x_\nu)}} & \text{else.} \end{cases}$$

So $\partial_\nu$ moves the derivatives around in the vector. Suppose a vector contains the derivatives of the function $f$, then applying $\partial_\nu$ to it one obtains the derivatives of $\partial f/\partial x_\nu$ through one order less. With this third operation, $_nD_v$ becomes a so-called Differential Algebra (DA)[5].

While in $_1D_1$, $d = (0, 1)$ was an infinitely small quantity, here we have a whole variety of infinitely small quantities that have the property that high enough powers of them vanish. We give special names to the ones in components $I$ belonging to first order monomials, denoting them by $dM_I$. In the example of $_2D_2$, we have $dx = (0, 1, 0, 0, 0, 0)$ and $dy = (0, 0, 1, 0, 0, 0)$. It then follows that instead of eq. (5.15) we obtain

$$f(x + dx, y + dy) = \left( f, \frac{\partial f}{\partial x}, \frac{\partial f}{\partial y}, \frac{\partial^2 f}{\partial x^2}, \frac{\partial^2 f}{\partial x \partial y}, \frac{\partial^2 f}{\partial y^2} \right)(x, y).$$

In the general case of $v$ variables and order $n$, after evaluating $f$ in DA one obtains:

$$\frac{\partial^{i_1 + i_2 + \cdots + i_v} f}{\partial x_1^{i_1} \partial x_2^{i_2} \cdots \partial x_v^{i_v}} = c_{I_{\left( x_1^{i_1} \cdots x_v^{i_v} \right)}},$$

where $I_{\left( x_1^{i_1} \cdots x_v^{i_v} \right)}$ is the index of the monomial $x_1^{i_1} \cdots x_v^{i_v}$, as defined in the beginning of the section.

### 5.2.3 Functions on Differential Algebras

In this subsection we will generalize standard functions like exponentials, logarithmic and trigonometric functions to Differential Algebras. As we will see below, virtually all functions existing on a computer can be generalized in a straightforward way.

We start our discussion by noting that for any Differential Algebra (DA) vector of the form $(0, a_1, \ldots, a_N) \in {}_nD_v$, i.e., with a zero in the component belonging to the zeroth order monomial, we have the following property:

$$(0, a_1, \ldots, a_N)^i = (0, 0, \ldots, 0) \quad \text{for } i > n, \tag{5.20}$$

which follows directly from the definition of the multiplication in ${}_nD_v$ defined in eq. (5.17).

Let us begin our discussion of special functions with the exponential function $\exp(x)$. Assume we have to compute the exponential of a DA vector that has already been created by previous operations. First we note that the functional equation

$$\exp(x + y) = \exp(x) \cdot \exp(y)$$

also holds for DA vectors. As we will see, this facilitates the computation of the exponential considerably.

$$\exp[(a_0, a_1, a_2, \ldots, a_N)] = \exp(a_0) \cdot \exp[(0, a_1, a_2, \ldots, a_N)]$$
$$= \exp(a_0) \cdot \sum_{i=0}^{\infty} \frac{(0, a_1, a_2, \ldots, a_N)^i}{i!} = \exp(a_0) \cdot \sum_{i=0}^{n} \frac{(0, a_1, a_2, \ldots, a_N)^i}{i!}.$$

In the last step use has been made of eq. (5.20) which entails that the sum has to be taken only through order $n$ and thus allows the computation of the root in finitely many steps. Hence the evaluation of the real number exponential $\exp(a_0)$, which internally on a computer requires a power series summation and hence cannot be done accurately, is more subtle than the rest of the operations in Differential Algebra.

A logarithm of a DA vector exists if and only if $a_0 > 0$. In this case one obtains

$$\log[(a_0, a_1, a_2, \ldots, a_N)] = \log\left\{ a_0 \cdot \left[ 1 + \left( 0, \frac{a_1}{a_0}, \frac{a_2}{a_0}, \cdots, \frac{a_N}{a_0} \right) \right] \right\}$$
$$= (\log(a_0), 0, \ldots, 0) + \sum_{i=1}^{\infty} (-1)^{i+1} \frac{1}{i} \left( 0, \frac{a_1}{a_0}, \frac{a_2}{a_0}, \cdots, \frac{a_N}{a_0} \right)^i$$
$$= (\log(a_0), 0, \ldots, 0) + \sum_{i=1}^{n} (-1)^{i+1} \frac{1}{i} \left( 0, \frac{a_1}{a_0}, \frac{a_2}{a_0}, \cdots, \frac{a_N}{a_0} \right)^i.$$

Again use has been made of the fundamental property of the logarithm

$$\log(x \cdot y) = \log(x) + \log(y)$$

which also holds for DA numbers and leads to simplifications by virtue of eq. (5.20).

As the last example, we will derive a formula for the root function. Even though there is a direct method to compute roots by solving a set of linear equations for the coefficients of the root, we present here a technique based on power series following an approach similar to the exponential and logarithm. The root has the following power series expansion:

$$\sqrt{1+x} = \sum_{i=0}^{\infty} (-1)^i \frac{1 \cdot 3 \cdots (2i-3)}{2 \cdot 4 \cdots (2i)} \cdot x^i.$$

Using this formula and the definitions of addition and multiplication in (5.17), one directly obtains for the square root of a DA vector:

$$\sqrt{(a_0, a_1, a_2, \ldots, a_N)} = \sqrt{a_0} \cdot \sqrt{1 + \left(0, \frac{a_1}{a_0}, \frac{a_2}{a_0}, \cdots, \frac{a_N}{a_0}\right)}$$

$$= \sqrt{a_0} \cdot \sum_{i=0}^{\infty} (-1)^i \frac{1 \cdot 3 \cdots (2i-3)}{2 \cdot 4 \cdots (2i)} \cdot \left(0, \frac{a_1}{a_0}, \frac{a_2}{a_0}, \cdots, \frac{a_N}{a_0}\right)^i$$

$$= \sqrt{a_0} \cdot \sum_{i=0}^{n} (-1)^i \frac{1 \cdot 3 \cdots (2i-3)}{2 \cdot 4 \cdots (2i)} \cdot \left(0, \frac{a_1}{a_0}, \frac{a_2}{a_0}, \cdots, \frac{a_N}{a_0}\right)^i.$$

Using the addition theorems for sine and cosine, one obtains formulas with finite sums in a quite similar way; in general, suppose a function $f$ has an addition theorem of the form

$$f(a+b) = g_a(b),$$

and $g_a(b)$ can be written in a power series, then by the same reasoning its Differential Algebraic extension is computable exactly in only finitely many steps.

In the meantime, there are numerous codes built on the ideas of Differential Algebraic methods, including the code COSY INFINITY [7, 50].

---

## 5.3    The Computation of Transfer Maps

### 5.3.1    An Illustrative Example

Differential Algebras (DA) can be used very efficiently to compute the transfer map of eq. (2.3) of particle optical systems in its Taylor series representation.

To illustrate this, let us start the discussion with a very simple example, the midplane motion in a 90° homogeneous bending magnet. Let $x_i$ and

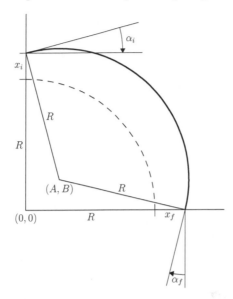

**FIGURE 5.2**: Motion in a 90° homogeneous dipole magnet. The dashed arc is the reference orbit.

$a_i = \sin(\alpha_i)$ denote the initial distance and scaled transverse momentum relative to the reference trajectory (see Fig. 5.2). Then we are interested in the values $x_f$ and $a_f = \sin(\alpha_f)$. Since the trajectories in the magnet are circles, we can readily read from Fig. 5.2:

$$A = R\sin(\alpha_i) = Ra_i,$$

$$B = R(1 - \cos(\alpha_i)) + x_i = R\left(1 - \sqrt{1 - a_i^2}\right) + x_i,$$

$$a_f = \sin(\alpha_f) = -\frac{B}{R},$$

$$x_f = A - R(1 - \cos(\alpha_f)) = A - R\left(1 - \sqrt{1 - a_f^2}\right).$$

These equations allow the computation of the final coordinates $x_f$ and $a_f$ in terms of the initial coordinates $x_i$ and $a_i$. However, taking these equations and performing all operations in Differential Algebra allows us to even obtain all derivatives of $x_f$ and $a_f$ with respect to $x_i$ and $a_i$. These so obtained derivatives, evaluated at $x_i = 0$, $a_i = 0$, are then the expansion coefficients of the map in eq. (2.3). For the sake of clarity, let us explicitly show how $x_f$ and $a_f$ are computed.

Using the ordering in (5.19) and identifying the variable $a$ with $y$, we obtain using the arithmetic defined in eqs. (5.17),

$$x_i = (0, 1, 0, 0, 0, 0), \qquad a_i = (0, 0, 1, 0, 0, 0),$$
$$A = (0, 0, R, 0, 0, 0), \qquad B = (0, 1, 0, 0, 0, R),$$
$$a_f = \left(0, -\frac{1}{R}, 0, 0, 0, -1\right), \quad x_f = \left(0, 0, R, -\frac{1}{R}, 0, 0\right). \qquad (5.21)$$

Comparing the obtained result with any matrix code, we find complete agreement; as an example, the fact that the second component of $x_f$ is zero implies that $\partial x_f / \partial x_i = 0$ and hence $(x|x) = 0$, which is a well known property of 90° bends.

In case an additional particle optical element is to follow this bending magnet, one does not have to start all over evaluating this new element at $x_i = (0, 1, 0, 0, 0, 0)$, $a_i = (0, 0, 1, 0, 0, 0)$, but one can start already with $x_f$ and $a_f$ of eq. (5.21). This way one can save the usually quite involved concatenation process and increase performance significantly.

## 5.3.2   Generation of Maps Using Numerical Integration

In this subsection we will address the general case in which no closed solution of the problem exists. We will see that also in this case we are actually able to compute transfer maps of arbitrary order for arbitrary particle optical elements. Even though we do not have analytical formulas that relate the final coordinates to the initial coordinates, there is still a way to computationally relate the final coordinates to the initial coordinates, by numerical integration of the equations of motion.

In this case, the final coordinates are still computed from the initial coordinates using standard arithmetic and functions; however, the relations are more complex than in the case of the homogeneous sector. As in any conventional numerical method, a numerical integrator is used to solve for the final coordinates.

Now blindfoldedly performing all these operations in Differential Algebra automatically leads to all desired derivatives of the transfer function, regardless of the form of the equations of motion. In other words, all coordinates and fields at any step are power series instead of real numbers.

Differential Algebraic (DA) techniques have been implemented in many programs. They allow the computation of transfer maps of elements with a dependence on the independent variable for which an analytic solution cannot be obtained. One example is magnets with fringing fields. Another example is an electrostatic round lens where the electric field varies with $s$ throughout the entire lens. As long as the electromagnetic field can be expressed in a differentiable function, the transfer map up to any given order can be obtained using the DA technique.

## 5.4 Manipulation of Maps

In most cases, a beam optical system consists of more than one element and it is necessary to connect the maps of individual pieces. Often the inverse of a map is needed. Sometimes one part of our system is the reversion of another part; therefore, it is time saving if the map of the reversed part can be obtained directly from that of the other part. All these map manipulations can be done elegantly using DA techniques.

### 5.4.1 Composition of Maps

Whenever a system contains more than one element, which is virtually always true, we have to deal with the composition of maps. This is the foundation of almost all other map manipulations, as we will see later.

Let us define $\circ$ as the symbol for composition. Hence the map $\mathcal{M}$ of a system consisting of two parts is

$$\mathcal{M} = \mathcal{M}_2 \circ \mathcal{M}_1, \tag{5.22}$$

where $\mathcal{M}_1$ and $\mathcal{M}_2$ are the maps of parts 1 and 2, respectively. According to Taylor's theorem, $\mathcal{M}_2$ can be expressed as the sum of a Taylor series and a remainder:

$$\mathcal{M}_2 = \mathcal{T}_n + \mathcal{R}_n,$$

where $\mathcal{R}_n$ is of order $n+1$. With the assumption that $\mathcal{M}_1$ is origin preserving, we have

$$[\mathcal{M}]_n = [\mathcal{M}_2 \circ \mathcal{M}_1]_n = [(\mathcal{T}_n + \mathcal{R}_n) \circ \mathcal{M}_1]_n = [\mathcal{T}_n \circ \mathcal{M}_1]_n + [\mathcal{R}_n \circ \mathcal{M}_1]_n$$
$$= [\mathcal{T}_n \circ \mathcal{M}_1]_n = \mathcal{T}_n([\mathcal{M}_1]_n).$$

Thus $[\mathcal{M}]_n$ can be obtained by composing two polynomials, which is called concatenation.

Concatenation is the most frequently used tool in DA calculations. It is extremely efficient when a given optical element appears multiple times in a system. Instead of computing the map every time it appears, the map of the element can be applied to the system using concatenation after the first appearance.

When the exact formula of $\mathcal{M}_2$ is known and it is relatively simple, eq. (5.22) can be used directly to compute $\mathcal{M}$, spending only a small fraction of the time required for concatenation. The saving comes from the fact that $\mathcal{M}_2$ is not expanded into a Taylor series. In fact, this method has been used whenever a element is a drift or a homogeneous dipole because their maps can be obtained from simple geometry.

Another application of the DA concatenator is the transformation of coordinates among different codes for the study of the dynamics of beams. For instance, while many codes work in the above discussed curvilinear canonical coordinates, other codes use the slope instead of the normalized momentum. In certain cases, Cartesian coordinates are used, which may be a better choice for certain elements, for example, wigglers, but are usually not very well suited for a discussion of the properties of beamlines. The transformation of a transfer map to a different set of coordinates, which can be expressed as

$$\mathcal{M}_T = \mathcal{C}^{-1} \circ \mathcal{M}_C \circ \mathcal{C},$$

is quite straightforward using Differential Algebra. One simply composes the transformation formulas between the two sets of coordinates, the map in one set of coordinates and the inverse transformation in Differential Algebra. Thus, one automatically obtains the map of the other set of coordinates to arbitrary orders.

### 5.4.2 Inversion of Maps

In this subsection, another kind of manipulation, the inversion, will be studied. The first step of inverting a Taylor map is to invert the linear part. If it exists, it can be done with any standard linear algebraic package. For the nonlinear part the inversion is done in an order-by-order fashion. To the second order we can write the map and its inverse as

$$\mathcal{M} =_2 \mathcal{M}_1 + \mathcal{M}_2, \quad \mathcal{M}^{-1} =_2 \mathcal{M}_1^{-1} + \mathcal{M}_2^{-1}.$$

Note that the subscript denotes the order of the map. Concatenating those two, we obtain

$$\mathcal{M} \circ \mathcal{M}^{-1} =_2 (\mathcal{M}_1 + \mathcal{M}_2) \circ (\mathcal{M}_1^{-1} + \mathcal{M}_2^{-1}) =_2 \mathcal{I} + \mathcal{M}_2 \circ \mathcal{M}_1^{-1} + \mathcal{M}_1 \circ \mathcal{M}_2^{-1} =_2 \mathcal{I},$$

where the term $\mathcal{M}_2 \circ \mathcal{M}_2^{-1}$ is dropped due to the fact that it contains terms of the fourth order. As a result, the second part of the inverse is

$$\mathcal{M}_2^{-1} =_2 -\mathcal{M}_1^{-1} \circ \mathcal{M}_2 \circ \mathcal{M}_1^{-1}.$$

Now let us assume that we already inverted the map up to the $(n-1)$th order; we write the map and its inverse up to the $n$th order as

$$\mathcal{M} =_n \mathcal{M}_1 + \mathcal{M}_n, \quad \mathcal{M}^{-1} =_n \mathcal{M}_1^{-1} + \mathcal{M}_n^{-1}.$$

Hence, we have

$$\mathcal{M} \circ \mathcal{M}^{-1} =_n (\mathcal{M}_1 + \mathcal{M}_n) \circ (\mathcal{M}_1^{-1} + \mathcal{M}_n^{-1})$$
$$=_n \mathcal{I} + \mathcal{M}_n \circ \mathcal{M}_1^{-1} + \mathcal{M}_1 \circ \mathcal{M}_n^{-1} + \mathcal{M}_n \circ \mathcal{M}_n^{-1} =_n \mathcal{I}.$$

Since $\mathcal{M}_n$ is of the second order and higher, only those terms of the order $[n/2]$ or lower in $\mathcal{M}_n^{-1}$ contribute. The result is

$$\mathcal{M}_n^{-1} =_n -\mathcal{M}_1^{-1} \circ \mathcal{M}_n \circ (\mathcal{M}_1^{-1} + \mathcal{M}_m^{-1}),$$

where $m = [n/2]$ and $m > 1$. It is clear that the inverse of a Taylor map exists as long as the inverse of its linear part exists. In later chapters, we will encounter a number of problems that have to be solved through finding the inverse of a map.

## 5.4.3 Reversion of Maps

Throughout the development of beam optical systems, mirror symmetry has been frequently used. Recently, mirror symmetric systems are being studied in great detail in the search for high order achromats. Among the various symmetry arrangements, reversion is the most commonly used. For a system which is the reversion of another one, the transfer map is the map obtained by going through the system in the reverse direction. The reversed motion can be described by first reversing time, i.e., switching all the signs of $p_x$ and $p_y$, then going through the inverse map, and finally re-reversing time. The time reversal operation can be performed easily using Differential Algebra, and the inversion of the transfer map is done as described in Section 5.4.2.

Specifically, reversion entails that, if a particle enters and exits the forward system at an initial point $(x_i, a_i, y_i, b_i, l_i, \delta_i)$ and a final point $(x_f, a_f, y_f, b_f, l_f, \delta_f)$, respectively, it will exit the reversed system at $(x_i, -a_i, y_i, -b_i, -l_i, \delta_i)$ after entering this system at $(x_f, -a_f, y_f, -b_f, -l_f, \delta_f)$. This determines the reversion transformation:

$$\hat{R} = \begin{pmatrix} 1 & 0 & 0 & 0 & 0 & 0 \\ 0 & -1 & 0 & 0 & 0 & 0 \\ 0 & 0 & 1 & 0 & 0 & 0 \\ 0 & 0 & 0 & -1 & 0 & 0 \\ 0 & 0 & 0 & 0 & -1 & 0 \\ 0 & 0 & 0 & 0 & 0 & 1 \end{pmatrix},$$

and hence the map of a reversed system:

$$\mathcal{M}^R = (\hat{R}) \circ \mathcal{M}^{-1} \circ (\hat{R}^{-1}).$$

In DA representation, $\mathcal{M}_n^R$ is computed through concatenation, where

$$\mathcal{M}_n^R = (\hat{R}) \circ \mathcal{M}_n^{-1} \circ (\hat{R}^{-1}).$$

In fact, the second composition $(\hat{R}) \circ \mathcal{M}_n^{-1}$ can be done by simply changing the signs of the rows for $a$, $b$ and $l$.

An interesting point worth noting is that $\hat{R}$ is not symplectic. In fact, it satisfies the following relation

$$\hat{R}^T \hat{J} \hat{R} = -\hat{J},$$

which we call an anti-symplectic relation.

# Chapter 6

## Linear Phase Space Motion

In this chapter, we want to study the action of transfer matrices on particles by looking in detail to what happens to **entire regions of phase space** as they are transported. This is important because the beam in an accelerator is just such a region, and of course we want to make sure that at any time, this region is within the beam pipe.

Let us begin by collecting several observations about two-dimensional transfer maps $\mathcal{M}$.

1. $\mathcal{M}$ preserves areas.

2. Different initial points have different final points.

3. Continuous curves stay continuous curves.

4. Closed curves stay closed curves.

5. A point inside a closed curve will stay inside of the closed curve.

Fig. 6.1 illustrates the second item, and Fig. 6.2 illustrates the other items.

The first item led us to giving a name, namely "emittance," to the preserved area. The last two items are particularly important, as they tell us that if we can enclose our beam within any closed boundary curve, then it is sufficient to study the dynamics of this boundary curve alone. It is interesting to note that while in the two-dimensional case, closed curves always stay closed curves, it

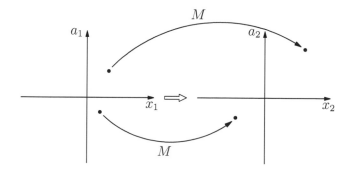

**FIGURE 6.1:** Mapping of individual points in phase space.

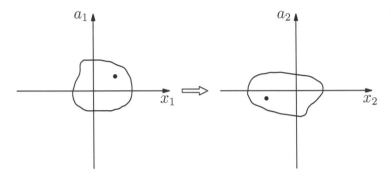

**FIGURE 6.2**: Mapping of a closed curve in phase space.

is not generally true that in higher dimensions, closed surfaces stay closed surfaces. While this is true for linear higher-dimensional transformation, nonlinear maps can produce some holes in the surfaces through which particles that were initially trapped inside the surface may find a way to escape, which is a very important mechanism that can lead to instability.

If in particular $\mathcal{M}$ is linear, then we also have the following observations.

1. Straight lines stay straight lines.

2. Ellipses stay ellipses.

Since straight lines stay straight lines, we may manufacture such a boundary curve as a **polygon**, and to study its motion it is completely sufficient to move only the corner points. Alternatively, we may try to enclose the beam by an **ellipse**. Before we follow these ideas, let us first study the action in phase space of some simple devices.

---

## 6.1 Phase Space Action

### 6.1.1 Drifts and Lenses

As seen in Section 2.2, the transfer matrix of a drift is given by

$$\hat{M} = \begin{pmatrix} 1 & l \\ 0 & 1 \end{pmatrix}.$$

This matrix leaves $a$ constant and moves $x$ by an amount proportional to $a$; hence it performs a **horizontal shearing** in phase space as shown in Fig. 6.3.

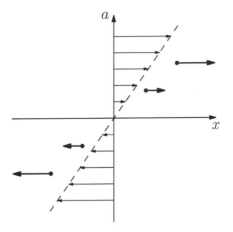

**FIGURE 6.3**: Action of a drift in phase space.

A lens has the transfer matrix

$$\hat{M} = \begin{pmatrix} 1 & 0 \\ -1/f & 1 \end{pmatrix},$$

and it leaves $x$ invariant and changes $a$ by a value proportional to $x$; hence it performs a **vertical shearing** as shown in Fig. 6.4.

## 6.1.2 Quadrupoles and Dipoles

In the case of quadrupoles and dipoles as seen in Sections 4.2 and 4.3, the matrices have the following form:

$$\hat{M} \propto \begin{pmatrix} \cos\phi & k\sin\phi \\ -(1/k)\sin\phi & \cos\phi \end{pmatrix}.$$

This corresponds roughly to a rotation, except that the $x$ and $a$ coordinates are also stretched or compressed; the result is a motion on an ellipse as shown in Fig. 6.5. In fact, computing the invariant ellipse of the motion following the procedure described in Section 8.1.2, we obtain

$$\alpha_i = 0, \quad \beta_i = k, \quad \gamma_i = \frac{1}{k}.$$

Applying to eq. (6.1), we see from $\alpha_i = 0$ that the ellipse is even upright.

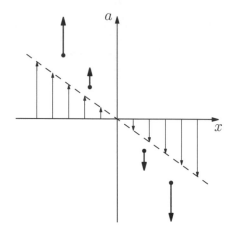

**FIGURE 6.4**: Action of a thin lens in phase space.

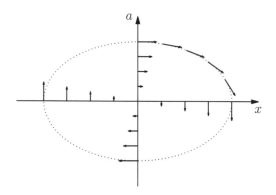

**FIGURE 6.5**: Action of a quadrupole or dipole in phase space.

## 6.2 Polygon–like Phase Space

In order to study the motion of ensembles of particles under linear transformations, it is useful to characterize them by certain simple geometric forms in which the particles are contained and requiring only few parameters. The two most useful such forms are the **polygon** and the **ellipse**.

A polygon in phase space is uniquely defined by its corner points; and since straight lines stay straight lines, it is sufficient to study just the motion of the corner points. Fig. 6.6 shows the motion of a polygon under a linear transformation.

Frequently a polygon with just four points is chosen; if its lines are initially symmetrically arranged around the origin, they will stay symmetrically ar-

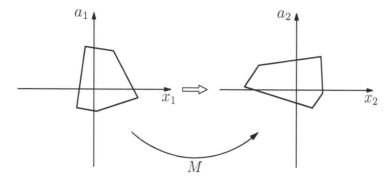

**FIGURE 6.6:** Mapping of a polygon in phase space.

ranged. Furthermore, a four-point polygon with symmetry around the origin is a parallelogram, and so parallelograms always stay parallelograms.

In many cases it is worthwhile to study how the actual beam width changes as a function of the $s$-position along the beamline. The beam width is apparently determined by the maximum of the horizontal positions of the corner points. In the special case in which we consider motion through a drift, each of the corner points moves on a straight line. Furthermore, the corner point that is furthest out will stay furthest out until it is possibly overtaken by another corner point; during the time it determines the beam width, it entails that the beam width changes linearly with $s$. Since the outermost corner point can change from time to time, the resulting beam width is **piecewise linear** as a function of $s$.

## 6.3 Elliptic Phase Space

The other choice that is worth considering is that of an elliptic phase space as shown in Fig. 6.7. In this case, the boundary of the phase space satisfies the **ellipse condition**

$$\gamma x^2 + 2\alpha x a + \beta a^2 = \varepsilon, \tag{6.1}$$

which can be written in matrix form using a symmetric matrix as

$$(x, a) \cdot \begin{pmatrix} \gamma & \alpha \\ \alpha & \beta \end{pmatrix} \cdot \begin{pmatrix} x \\ a \end{pmatrix} = \varepsilon.$$

For future simplicity, we denote the matrix describing the ellipse by $\hat{\sigma}$.

We first note that there is a redundancy in the description of the ellipse: obviously, doubling the values of $\alpha$, $\beta$, $\gamma$ as well as $\varepsilon$ simultaneously leads to

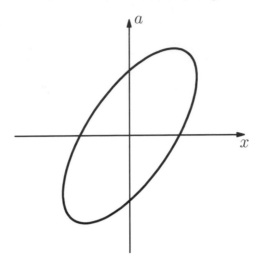

**FIGURE 6.7:** An ellipse in phase space.

the same ellipse. In order to eliminate this redundancy, we demand that the determinant of the ellipse be unity, i.e.,

$$\beta\gamma - \alpha^2 = 1.$$

With this choice of the matrix, the quantity $\varepsilon$ is a unique measure of its area, called the **emittance**. The four quantities $\alpha, \beta, \gamma$ and $\varepsilon$ are called the **Twiss parameters** of the matrix.

Now we are ready to study the question how the phase space ellipse changes as we pass through a system. Let $\hat{M}$ be the transfer matrix of the system; then the coordinates $x_1, a_1$ are transformed to $x_2, a_2$ via

$$\begin{pmatrix} x_2 \\ a_2 \end{pmatrix} = \hat{M} \cdot \begin{pmatrix} x_1 \\ a_1 \end{pmatrix};$$

and we also have

$$\begin{pmatrix} x_1 \\ a_1 \end{pmatrix} = \hat{M}^{-1} \cdot \begin{pmatrix} x_2 \\ a_2 \end{pmatrix}.$$

The new ellipse after the system characterized by $\hat{M}$ must obviously satisfy

$$(x_2, a_2) \cdot \hat{\sigma}_2 \cdot \begin{pmatrix} x_2 \\ a_2 \end{pmatrix} = \varepsilon. \qquad (6.2)$$

Observe that if we demand $\det(\hat{\sigma}_2) = 1$, even the measure for the occupied area $\varepsilon$ must be the same as before since we know the transfer map preserves area. We recall that the old coordinates satisfy

$$(x_1, a_1) \cdot \hat{\sigma}_1 \cdot \begin{pmatrix} x_1 \\ a_1 \end{pmatrix} = \varepsilon.$$

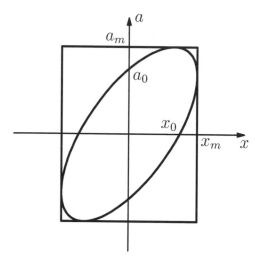

**FIGURE 6.8**: Characteristic points of an ellipse in phase space.

Expressing $x_1$, $a_1$ in terms of $x_2$, $a_2$, which is accomplished by the inverse matrix, we obtain

$$(x_2, a_2) \cdot \left( \left( \hat{M}^{-1} \right)^T \cdot \hat{\sigma}_1 \cdot \hat{M}^{-1} \right) \cdot \begin{pmatrix} x_2 \\ a_2 \end{pmatrix} = \varepsilon. \tag{6.3}$$

It is not difficult to show that $(\hat{M}^{-1})^T \cdot \hat{\sigma}_1 \cdot \hat{M}^{-1}$ is a symmetric matrix, thus we conclude that the resulting object is again an ellipse. So ellipses are indeed preserved under linear transformation. Furthermore, since the determinant of the matrix is unity (see Section 5.1.4), such a representation of an ellipse by a symmetric matrix of unity determinant is unique, and because eqs. (6.2) and (6.3) hold at the same time, we must conclude that

$$\hat{\sigma}_2 = \left( \hat{M}^{-1} \right)^T \cdot \hat{\sigma}_1 \cdot \hat{M}^{-1}. \tag{6.4}$$

This equation describes the **transformation of the ellipse** in phase space under the linear transformation.

### 6.3.1 The Practical Meaning of $\alpha$, $\beta$ and $\gamma$

As we propagate the beam through a system, the value of $\hat{\sigma}$ changes with $s$, and so do its three characteristic quantities $\alpha$, $\beta$ and $\gamma$. It is important to study how the three quantities $\alpha$, $\beta$ and $\gamma$ describe important characteristics of the beam. Another important question relates to the shape and degree of deformation of the ellipse. Together with the widths, this is characterized by the points at which the ellipse intersects the axes as shown in Fig. 6.8.

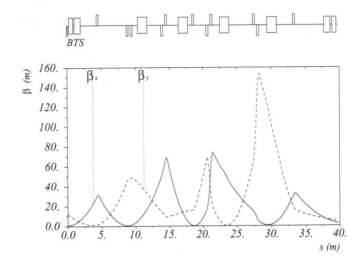

**FIGURE 6.9:**   Sketch of horizontal (solid) and vertical (dashed) $\beta$ functions of a beamline.

The question of **axes intersection** can be answered readily. In

$$\gamma x^2 + 2\alpha x a + \beta a^2 = \varepsilon,$$

we just set $a$ and $x$ to zero, and obtain

$$x_0 = \sqrt{\frac{\varepsilon}{\gamma}}, \quad a_0 = \sqrt{\frac{\varepsilon}{\beta}}.$$

Now we address the calculation of the **maximal points** $x_m$ and $a_m$, which characterize the width as well as the maximum angle in the ellipse. To this end, we view the elliptic shape as the contour line of a function, and remember that the gradient is always perpendicular to the contour lines. Hence the maximum position occurs where the angular component of the gradient vanishes, and the maximum angle occurs where the positional component of the gradient disappears. For the function

$$f(x, a) = \gamma x^2 + 2\alpha x a + \beta a^2,$$

we have

$$\vec{\nabla} f = (2\gamma x + 2\alpha a, 2\alpha x + 2\beta a),$$

and we infer that for the maximum position, we must have $ax = -\beta a$, namely $a = -\alpha/\beta \cdot x$. Inserting this into the ellipse yields

$$\gamma x^2 + 2\alpha x \left(-\frac{\alpha}{\beta} x\right) + \beta \left(-\frac{\alpha}{\beta} x\right)^2 = \varepsilon \quad \Longrightarrow \quad (\beta\gamma - \alpha^2) x^2 = \varepsilon\beta,$$

thus

$$x_m = \sqrt{\varepsilon\beta}.$$

Because of the symmetry of the equations with respect to interchange of $x$ and $a$, we see that also

$$a_m = \sqrt{\varepsilon\gamma}.$$

So the **maximal width** in $x$ direction is determined by the area of phase space $\varepsilon$ as well as the function $\beta$. Thus, $\beta$ plays an eminent role, as it immediately tells the width of a beam at a given point; and plots of its value for different positions around the accelerator are very commonly studied. An example of such a plot showing the $\beta$ functions for horizontal ($x$) and vertical ($y$) motion of a beamline at the Advanced Light Source (ALS) at Lawrence Berkeley National Laboratory (LBNL, LBL), California, USA, is shown in Fig. 6.9 [54, 21].

## 6.3.2 The Algebraic Relations among the Twiss Parameters

In this section, we attempt to introduce concept of the **phase advance**, which is the difference in phase between two points on the $s$-axis, and obtain the relations among the Twiss parameters. Let $\hat{M}(s)$ be the transfer matrix of a beamline, which may or may not be part of a periodic transport system, from $s_1 = 0$ to $s_2 = s$. Let $\alpha$, $\beta$ and $\gamma$ be the Twiss parameters at $s_1 = 0$ and $\alpha(s)$, $\beta(s)$ and $\gamma(s)$ be the ones at $s_2 = s$. From eq. (6.4), we obtain that

$$\begin{pmatrix} \gamma(s) & \alpha(s) \\ \alpha(s) & \beta(s) \end{pmatrix} = \left( \left( \hat{M}(s) \right)^{-1} \right)^T \begin{pmatrix} \gamma & \alpha \\ \alpha & \beta \end{pmatrix} \left( \hat{M}(s) \right)^{-1},$$

or in another form

$$\left( \hat{M}(s) \right)^T \begin{pmatrix} \gamma(s) & \alpha(s) \\ \alpha(s) & \beta(s) \end{pmatrix} \hat{M}(s) = \begin{pmatrix} \gamma & \alpha \\ \alpha & \beta \end{pmatrix}. \tag{6.5}$$

On the other hand, it is straightforward to show that

$$\begin{pmatrix} \sqrt{\beta} & -\alpha/\sqrt{\beta} \\ 0 & 1/\sqrt{\beta} \end{pmatrix} \begin{pmatrix} \gamma & \alpha \\ \alpha & \beta \end{pmatrix} \begin{pmatrix} \sqrt{\beta} & 0 \\ -\alpha/\sqrt{\beta} & 1/\sqrt{\beta} \end{pmatrix} = \hat{I},$$

which means that the matrix

$$A = \begin{pmatrix} \sqrt{\beta} & 0 \\ -\alpha/\sqrt{\beta} & 1/\sqrt{\beta} \end{pmatrix}$$

transforms the ellipse into a circle. The new coordinates are sometimes called the normal coordinates. This equation entails that

$$\begin{pmatrix} \gamma & \alpha \\ \alpha & \beta \end{pmatrix} = \begin{pmatrix} 1/\sqrt{\beta} & \alpha/\sqrt{\beta} \\ 0 & \sqrt{\beta} \end{pmatrix} \begin{pmatrix} 1/\sqrt{\beta} & 0 \\ \alpha/\sqrt{\beta} & \sqrt{\beta} \end{pmatrix}. \tag{6.6}$$

Plugging eq. (6.6) into eq. (6.5), we obtain

$$\left(\hat{M}(s)\right)^T \begin{pmatrix} 1/\sqrt{\beta(s)} & \alpha(s)/\sqrt{\beta(s)} \\ 0 & \sqrt{\beta(s)} \end{pmatrix} \begin{pmatrix} 1/\sqrt{\beta(s)} & 0 \\ \alpha(s)/\sqrt{\beta(s)} & \sqrt{\beta(s)} \end{pmatrix} \hat{M}(s)$$

$$= \begin{pmatrix} 1/\sqrt{\beta} & \alpha/\sqrt{\beta} \\ 0 & \sqrt{\beta} \end{pmatrix} \begin{pmatrix} 1/\sqrt{\beta} & 0 \\ \alpha/\sqrt{\beta} & \sqrt{\beta} \end{pmatrix}.$$

Hence, we have the following relation

$$\begin{pmatrix} \sqrt{\beta} & -\alpha/\sqrt{\beta} \\ 0 & 1/\sqrt{\beta} \end{pmatrix} \left(\hat{M}(s)\right)^T \begin{pmatrix} 1/\sqrt{\beta(s)} & \alpha(s)/\sqrt{\beta(s)} \\ 0 & \sqrt{\beta(s)} \end{pmatrix}$$

$$\cdot \begin{pmatrix} 1/\sqrt{\beta(s)} & 0 \\ \alpha(s)/\sqrt{\beta(s)} & \sqrt{\beta(s)} \end{pmatrix} \hat{M}(s) \begin{pmatrix} \sqrt{\beta} & 0 \\ -\alpha/\sqrt{\beta} & 1/\sqrt{\beta} \end{pmatrix} = \hat{I}.$$

Defining

$$\hat{R}(s) = \begin{pmatrix} 1/\sqrt{\beta(s)} & 0 \\ \alpha(s)/\sqrt{\beta(s)} & \sqrt{\beta(s)} \end{pmatrix} \cdot \hat{M}(s) \cdot \begin{pmatrix} \sqrt{\beta} & 0 \\ -\alpha/\sqrt{\beta} & 1/\sqrt{\beta} \end{pmatrix}, \qquad (6.7)$$

we immediately have

$$\left(\hat{R}(s)\right)^T \cdot \hat{R}(s) = \hat{I},$$

or equivalently

$$\left(\hat{R}(s)\right)^T = \left(\hat{R}(s)\right)^{-1}.$$

The matrix $\hat{R}(s)$ can be expressed in the extended form

$$\hat{R}(s) = \begin{pmatrix} R_{11}(s) & R_{12}(s) \\ R_{21}(s) & R_{22}(s) \end{pmatrix},$$

which entails that

$$\left(\hat{R}(s)\right)^T = \begin{pmatrix} R_{11}(s) & R_{21}(s) \\ R_{12}(s) & R_{22}(s) \end{pmatrix},$$

$$\left(\hat{R}(s)\right)^{-1} = \begin{pmatrix} R_{22}(s) & -R_{12}(s) \\ -R_{21}(s) & R_{11}(s) \end{pmatrix};$$

the latter holds due to the fact that $\det(\hat{R}(s)) = 1$. As a result, we have

$$\begin{pmatrix} R_{11}(s) & R_{21}(s) \\ R_{12}(s) & R_{22}(s) \end{pmatrix} = \begin{pmatrix} R_{22}(s) & -R_{12}(s) \\ -R_{21}(s) & R_{11}(s) \end{pmatrix},$$

which leads to the relations

$$R_{11}(s) = R_{22}(s), \quad R_{12}(s) = -R_{21}(s).$$

Plugging them into the equation

$$R_{11}(s)R_{22}(s) - R_{12}(s)R_{21}(s) = 1,$$

we obtain

$$R_{11}^2(s) + R_{12}^2(s) = 1.$$

Therefore, the matrix elements can be expressed as trigonometric functions of a single variable $\phi(s)$ and the matrix $\hat{R}(s)$ takes the form

$$\hat{R}(s) = \begin{pmatrix} \cos\phi(s) & \sin\phi(s) \\ -\sin\phi(s) & \cos\phi(s) \end{pmatrix}. \tag{6.8}$$

Since $\hat{R}(0) = \hat{I}$, $\phi(0) = 0$. Plugging eq. (6.8) into eq. (6.7), we obtain the explicit form of the transfer matrix, which is

$$
\begin{aligned}
\hat{M}(s) &= \begin{pmatrix} \sqrt{\beta(s)} & 0 \\ -\alpha(s)/\sqrt{\beta(s)} & 1/\sqrt{\beta(s)} \end{pmatrix} \hat{R}(s) \begin{pmatrix} 1/\sqrt{\beta} & 0 \\ \alpha/\sqrt{\beta} & \sqrt{\beta} \end{pmatrix} \\
&= \begin{pmatrix} \sqrt{\beta(s)} & 0 \\ -\alpha(s)/\sqrt{\beta(s)} & 1/\sqrt{\beta(s)} \end{pmatrix} \begin{pmatrix} \cos\phi(s) & \sin\phi(s) \\ -\sin\phi(s) & \cos\phi(s) \end{pmatrix} \begin{pmatrix} 1/\sqrt{\beta} & 0 \\ \alpha/\sqrt{\beta} & \sqrt{\beta} \end{pmatrix}
\end{aligned} \tag{6.9}
$$

$$= \begin{pmatrix} \sqrt{\beta_s/\beta}\,(\cos\phi_s + \alpha\sin\phi_s) & \sqrt{\beta_s\beta}\,\sin\phi_s \\ m_{21}(s) & \sqrt{\beta/\beta_s}\,(\cos\phi_s - \alpha_s\sin\phi_s) \end{pmatrix}, \tag{6.10}$$

where

$$m_{21}(s) = -\frac{1}{\sqrt{\beta_s\beta}}\left[(\alpha_s - \alpha)\cos\phi_s + (1 + \alpha\alpha_s)\sin\phi_s\right]$$

and $\alpha_s$, $\beta_s$ and $\phi_s$ represent $\alpha(s)$, $\beta(s)$ and $\phi(s)$, respectively. Denoting

$$\hat{M}(s) = \begin{pmatrix} (x|x) & (x|a) \\ (a|x) & (a|a) \end{pmatrix},$$

where each element is a function of $s$, it is easy to show that

$$\tan\phi_s = \frac{(x|a)}{\beta(x|x) - \alpha(x|a)}$$

when the values of $(x|x)$ and $(x|a)$ from eq. (6.10) are plugged in.

From the coordinate transformation point of view, eq. (6.9) illustrates the relation between the physical coordinates and the normal coordinates. The right most matrix transforms the physical coordinates into normal coordinates where the motion in the normalized phase space is a rotation, represented by the middle matrix. At end of the transport, the normalized coordinates are transformed back to the physical coordinates.

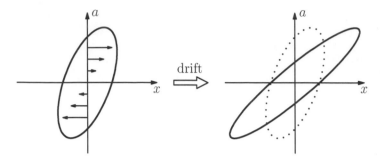

**FIGURE 6.10:** Transformation of an ellipse under a drift.

For many practical applications, it is useful to explicitly study the transformation of the ellipse (6.4) through the influence of the matrix $\hat{M}$. We first observe that if

$$\hat{M} = \begin{pmatrix} (x|x) & (x|a) \\ (a|x) & (a|a) \end{pmatrix},$$

then

$$\hat{M}^{-1} = \begin{pmatrix} (a|a) & -(x|a) \\ -(a|x) & (x|x) \end{pmatrix},$$

as simple arithmetic shows. So we have

$$\hat{\sigma}_2 = \left(\hat{M}^{-1}\right)^T \cdot \begin{pmatrix} \gamma_1 & \alpha_1 \\ \alpha_1 & \beta_1 \end{pmatrix} \cdot \left(\hat{M}^{-1}\right) = \begin{pmatrix} \gamma_2 & \alpha_2 \\ \alpha_2 & \beta_2 \end{pmatrix}$$

$$= \begin{pmatrix} (a|a) & -(a|x) \\ -(x|a) & (x|x) \end{pmatrix} \cdot \begin{pmatrix} \gamma_1 & \alpha_1 \\ \alpha_1 & \beta_1 \end{pmatrix} \cdot \begin{pmatrix} (a|a) & -(x|a) \\ -(a|x) & (x|x) \end{pmatrix}.$$

Performing the calculations, we see first of all that $\alpha_2$, $\beta_2$ and $\gamma_2$ depend **linearly** on $\alpha_1$, $\beta_1$ and $\gamma_1$, and hence the relationship can be written in matrix form. Explicitly, we have

$$\begin{pmatrix} \beta_2 \\ \alpha_2 \\ \gamma_2 \end{pmatrix} = \begin{pmatrix} (x|x)^2 & -2(x|x)(x|a) & (x|a)^2 \\ -(x|x)(a|x) & (x|x)(a|a)+(x|a)(a|x) & -(x|a)(a|a) \\ (a|x)^2 & -2(a|x)(a|a) & (a|a)^2 \end{pmatrix} \begin{pmatrix} \beta_1 \\ \alpha_1 \\ \gamma_1 \end{pmatrix}.$$

$$(6.11)$$

One particularly interesting case is the one where we let an ellipse evolve under the action of a drift, as shown in Fig. 6.10.

If we are interested in the way in which the width of the beam changes, we must look at the function $\beta(s)$. For the special case of the drift matrix with $(x|x) = (a|a) = 1$, $(a|x) = 0$ and $(x|a) = L$, we have

$$\beta(L) = (x|x)^2 \beta_1 - 2(x|x)(x|a)\alpha_1 + (x|a)^2 \gamma_1 = \beta_1 - 2L\alpha_1 + L^2\gamma_1$$

$$= \gamma_1 \left(L - \frac{\alpha_1}{\gamma_1}\right)^2 - \frac{\alpha_1^2}{\gamma_1} + \beta_1 = \gamma_1 \left(L - \frac{\alpha_1}{\gamma_1}\right)^2 + \frac{1}{\gamma_1}. \qquad (6.12)$$

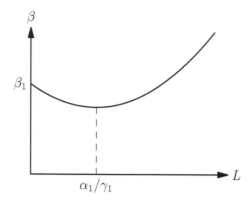

**FIGURE 6.11:** Plot of the $\beta$ function in a drift.

So as a function of $L$, $\beta(L)$ changes quadratically. We also see readily that at the point where

$$L = \frac{\alpha_1}{\gamma_1},$$

the beam has minimum width, and we have what is called a **waist** (see Fig. 6.11). On the other hand, we obtain from eq. (6.11) that

$$\gamma(L) = \gamma_1,$$

which reflects the fact that the divergence of the beam is not changed in a drift. As a result, we have

$$\beta(L) = \frac{1}{\gamma(L)},$$

which entails that

$$\alpha(L) = 0$$

at the waist. Meanwhile the behavior of $\alpha(L)$, which is

$$\alpha(L) = -\frac{1}{2}\frac{d\beta}{dL} = \alpha_1 - L\gamma_1,$$

vanishes at the waist. Finally, eq. (6.12) can be re-formulated as

$$\beta(s) = \beta^* + \frac{s^2}{\beta^*}, \tag{6.13}$$

where $\beta^*$ is the value at the waist and $s$ is the longitudinal distance from the waist.

### 6.3.3 The Differential Relations among the Twiss Parameters

Now let us consider the case that $s$ is small and to the first order $\beta(s) =_1 \beta + \beta' s$, $\alpha(s) =_1 \alpha + \alpha' s$ and $\phi(s) =_1 \phi' s$. Similarly, the transfer matrix becomes

$$\hat{M}(s) =_1 \begin{pmatrix} 1 + \left(\beta'/2\beta + \alpha\phi'\right) s & \beta\phi' s \\ -(1/\beta)\left[\alpha' + (1+\alpha^2)\phi'\right] s & 1 - \left(\beta'/2\beta + \alpha\phi'\right) s \end{pmatrix}, \quad (6.14)$$

where each element $M_{ij}$, the $i,j$-th element, is computed as

$$M_{11}(s) =_1 \sqrt{\frac{\beta+\beta's}{\beta}} \left[\cos\left(\phi's\right) + \alpha\sin\left(\phi's\right)\right] =_1 1 + \left(\frac{\beta'}{2\beta} + \alpha\phi'\right) s,$$

$$M_{12}(s) =_1 \sqrt{(\beta+\beta's)\beta}\sin\left(\phi's\right) =_1 \beta\phi's,$$

$$M_{21}(s) =_1 -\sqrt{\frac{1}{(\beta+\beta's)\beta}} \left[\alpha's\cos\left(\phi's\right) + \left(1+\alpha\left(\alpha+\alpha's\right)\right)\sin\left(\phi's\right)\right]$$

$$=_1 -\frac{1}{\beta}\left[\alpha' + (1+\alpha^2)\phi'\right] s,$$

$$M_{22}(s) =_1 \sqrt{\frac{\beta}{\beta+\beta's}} \left[\cos\left(\phi's\right) - \left(\alpha+\alpha's\right)\sin\left(\phi's\right)\right]$$

$$=_1 1 - \left(\frac{\beta'}{2\beta} + \alpha\phi'\right) s.$$

On the other hand, the transfer matrix to the first order of $s$ can be solved directly from the equations of motion, which is

$$\hat{M}(s) =_1 \begin{pmatrix} 1 & s \\ -ks & 1 \end{pmatrix}, \quad (6.15)$$

where $k$ is the focusing strength at $s_1 = 0$. Equating the corresponding terms in eqs. (6.14) and (6.15) yields

$$\beta\phi' = 1, \quad \frac{\beta'}{2\beta} + \alpha\phi' = 0, \quad -\frac{1}{\beta}\left[\alpha' + (1+\alpha^2)\phi'\right] = -k,$$

which can be simplified to the familiar form

$$\phi' = \frac{1}{\beta},$$

$$\alpha' = k\beta - \gamma,$$

$$\beta' = -2\alpha,$$

$$\gamma' = 2k\alpha.$$

The behavior of $\phi$, $\alpha$, $\beta$ and $\gamma$ are determined by the above set of differential equations.

## 6.4 *Edwards-Teng Parametrization

Here we have shown an alternative way of propagating the Twiss parameters through a beamline or a ring. The advantage is that only matrix multiplication and inner products of vectors (rows of a matrix) are used. In the following we will show that this way of tracking the Twiss parameters can be readily extended to coupled $x$ and $y$ motions.

The approach is based on the Edwards-Teng parametrization of a four-dimensional symplectic matrix. From the relation

$$\hat{M}_4 = \begin{pmatrix} \hat{M} & \hat{n} \\ \hat{m} & \hat{N} \end{pmatrix} = \begin{pmatrix} \hat{I}\cos\varphi & \hat{D}^{-1}\sin\varphi \\ -\hat{D}\sin\varphi & \hat{I}\cos\varphi \end{pmatrix} \begin{pmatrix} \hat{A} & \hat{0} \\ \hat{0} & \hat{B} \end{pmatrix} \begin{pmatrix} \hat{I}\cos\varphi & -\hat{D}^{-1}\sin\varphi \\ \hat{D}\sin\varphi & \hat{I}\cos\varphi \end{pmatrix},$$

we obtain

$$\hat{M} = \hat{A}\cos^2\varphi + \hat{D}^{-1}\hat{B}\hat{D}\sin^2\varphi,$$
$$\hat{N} = \hat{B}\cos^2\varphi + \hat{D}\hat{A}\hat{D}^{-1}\sin^2\varphi,$$
$$\hat{m} = -\left(\hat{D}\hat{A} - \hat{B}\hat{D}\right)\sin\varphi\cos\varphi,$$
$$\hat{n} = -\left(\hat{A}\hat{D}^{-1} - \hat{D}^{-1}\hat{B}\right)\sin\varphi\cos\varphi, \qquad (6.16)$$

where $\hat{A}$, $\hat{B}$ and $\hat{D}$ are symplectic. It is straightforward to obtain

$$\hat{A} = \hat{M} - \hat{D}^{-1}\hat{m}\tan\varphi, \quad \hat{B} = \hat{N} + \hat{D}\hat{n}\tan\varphi.$$

Subtracting the second equation from the first equation of (6.16) and taking its trace, we have

$$\begin{aligned}
\text{tr}(\hat{M} - \hat{N}) &= \text{tr}\left[\left(\hat{A} - \hat{B}\right)\cos^2\varphi + \left(\hat{D}^{-1}\hat{B}\hat{D} - \hat{D}\hat{A}\hat{D}^{-1}\right)\sin^2\varphi\right] \\
&= \left(\text{tr}\,\hat{A} - \text{tr}\,\hat{B}\right)\cos^2\varphi + \left[\text{tr}\left(\hat{D}^{-1}\hat{B}\hat{D}\right) - \text{tr}\left(\hat{D}\hat{A}\hat{D}^{-1}\right)\right]\sin^2\varphi \\
&= \left(\text{tr}\,\hat{A} - \text{tr}\,\hat{B}\right)\left(\cos^2\varphi - \sin^2\varphi\right) \\
&= 2\left(\cos\mu_1 - \cos\mu_2\right)\cos(2\varphi). \qquad (6.17)
\end{aligned}$$

In the last step, the assumption that $|\,\text{tr}\,\hat{A}\,| < 2$ and $|\,\text{tr}\,\hat{B}\,| < 2$ is adopted following [25], where the matrix $\hat{M}_4$ is the one turn matrix of a ring. For a generic symplectic matrix, all the conclusions hold after replacing $\cos \mu_1 - \cos \mu_2$ with $(\operatorname{tr} \hat{A} - \operatorname{tr} \hat{B})/2$. As a result, we have

$$\cos (2\varphi) = \frac{\operatorname{tr}(\hat{M} - \hat{N})}{2 \left(\cos \mu_1 - \cos \mu_2\right)}.$$

From the fourth equation of (6.16), on the other hand, we have

$$\hat{n} = - \left(\hat{A}\hat{D}^{-1} - \hat{D}^{-1}\hat{B}\right) \sin \varphi \cos \varphi = \left(\hat{A}\hat{J}\hat{D}^T\hat{J} - \hat{J}\hat{D}^T\hat{J}\hat{B}\right) \sin \varphi \cos \varphi,$$

and hence

$$\hat{J}\hat{n}^T \hat{J}^T = \hat{J} \left(\hat{J}^T \hat{D}\hat{J}\hat{A}^T - \hat{B}^T \hat{J}^T \hat{D}\hat{J}\right) \hat{J}^T \sin \varphi \cos \varphi$$
$$= - \left(\hat{D}\hat{A}^{-1} - \hat{B}^{-1}\hat{D}\right) \sin \varphi \cos \varphi.$$

Adding the third equation of (6.16) on both sides, we obtain

$$\hat{m} + \hat{J}\hat{n}^T \hat{J}^T = - \left[\hat{D} \left(\hat{A} + \hat{A}^{-1}\right) - \left(\hat{B} + \hat{B}^{-1}\right) \hat{D}\right] \sin \varphi \cos \varphi$$
$$= - \left[\hat{D}(\operatorname{tr} \hat{A}) - (\operatorname{tr} \hat{B})\hat{D}\right] \sin \varphi \cos \varphi$$
$$= - \left(\cos \mu_1 - \cos \mu_2\right) \sin (2\varphi) \hat{D}. \qquad (6.18)$$

Finally, we obtain

$$\hat{D} = -\frac{\hat{m} + \hat{J}\hat{n}^T \hat{J}^T}{\left(\cos \mu_1 - \cos \mu_2\right) \sin (2\varphi)}.$$

Now, the only unknown quantity is $\cos \mu_1 - \cos \mu_2$, which can be obtained below. From eq. (6.18), we have

$$\det \left(\hat{m} + \hat{J}\hat{n}^T \hat{J}^T\right) = \left(\cos \mu_1 - \cos \mu_2\right)^2 \sin^2 (2\varphi).$$

Adding the square of eq. (6.17), we obtain

$$\cos \mu_1 - \cos \mu_2 = \frac{1}{2} \operatorname{tr}(\hat{M} - \hat{N}) \sqrt{1 + \frac{\det(\hat{m} + \hat{J}\hat{n}^T \hat{J}^T)}{\left[\operatorname{tr}(\hat{M} - \hat{N})/2\right]^2}}.$$

Now let us consider the case that $\hat{M}_4$ is symmetric, i.e., $\hat{M}^T = \hat{M}$, $\hat{N}^T = \hat{N}$ and $\hat{n}^T = \hat{m}$. As a result, we have

$$\hat{D} = -\frac{\hat{m} + \hat{J}\hat{m}\hat{J}^T}{\left(\cos \mu_1 - \cos \mu_2\right) \sin (2\varphi)},$$

and so

$$\hat{D}^{-1} = \hat{J}\hat{D}^T\hat{J}^T = -\frac{\hat{J}\hat{m}^T\hat{J}^T + \hat{J}(\hat{J}\hat{m}^T\hat{J}^T)\hat{J}^T}{(\cos\mu_1 - \cos\mu_2)\sin(2\varphi)}$$

$$= -\frac{\hat{J}\hat{m}^T\hat{J}^T + \hat{m}^T}{(\cos\mu_1 - \cos\mu_2)\sin(2\varphi)} = \hat{D}^T.$$

Let us define

$$\hat{D} = \begin{pmatrix} a & b \\ c & d \end{pmatrix}.$$

Since $\det\hat{D} = 1$, we have

$$\hat{D}^{-1} = \begin{pmatrix} d & -b \\ -c & a \end{pmatrix}.$$

The relation $\hat{D}^T = \hat{D}^{-1}$ entails that $d = a$ and $c = -b$. Combining with $\det\hat{D} = 1$, we obtain $a^2 + b^2 = 1$, which means that $\hat{D}$ is a rotation. Furthermore, we have

$$\hat{D}^{-1}\hat{m} = -\frac{(\hat{J}\hat{m}^T\hat{J}^T + \hat{m}^T)\hat{m}}{(\cos\mu_1 - \cos\mu_2)\sin(2\varphi)} = -\frac{(\det\hat{m})\hat{I} + \hat{m}^T\hat{m}}{(\cos\mu_1 - \cos\mu_2)\sin(2\varphi)},$$

and

$$\hat{D}\hat{n} = -\frac{(\hat{m} + \hat{J}\hat{m}\hat{J}^T)\hat{m}^T}{(\cos\mu_1 - \cos\mu_2)\sin(2\varphi)} = -\frac{\hat{m}\hat{m}^T + (\det\hat{m})\hat{I}}{(\cos\mu_1 - \cos\mu_2)\sin(2\varphi)}.$$

Therefore both $\hat{D}^{-1}\hat{m}$ and $\hat{D}\hat{n}$ are symmetric and, as a result, $\hat{A}$ and $\hat{B}$ are symmetric as well.

### 6.4.1 The Algebraic Relations with Coupling

Let us start by noting that the $4 \times 4$ matrix that describes the beam ellipse of the coupled transverse motion is symmetric and symplectic, which is a result of the fact that the motion of the particles is symplectic. From the argument above, we have

$$\hat{M}_4 = \begin{pmatrix} \hat{M} & \hat{n} \\ \hat{m} & \hat{N} \end{pmatrix} = \begin{pmatrix} \hat{I}\cos\varphi & \hat{D}^{-1}\sin\varphi \\ -\hat{D}\sin\varphi & \hat{I}\cos\varphi \end{pmatrix} \begin{pmatrix} \hat{A} & \hat{0} \\ \hat{0} & \hat{B} \end{pmatrix} \begin{pmatrix} \hat{I}\cos\varphi & -\hat{D}^{-1}\sin\varphi \\ \hat{D}\sin\varphi & \hat{I}\cos\varphi \end{pmatrix},$$

where $\det\hat{A} = \det\hat{B} = \det\hat{D} = 1$. For a symmetric matrix characterizing the four-dimensional beam ellipsoid, the parametrization can be written as

$$\hat{T} = \begin{pmatrix} \hat{I}\cos\varphi & \hat{D}^{-1}\sin\varphi \\ -\hat{D}\sin\varphi & \hat{I}\cos\varphi \end{pmatrix} \begin{pmatrix} \hat{T}_x & \hat{0} \\ \hat{0} & \hat{T}_y \end{pmatrix} \begin{pmatrix} \hat{I}\cos\varphi & -\hat{D}^{-1}\sin\varphi \\ \hat{D}\sin\varphi & \hat{I}\cos\varphi \end{pmatrix},$$

where $\hat{T}_x^T = \hat{T}_x$, $\hat{T}_y^T = \hat{T}_y$ and $\hat{D}^T = \hat{D}^{-1}$. Using the relation $\hat{D}^T = \hat{D}^{-1}$, we obtain immediately

$$
\begin{pmatrix} \hat{I}\cos\varphi & \hat{D}^{-1}\sin\varphi \\ -\hat{D}\sin\varphi & \hat{I}\cos\varphi \end{pmatrix} = \begin{pmatrix} \hat{I}\cos\varphi & \hat{D}^T\sin\varphi \\ -\hat{D}\sin\varphi & \hat{I}\cos\varphi \end{pmatrix} = \begin{pmatrix} \hat{I}\cos\varphi & -\hat{D}^{-1}\sin\varphi \\ \hat{D}\sin\varphi & \hat{I}\cos\varphi \end{pmatrix}^T.
$$

As a result, we have

$$
\hat{T} = \begin{pmatrix} \hat{I}\cos\varphi & -\hat{D}^{-1}\sin\varphi \\ \hat{D}\sin\varphi & \hat{I}\cos\varphi \end{pmatrix}^T \begin{pmatrix} \hat{T}_x & \hat{0} \\ \hat{0} & \hat{T}_y \end{pmatrix} \begin{pmatrix} \hat{I}\cos\varphi & -\hat{D}^{-1}\sin\varphi \\ \hat{D}\sin\varphi & \hat{I}\cos\varphi \end{pmatrix},
$$

which means that the coordinate change that block diagonalizes $\hat{T}$ in the linear form also block diagonalizes $\hat{T}$ in the bilinear form. The matrices $\hat{T}_x$ and $\hat{T}_y$ can be decomposed the same way as eq. (6.6) and we obtain

$$
\begin{pmatrix} \hat{A}_x^T & \hat{0} \\ \hat{0} & \hat{A}_y^T \end{pmatrix} \begin{pmatrix} \hat{I}\cos\varphi & \hat{D}^{-1}\sin\varphi \\ -\hat{D}\sin\varphi & \hat{I}\cos\varphi \end{pmatrix}^T \hat{T} \begin{pmatrix} \hat{I}\cos\varphi & \hat{D}^{-1}\sin\varphi \\ -\hat{D}\sin\varphi & \hat{I}\cos\varphi \end{pmatrix} \begin{pmatrix} \hat{A}_x & \hat{0} \\ \hat{0} & \hat{A}_y \end{pmatrix} = \begin{pmatrix} \hat{I} & \hat{0} \\ \hat{0} & \hat{I} \end{pmatrix},
$$

$$\tag{6.19}$$

where

$$
\hat{A}_{x,y} = \begin{pmatrix} \sqrt{\beta_{x,y}} & 0 \\ -\alpha_{x,y}/\sqrt{\beta_{x,y}} & 1/\sqrt{\beta_{x,y}} \end{pmatrix}.
$$

Similar to eq. (6.9), this equation shows that the matrix

$$
\hat{A}_4 = \begin{pmatrix} \hat{I}\cos\varphi & \hat{D}^{-1}\sin\varphi \\ -\hat{D}\sin\varphi & \hat{I}\cos\varphi \end{pmatrix} \begin{pmatrix} \hat{A}_x & \hat{0} \\ \hat{0} & \hat{A}_y \end{pmatrix}
$$

transforms the four-dimensional ellipse into two decoupled circles and the transfer matrix can be written as

$$
\hat{M}_4 = \begin{pmatrix} \hat{I}\cos\varphi_2 & \hat{D}_2^{-1}\sin\varphi_2 \\ -\hat{D}_2\sin\varphi_2 & \hat{I}\cos\varphi_2 \end{pmatrix} \begin{pmatrix} \hat{A}_{x2} & \hat{0} \\ \hat{0} & \hat{A}_{y2} \end{pmatrix} \begin{pmatrix} \hat{R}_x & \hat{0} \\ \hat{0} & \hat{R}_y \end{pmatrix}
$$
$$
\cdot \begin{pmatrix} \hat{A}_{x1}^{-1} & \hat{0} \\ \hat{0} & \hat{A}_{y1}^{-1} \end{pmatrix} \begin{pmatrix} \hat{I}\cos\varphi_1 & -\hat{D}_1^{-1}\sin\varphi_1 \\ \hat{D}_1\sin\varphi & \hat{I}\cos\varphi_1 \end{pmatrix},
$$

where

$$
\hat{R}_{x,y} = \begin{pmatrix} \cos\phi_{x,y} & \sin\phi_{x,y} \\ -\sin\phi_{x,y} & \cos\phi_{x,y} \end{pmatrix}.
$$

Similar to the two-dimensional case, we have

$$\begin{pmatrix} \hat{I}\cos\varphi_2 & \hat{D}_2^{-1}\sin\varphi_2 \\ -\hat{D}_2\sin\varphi_2 & \hat{I}\cos\varphi_2 \end{pmatrix}\begin{pmatrix} \hat{A}_{x2} & \hat{0} \\ \hat{0} & \hat{A}_{y2} \end{pmatrix}$$

$$= \hat{M}_4 \begin{pmatrix} \hat{I}\cos\varphi_1 & \hat{D}_1^{-1}\sin\varphi_1 \\ -\hat{D}_1\sin\varphi & \hat{I}\cos\varphi_1 \end{pmatrix}\begin{pmatrix} \hat{A}_{x1} & \hat{0} \\ \hat{0} & \hat{A}_{y1} \end{pmatrix}\begin{pmatrix} \hat{R}_x^{-1} & \hat{0} \\ \hat{0} & \hat{R}_y^{-1} \end{pmatrix}$$

$$= \begin{pmatrix} \hat{M} & \hat{n} \\ \hat{m} & \hat{N} \end{pmatrix}\begin{pmatrix} \hat{I}\cos\varphi_1 & \hat{D}_1^{-1}\sin\varphi_1 \\ -\hat{D}_1\sin\varphi & \hat{I}\cos\varphi_1 \end{pmatrix}\begin{pmatrix} \hat{A}_{x1} & \hat{0} \\ \hat{0} & \hat{A}_{y1} \end{pmatrix}\begin{pmatrix} \hat{R}_x^{-1} & \hat{0} \\ \hat{0} & \hat{R}_y^{-1} \end{pmatrix}$$

$$= \begin{pmatrix} \hat{M}\hat{A}_{x1}\cos\varphi_1 - \hat{n}\hat{D}_1\hat{A}_{x1}\sin\varphi_1 & \hat{M}\hat{D}_1^{-1}\hat{A}_{y1}\sin\varphi_1 + \hat{n}\hat{A}_{y1}\cos\varphi_1 \\ \hat{m}\hat{A}_{x1}\cos\varphi_1 - \hat{N}\hat{D}_1\hat{A}_{x1}\sin\varphi_1 & \hat{m}\hat{D}_1^{-1}\hat{A}_{y1}\sin\varphi_1 + \hat{N}\hat{A}_{y1}\cos\varphi_1 \end{pmatrix}\begin{pmatrix} \hat{R}_x^{-1} & \hat{0} \\ \hat{0} & \hat{R}_y^{-1} \end{pmatrix}.$$

Defining

$$\hat{M}\hat{A}_{x1}\cos\varphi_1 - \hat{n}\hat{D}_1\hat{A}_{x1}\sin\varphi_1 = \begin{pmatrix} \tilde{m}_{11} & \tilde{m}_{12} \\ \tilde{m}_{21} & \tilde{m}_{22} \end{pmatrix},$$

we obtain that

$$\begin{pmatrix} \sqrt{\beta_{x2}} & 0 \\ -\alpha_{x2}/\sqrt{\beta_{x2}} & 1/\sqrt{\beta_{x2}} \end{pmatrix}\cos\varphi_2 = \begin{pmatrix} \tilde{m}_{11} & \tilde{m}_{12} \\ \tilde{m}_{21} & \tilde{m}_{22} \end{pmatrix}\begin{pmatrix} \cos\phi_{x2} & -\sin\phi_{x2} \\ \sin\phi_{x2} & \cos\phi_{x2} \end{pmatrix}$$

$$= \begin{pmatrix} \tilde{m}_{11}\cos\phi_{x2} + \tilde{m}_{12}\sin\phi_{x2} & -\tilde{m}_{11}\sin\phi_{x2} + \tilde{m}_{12}\cos\phi_{x2} \\ \tilde{m}_{21}\cos\phi_{x2} + \tilde{m}_{22}\sin\phi_{x2} & -\tilde{m}_{21}\sin\phi_{x2} + \tilde{m}_{22}\cos\phi_{x2} \end{pmatrix}. \tag{6.20}$$

Similar to the two-dimensional case, we obtain the relations

$$\tan\phi_{x2} = \frac{\tilde{m}_{12}}{\tilde{m}_{11}}, \quad \cos\varphi_2 = \sqrt{\tilde{m}_{11}\tilde{m}_{22} - \tilde{m}_{12}\tilde{m}_{21}},$$

$$\beta_{x2} = \frac{\tilde{m}_{11}^2 + \tilde{m}_{12}^2}{\tilde{m}_{11}\tilde{m}_{22} - \tilde{m}_{12}\tilde{m}_{21}}, \quad \alpha_{x2} = -\frac{\tilde{m}_{21}\tilde{m}_{11} + \tilde{m}_{22}\tilde{m}_{12}}{\tilde{m}_{11}\tilde{m}_{22} - \tilde{m}_{12}\tilde{m}_{21}}.$$

The relations in the vertical plane are the same. Note that both planes give the same tilt angle $\varphi_2$ due to the fact that

$$\det\left(\hat{M}\hat{A}_{x1}\cos\varphi_1 - \hat{n}\hat{D}_1\hat{A}_{x1}\sin\varphi_1\right) = \det\left(\hat{m}\hat{D}_1^{-1}\hat{A}_{y1}\sin\varphi_1 + \hat{N}\hat{A}_{y1}\cos\varphi_1\right).$$

The four-dimensional case reduces to the two-dimensional case when $\varphi_2 = 0$. At this point, the advantage of this procedure based on coordinate transformation is rather clear since it avoids the coding of complicated formulas which are prone to errors. Instead, it divides the task into a few simple and standard steps which are finding the normal coordinates of the initial ellipsoid, tracking the transformation matrix to the point of interest and finding the Twiss parameters at the point of interest. With the help of the Differential Algebraic (DA) technique, it is straightforward to include parameter dependence of the Twiss parameters, which can introduce **beating** due to momentum deviation or quadrupole errors.

# Chapter 7

## Imaging Devices

In the following chapters, we will discuss what specifically has to be done to the map of a system to make the system useful for a specific task. In many cases, this requires that certain matrix elements vanish, or sometimes assume specific values. The most important device is probably the imaging device, in which **final positions are independent of initial angles**, as shown schematically in Fig. 7.1. Represented in the language of the transfer map, this entails that

$$(x|a) = 0, \quad (y|b) = 0.$$

It is apparent the final angles $a_f$ and $b_f$ are unimportant since it does not matter at what angle the rays strike at the image position; so all terms of the form $(a|\ldots)$ or $(b|\ldots)$ are insignificant. Additional requirements usually exist for the various subclasses of imaging systems.

There are many types of devices that form images of charged particles. The following are a few types that are widely used.

---

## 7.1 The Cathode Ray Tube (CRT)

The cathode ray tubes (CRT) are a class of imaging that have seen wide use in electronic displays, such as the **television tube** and the **oscilloscope**. As far as practical use, impact on society, and revenues are concerned, the TV

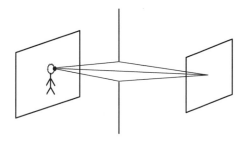

**FIGURE 7.1**:  Sketch of an imaging system.

tube was until recently the most important application of particle optics. In this case, for each color an electron beam is deflected vertically and horizontally by two simple magnetic deflectors in order to sweep over the screen area, and the intensity of each beam is adjusted according to the color saturation at the respective point. The cathode ray tubes used in the oscilloscopes use electrostatic deflection plates to achieve high frequency. Yet the limit of a single pair of plates is around 150 MHz, above which the single pair is replaced with segmented pairs of plates. This type of deflector can reach a frequency of 300 MHz. To reach even higher frequency, a double helix line is used to deflect the beam, which has reached 10 to 20 GHz. Nowadays, 33 GHz oscilloscopes are commercially available.

At any given point on the screen, the resulting spot should not be wider than the distance between two pixels, so whatever size the beam had initially should not be amplified very much; so

$$(x|x) \quad \text{and} \quad (y|y)$$

**should not be large**.

The requirements for aberrations are usually benign as the phase space volume of the beam is small. Yet those aberrations do limit performance and over the decades various ways have been developed to minimize them. The advent of plasma and liquid crystal displays (LCD) for TV on one hand and digital oscilloscopes on the other has caused a precipitous decline of the use of cathode ray tubes, which are preferred now only in specialized markets.

---

## 7.2 The Camera and the Microscope

The purpose of a **camera** and an **electron microscope** is to create an image of an object formed by light or particle rays. The quantities

$$(x|x) \quad \text{and} \quad (y|y)$$

are **magnifications**, and in most cases it is desirable to have them equal. The electron microscope is just a special case in which both of them are made to be very large to increase the resolution.

If a true image is desired, it is important that the relationship between final and initial coordinates is really linear, which requires that all higher order position dependent matrix elements vanish, and so

$$(x|xx) = 0, \quad (y|yy) = 0, \quad (x|xxx) = 0, \ldots.$$

In reality, of course, it is difficult to achieve this to higher orders, hence some distortion remains. In the case of an electron microscope, this is often

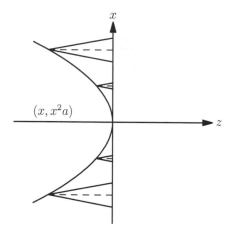

$(x, x^2 a)$

**FIGURE 7.2**:  Curvature of the image.

not detrimental as one can retroactively correct the effects by calculation and the resolution is not affected. The effects that appear usually have the consequence that rectangles are distorted into either the shape of a **pincushion** or into the shape of a **barrel**; these effects are due to

$$(x|xyy) \quad \text{and} \quad (y|yxx),$$

which entail that rays that simultaneously have $x$ and $y$ coordinates are either pushed out from the center (pincushion) or pulled in (barrel). Higher order terms in $x$ and $y$ produce similar effects.

There should be no effect of energy on position, so

$$(x|\delta) = 0 \quad \text{and} \quad (y|\delta) = 0$$

should be maintained. Similarly, all higher order aberrations involving $\delta$ should vanish; if this is not the case, some color dependent blurring called **chromatic aberration** may occur, in particular for larger values of $x$ and $y$, an effect that can be easily observed in the case of less expensive **binoculars**.

There should also be no effects of position on initial angles to higher order; so it is necessary that

$$(x|a^{i_a} b^{i_b}) = (y|a^{i_a} b^{i_b}) = 0,$$

and since the range of accepted angles corresponding to $a$ and $b$ is often rather large, to correct these terms is often very important. If any of them prevail, they will entail a color independent fuzziness; in case the order of the coordinates $a$ and $b$ is even, the fuzz will be oriented toward one side like the **coma** of a comet; if the powers are odd, it will lead to a uniformly distributed fuzziness.

Similarly, all aberrations involving positions and angles simultaneously have to vanish, and hence it is necessary to have

$$(x|x^{i_x}y^{i_y}a^{i_a}b^{i_b}) = (y|x^{i_x}y^{i_y}a^{i_a}b^{i_b}) = 0;$$

if any of them prevail, they will entail a position dependent fuzziness that becomes stronger with an increase of the positions $x$ and $y$.

Interestingly enough, all higher order aberrations depending on $a$ and $b$ only linearly can be corrected by a reshaping of the focal plane; in fact, $(x|xa)$, etc., produce a tilt of the image, and $(x|xxa)$, etc., produce a curvature of the image. Fig. 7.2 shows how the matrix element $(x|xxa)$ can be corrected by shaping the image position parabolically.

At any given position, due to the matrix element $(x|xxa)$, any ray with a given $a$ is moved up or down in proportion to $a$, where the amount of deflection depends quadratically on $x$; so the rays arrive at the $x$ plane as shown. However, tracing the rays backwards shows that they in fact all intersect before the plane, and the point where this happens depends quadratically on $x$. In similar ways, $(x|x^4a)$, etc., can be corrected.

---

## 7.3 Spectrometers and Spectrographs

Spectrometers and spectrographs are devices for the purpose of **measuring momentum, energy**, or **mass** of charged particles. **Momentum spectrometers** are mainly used in nuclear physics for the determination of the momentum distribution of nuclear reaction products. Most of the momentum spectrometers known are magnetic because of the fact that the energies that need to be analyzed are too high to allow sufficient deflection by electric fields. In addition, magnets have two more advantages. They automatically preserve the momentum, and can be built big enough to achieve large acceptance.

The **mass spectrometer** is mainly used for the analysis of masses of molecules, and they can be operated at much lower energies. They have a long history, and their applications pervade many disciplines from physics and chemistry to biology, environmental sciences, etc. Mass spectrometers are also more diverse; among the major types, there are sector field, quadrupole, accelerator, energy loss, time-of-flight, Fourier transform ion cyclotron resonance and ion trap mass spectrometers.

For all the different types of spectrometers, the goal is to achieve **high resolution**, and in many cases **large acceptance** at the same time. As resolution improves, the need for better understanding and correction of high order aberrations increases. In the following, the linear theory of various types of spectrometers will be discussed, followed by the studies of aberrations and their correction.

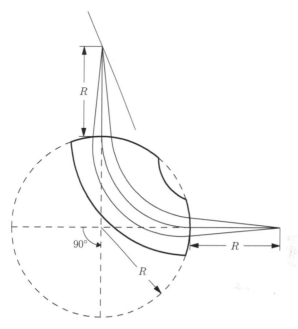

**FIGURE 7.3:** The Browne-Buechner spectrograph.

As alluded to before, in spectrometers the **final position** is used as a measure of momentum, energy, or mass of the particle. This requires that the final position be independent of other quantities, which in particular requires that the device be focusing such that

$$(x|a) = 0.$$

Due to Liouville's theorem, it is impossible to obtain focusing and zero image size simultaneously, and hence the **initial spot width** has to be minimized to ensure that the final image is narrow enough. Furthermore, the dependence on the spectroscopic quantity of interest $\delta$, the so-called **dispersion**

$$(x|\delta),$$

**should be large**. Finally, in a mass spectrometer where particles of different energies are present, the dependence on energy $(x|\delta)$ should vanish. In the map picture, the linear behavior is thus given by the transfer matrix of the horizontal motion

$$\hat{M} = \begin{pmatrix} (x|x) & 0 & (x|\delta) \\ (a|x) & (a|a) & (a|\delta) \\ 0 & 0 & 1 \end{pmatrix}. \tag{7.1}$$

Let $2D_i$ be the width of the source. From eq. (7.1), it is clear that the particles to be detected focus around a spot at $(x|\delta)\delta$ with a width of $|2(x|x)D_i|$.

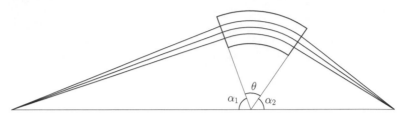

**FIGURE 7.4**:   Illustration of the imaging condition arising from Barber's rule.

Hence, the distance between the centers of particles of different energies must be larger than the width, i.e.,

$$|(x|\delta)\delta| > |2(x|x)D_i|.$$

This sets an upper bound for $1/\delta$, which we call the linear resolving power (or linear resolution),

$$R_l = \left(\frac{1}{\delta}\right)_{max} = \left|\frac{(x|\delta)}{2(x|x)D_i}\right|.$$

Hence in order to increase the resolution, it is necessary to increase $|(x|\delta)|$ and/or decrease $|D_i|$.

As an example, let us study the first broad range momentum spectrometer, the **Browne-Buechner** spectrometer. It contains only a homogeneous dipole with 90° bending and circular pole boundaries. The layout is depicted in Fig. 7.3, and it is applicable for particles with energies up to 25 MeV/u. As it turns out, there is a simple condition known as **Barber's rule** that assures that the system is $x$-focusing. This is in fact the case whenever the source location, the center of deflection of the magnet, and the image location lie on a straight line, as shown in Fig. 7.4. To prove Barber's rule, we first write down the transfer matrix of the horizontal plane, which is

$$
\hat{M}_x = \begin{pmatrix} 1 & l_2 \\ 0 & 1 \end{pmatrix} \begin{pmatrix} \cos\theta & R\sin\theta \\ -(1/R)\sin\theta & \cos\theta \end{pmatrix} \begin{pmatrix} 1 & l_1 \\ 0 & 1 \end{pmatrix}
$$

$$
= \begin{pmatrix} \cos\theta - (l_2/R)\sin\theta & R\sin\theta + l_2\cos\theta \\ -(1/R)\sin\theta & \cos\theta \end{pmatrix} \begin{pmatrix} 1 & l_1 \\ 0 & 1 \end{pmatrix}
$$

$$
= \begin{pmatrix} \cos\theta - (l_2/R)\sin\theta & (l_1 + l_2)\cos\theta + (R - l_1 l_2/R)\sin\theta \\ -(1/R)\sin\theta & \cos\theta - (l_1/R)\sin\theta \end{pmatrix},
$$

where the angles $\theta$, $\alpha_1$ and $\alpha_2$ are shown in Fig. 7.4 and the quantities $R$, $l_1$ and $l_2$ are the bending radius and the drifts before and after the dipole magnet, respectively. Using the relations $l_1 = R\tan\alpha_1$, $l_2 = R\tan\alpha_2$ and $\alpha_1 + \alpha_2 + \theta = \pi$, as well as $\tan A + \tan B = (1 - \tan A \tan B) \cdot \tan(A + B)$,

**TABLE 7.1:** The first order map of the Browne-Buechner spectrograph (Exponents in the initial variables $x, a, y, b, l, \delta$)

| $x_f$ | $a_f$ | $y_f$ | $b_f$ | exponents |
|---|---|---|---|---|
| -1.000000 | -1.950458 | 0 | 0 | 100000 |
| 0 | -1.000000 | 0 | 0 | 010000 |
| 0 | 0 | 1.000000 | 0 | 001000 |
| 0 | 0 | 1.830747 | 1.000000 | 000100 |
| 0 | 0 | 0 | 0 | 000010 |
| 0.519441 | 0.506574 | 0 | 0 | 000001 |

we have

$$m_{12} = (l_1 + l_2)\cos\theta + \left(R - \frac{l_1 l_2}{R}\right)\sin\theta$$
$$= R\left[(\tan\alpha_1 + \tan\alpha_2)\cos\theta + (1 - \tan\alpha_1\tan\alpha_2)\sin\theta\right]$$
$$= R(1 - \tan\alpha_1\tan\alpha_2)\cos\theta \cdot \left[\tan(\alpha_1 + \alpha_2) + \tan\theta\right]$$
$$= R(1 - \tan\alpha_1\tan\alpha_2)\cos\theta \cdot \left[\tan(\pi - \theta) + \tan\theta\right].$$

Because of $\tan(\pi - \theta) = -\tan\theta$, we obtain the desired result

$$m_{12} = 0.$$

In general, it is not hard to obtain the first order map by hand, and it is very easy to obtain it using a computer code; the result is shown in Table 7.1. With the typical assumption that the half width $D_i$ is 0.25 mm, the resulting linear energy resolution is

$$R_l = \frac{1}{\delta_{min}} = \left|\frac{(x|\delta)}{2(x|x)D_i}\right| \approx 1000.$$

Since all electric and magnetic devices produce nonlinear terms in the map called **aberrations**, their impact on the resolution has to be studied whenever necessary. The nonlinear effects are very important in the case of the momentum spectrometers due to their large angular and momentum acceptances. Considering the aberrations, the final width will be a new value $\Delta x_{ab}$ instead of $|(x|x)D_i|$, which has as an upper bound

$$\Delta x_{ab} = (2|(x|x)D_i| + |(x|x^2)|D_i^2 + |(x|xa)D_i A_i| + \cdots),$$

where $A_i$ is the half width of the spread in the quantity $a$. So, the actual **resolution** $R_{ab}$ is

$$R_{ab} = \frac{|(x|\delta)|}{\Delta x_{ab}}.$$

A parameter often used as a comprehensive quality indicator for a spectrometer is the so-called **Q value**

$$Q = \frac{\Omega \ln(p_{max}/p_{min})}{\ln 2},$$

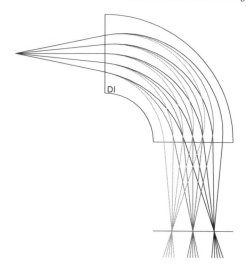

**FIGURE 7.5**: Sketch of a generic spectrograph consisting of a single dipole, including rays that show the imaging condition and the dispersion of the device.

where $\Omega$ is the nominal solid angle from which a reasonable resolution is expected. The $Q$-value shows both the **geometric** and **momentum acceptance** of a spectrometer. For example, the $\Omega$ and $p_{max}/p_{min}$ for the Browne-Buechner spectrometer are 0.4 msr and 1.5, respectively. Large $\Omega$ translates into high intensity, which is important for nuclear studies and other situations where the number of available particles are small. Large momentum acceptance can reduce the number of exposures to cover a certain momentum range.

As discussed before, the purpose of the spectrograph is to translate energy information into position information, and in order to have high resolution, the position should not depend on anything else if possible. Rays originate from a source, travel through the spectrograph, and finally reach the screen, as shown in Fig. 7.5.

It is possible to measure energies in terms of final positions by making the **dispersion**

$$(x|\delta)$$

**large**. In practice this requires the use of at least one bending element, because all other elements have vanishing $(x|\delta)$. The final position should not depend on anything else besides $\delta$, and since it is important to be able to accept rays covering a wide range of angles, it is necessary to have

$$(x|a) = (y|b) = 0.$$

So the spot size is limited by $(x|x) = 1/(a|a)$, which is usually kept small,

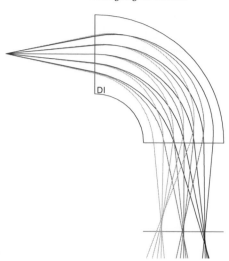

**FIGURE 7.6**: Sketch of a generic spectrograph as in the previous picture, but now subject to aberrations up to order seven.

and the size of the object, the $x$ size of which is usually kept in the range of fractions of mm.

Any contribution to the final position should be due to energy, and so aberrations depending on initial angle should be avoided. These aberrations are usually called **spherical**, since they historically first manifested themselves from the grinding of lens surfaces as spheres, which is much easier to achieve than other shapes. So if possible we want

$$(x|aa) = (x|aaa) = 0, \quad (x|bb) = (x|abb) = 0.$$

These conditions are not satisfied for the simple spectrograph shown in Fig. 7.5. Rather, when they are considered, the trajectories of the rays look like in Fig. 7.6, showing very noticeable broadening of the image due to aberrations.

The aberrations involving also $x$-positions are less significant as positions are kept small. The ones involving also $y$ positions are more important as $y$ is not necessarily kept small; but if $(y|y)$ is kept large enough, particles with significant initial $y$ reach the focal plane with significant final $y$; the interplay of $(y|y)$ and $(x|yy)$ then leads to a parabolic shape of the resulting image, but the sharpness of the parabola, which determines the resolution, is unaffected by $(x|yy)$.

It is also important to consider aberrations involving energy. Of these, the terms depending only on energy of the form

$$(x|\delta^{i_\delta})$$

do not necessarily have to be corrected as long as they are known, since they just turn the relationship of final $x$ and initial $\delta$ into a nonlinear one, which

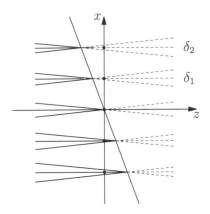

**FIGURE 7.7**: The effect of the aberration $(x|a\delta)$.

still allows an accurate measurement of $\delta$. The most important aberrations are usually those that involve initial angles and energies simultaneously, as both of these can be large. Of these, the lowest order aberration $(x|a\delta)$ can be corrected by a simple **tilt of the focal plane**: the final $x$ of a particle, which depends mostly on $\delta$, is moved up or down linearly depending on the value of $a$. As shown in Fig. 7.7, similar to before, all these rays with different values of $a$ go through a common point at a distance before or after the $x$ plane, where the effect of $(x|a\delta)$ does not manifest itself. The tilt of the focal plane is also very clearly visible in the actual example of Fig. 7.6.

In a similar way, spectrographs can also be used to measure **masses** of particles, and all previous arguments remain valid if the energy deviation $\delta$ is replaced by the mass deviation $\delta_m$. If mass resolution is to be achieved to very high precision and the initial energy is not uniform, then in addition to the above requirements, it is also important that the final position does not depend on $\delta$; this requires that

$$(x|\delta) = 0,$$

while of course at the same time trying to have

$$(x|\delta_m)$$

**large**. The simultaneous satisfaction of these conditions is not possible using only magnetic devices; for low energies, it is usually achieved by combining magnetic and electric deflectors.

### 7.3.1 Aberrations and Correction

For all the spectrographs mentioned above, nonlinear effects have always been a concern for the designers. Looking back to the Browne-Buechner spectrograph, the linear energy resolution obtained was $\sim 1000$. When aberrations

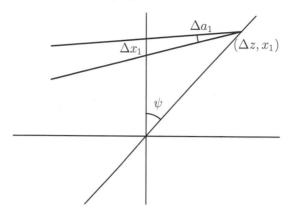

**FIGURE 7.8:** A magnified drawing of the effect of the aberration $(x|a\delta)$. $(\Delta z, x_1)$ is the image on the tilted focal plane caused by the term $(x|a\delta)$.

are considered, the resolution of the eighth order drops sharply to around 60, which is far below the actually achieved resolution. This shows the importance of the aberrations. They have to be studied carefully, and the prominent ones have to be corrected.

Since the entrance slit $D_i$ is usually small and the solid angle large, only the angle and dispersion aberrations are important. Both map calculations and geometric considerations show that all the terms $(x|x^m a^n)$ $(m + n$ even) vanish. Since $(x|b^2)$ is small (see Table 7.2), the only second order term that has a strong impact is $(x|a\delta)$ which can be as large as 8 mm when $a = 40$ mrad and $\delta = 20\%$. In fact, this is the most important factor that causes the decrease of the resolution. This becomes apparent when the resolution considering $(x|a\delta)$ is calculated:

$$R_{ab} = \frac{(x|\delta)}{2(x|a\delta)a_i\delta} = 63.$$

Fortunately $(x|a\delta)$ is easy to correct, because it only causes the tilt of the focal plane, which is illustrated in Fig. 7.8.

To prove the last statement, suppose a particle of energy $K_0(1 + \delta)$ starts from the origin with slope $a_0$ and goes through an angle focusing system. The final position and angle to first order are

$$x_1 = (x|\delta)\delta, \quad a_1 = (a|a)a_0 + (a|\delta)\delta,$$

respectively. Taking into account $(x|a\delta)$, the result becomes

$$\tilde{x}_1 = (x|\delta)\delta + (x|a\delta)a_0\delta, \quad \tilde{a}_1 = (a|a)a_0 + (a|\delta)\delta.$$

Consequently, the system is not focusing anymore. Now consider a second particle of the same energy starting from the same point but with a different

**TABLE 7.2:** Aberrations of the Browne-Buechner spectrograph at the straight focal plane that are 10 $\mu$m or larger (Parameters: $x_{max} = 0.23$ mm, $a_{max} = 40$ mrad, $y_{max} = 1$ mm, $b_{max} = 10$ mrad, $\delta_{max} = 20\%$. Exponents: in the initial variables $x, a, y, b, l, \delta$)

| # | Coefficient | Order | Exponents | | | | | |
|---|---|---|---|---|---|---|---|---|
| 1 | -0.2300000000000000e-3 | 1 | 1 0 | 0 0 | 0 0 |
| 2 | 0.1038881145421565 | 1 | 0 0 | 0 0 | 0 1 |
| 3 | 0.4660477149345817e-4 | 2 | 1 0 | 0 0 | 0 1 |
| 4 | 0.8311049163372523e-2 | 2 | 0 1 | 0 0 | 0 1 |
| 5 | -0.5127000000000000e-4 | 2 | 0 0 | 0 2 | 0 0 |
| 6 | -0.1025577239401090e-1 | 2 | 0 0 | 0 0 | 0 2 |
| 7 | -0.6562560000000000e-4 | 3 | 0 3 | 0 0 | 0 0 |
| 8 | 0.3324419665349009e-3 | 3 | 0 2 | 0 0 | 0 1 |
| 9 | -0.1873001321765092e-2 | 3 | 0 1 | 0 0 | 0 2 |
| 10 | 0.1038881145421565e-4 | 3 | 0 0 | 0 2 | 0 1 |
| 11 | 0.1544932687715805e-2 | 3 | 0 0 | 0 0 | 0 3 |
| 12 | 0.4654187531488613e-4 | 4 | 0 3 | 0 0 | 0 1 |
| 13 | -0.1338622665642800e-3 | 4 | 0 2 | 0 0 | 0 2 |
| 14 | 0.4380445064894873e-3 | 4 | 0 1 | 0 0 | 0 3 |
| 15 | -0.2577160791614789e-3 | 4 | 0 0 | 0 0 | 0 4 |
| 16 | 0.4622130002260186e-4 | 5 | 0 2 | 0 0 | 0 3 |
| 17 | -0.9970354551245065e-4 | 5 | 0 1 | 0 0 | 0 4 |
| 18 | 0.4511716453676560e-4 | 5 | 0 0 | 0 0 | 0 5 |
| 19 | 0.2219821460952441e-4 | 6 | 0 1 | 0 0 | 0 5 |

angle $a_0 + \Delta a_0$. The differences in final position and angle between the two particles are

$$\Delta x_1 = (x|a\delta)\Delta a_0 \delta, \quad \Delta a_1 = (a|a)\Delta a_0.$$

The fact that $\Delta x_1/\Delta a_1$ is independent of $\Delta a_0$ indicates that particles of energy $K_0(1 + \delta)$ are focusing at

$$\Delta z = -\frac{\Delta x_1}{\Delta a_1} = -\frac{(x|a\delta)\delta}{(a|a)},$$

which is proportional to $\delta$. So the tilting angle is

$$\tan \psi = \frac{\Delta z}{x_1} = -\frac{(x|a\delta)}{(a|a)(x|\delta)},$$

where $\psi$ is the angle between the normal to the focal plane and the $z$-axis.

Furthermore, the correction of $(x|a\delta)$ even increases the resolution under certain circumstances. When $\Delta x_{ab}$ is smaller than the detector resolution $\Delta x_d$, $\Delta x_d$ becomes the limitation of the momentum resolution and is independent of $\psi$. Since the distance between two peaks increases by a factor of

**TABLE 7.3:** Aberrations of the Browne-Buechner spectrograph at the tilted focal plane that are 10 $\mu$m or larger (Parameters: $x_{max} = 0.23$ mm, $a_{max} = 40$ mrad, $y_{max} = 1$ mm, $b_{max} = 10$ mrad, $\delta_{max} = 20\%$. Exponents: in the initial variables $x, a, y, b, l, \delta$)

| # | Coefficient | Order | Exponents | | | | | |
|---|---|---|---|---|---|---|---|---|
| 1 | -0.2300000000000000e-3 | 1 | 1 | 0 | 0 | 0 | 0 | 0 |
| 2 | 0.1038881145421565 | 1 | 0 | 0 | 0 | 0 | 0 | 1 |
| 3 | -0.9320954298691634e-4 | 2 | 1 | 0 | 0 | 0 | 0 | 1 |
| 4 | -0.5127000000000000e-4 | 2 | 0 | 0 | 0 | 2 | 0 | 0 |
| 5 | 0.1079501821087351e-1 | 2 | 0 | 0 | 0 | 0 | 0 | 2 |
| 6 | -0.6562560000000000e-4 | 3 | 0 | 3 | 0 | 0 | 0 | 0 |
| 7 | 0.3324419665349009e-3 | 3 | 0 | 2 | 0 | 0 | 0 | 1 |
| 8 | -0.2105079060488440e-3 | 3 | 0 | 1 | 0 | 0 | 0 | 2 |
| 9 | -0.1038881145421566e-4 | 3 | 0 | 0 | 0 | 2 | 0 | 1 |
| 10 | 0.1654199766297046e-2 | 3 | 0 | 0 | 0 | 0 | 0 | 3 |
| 11 | -0.1329767866139604e-4 | 4 | 0 | 3 | 0 | 0 | 0 | 1 |
| 12 | 0.5138469075870278e-4 | 4 | 0 | 2 | 0 | 0 | 0 | 2 |
| 13 | -0.2242022047923136e-4 | 4 | 0 | 1 | 0 | 0 | 0 | 3 |
| 14 | 0.1745846601755523e-3 | 4 | 0 | 0 | 0 | 0 | 0 | 4 |
| 15 | 0.1305877080464639e-4 | 5 | 0 | 2 | 0 | 0 | 0 | 3 |
| 16 | 0.2771149616039620e-4 | 5 | 0 | 0 | 0 | 0 | 0 | 5 |

$1/\cos\psi$ while $\Delta x_d$ remains unchanged, the resolution is

$$R_{ab} = \frac{(x|\delta)}{\Delta x_d \cos\psi},$$

which is greater than the linear resolution. Rigorous computation of the actual resolution requires that the aberrations on the tilted focal plane be calculated.

For the Browne-Buechner spectrograph, eighth order maps of both straight and tilted focal planes are computed. Table 7.2 shows the aberrations on the straight focal plane, where $(x|a\delta)$ is clearly the major contributor. Table 7.3 contains the aberrations on the tilted focal plane, where $(x|a\delta)$ vanishes and others are either reduced or unchanged. Both tables are taken from [5]. The resolution after the cancellation of $(x|a\delta)$ bounces back to 780 (or 1560 in momentum), which is quite close to the linear resolution. This entails that the remaining aberrations are weak enough to be ignored at the time when it was first designed.

When higher resolution is required, more aberrations have to be corrected. Usually $(x|a^2)$ and $(x|b^2)$ are corrected first. Then $(x|a^3)$, $(x|a^2b)$ and $(x|ab^2)$ are tackled. If necessary, fourth order terms like $(x|a^4)$ also have to be minimized. At least in one instance, eighth order terms, such as $(x|a^5b^3)$, are corrected, too. This is done in Quad-Dipole-Quad (QDQ) spectrographs. The pole faces of the last quadrupole are shaped to produce quadrupole, sextupole, octupole and decapole field components. Generally, corrections are done by placing magnetic multipoles of the same order into the system. They are either separate adjustable multipoles or combined fixed elements to dipoles or quadrupoles.

## 7.3.2 Energy Loss On–Line Isotope Separators

As part of the growing developments in the study of radioactive beams, **on–line isotope separation** is more and more widely performed. As in other mass spectrometers described above, different isotopes have to be laterally separated. Yet they can no longer be bent by electrostatic sectors anymore due to their high energy. Among the different methods, the energy loss method is a very interesting one. First, particles of the right rigidity are selected by a slit at the dispersive focal point. Second, the selected particles are sent through an energy degrader which creates new momentum spread according to the mass of the particles. And finally, a second slit picks up the desired nuclei. The best spatial separation can be achieved when the whole beamline is achromatic and the degrader preserves the achromaticity. This is due to the fact that nuclei of the same mass but different momentum are focused at the same spot. Hence it is important to study the transfer map of an energy degrader and the achromatic conditions.

In most of the cases, the degrader is thin enough for us to neglect the straggling effect from multiple scattering. So a degrader is the combination of a drift and an energy loss, which has the first order matrix

$$\hat{M}_d = \begin{pmatrix} 1 & d & 0 \\ 0 & 1 & 0 \\ (\delta|x)_d & (\delta|a)_d & (\delta|\delta)_d \end{pmatrix}.$$

Here $d$ is the thickness of the degrader. It is easy to show that the spatial part of $\hat{M}_d$ can be reduced to a unity matrix. After applying a negative drift behind the degrader, $\hat{M}_d$ becomes

$$\hat{M}_d = \begin{pmatrix} 1 & -d & 0 \\ 0 & 1 & 0 \\ 0 & 0 & 1 \end{pmatrix} \begin{pmatrix} 1 & d & 0 \\ 0 & 1 & 0 \\ (\delta|x)_d & (\delta|a)_d & (\delta|\delta)_d \end{pmatrix} = \begin{pmatrix} 1 & 0 & 0 \\ 0 & 1 & 0 \\ (\delta|x)_d & (\delta|a)_d & (\delta|\delta)_d \end{pmatrix}.$$

For heavy ions at the intermediate energy region of $\geq 10$ MeV/u, the energy at the exit can be described with the formula

$$K_d = K \left(1 - \frac{d}{R}\right)^{1/\gamma},$$

where $K$ is the energy at the entrance of the degrader and $R$ is the range. Furthermore,

$$R = kA^{1-\gamma}K^\gamma/Z^2,$$

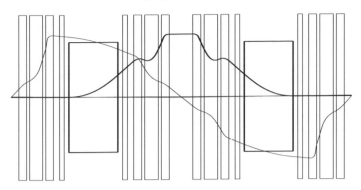

**FIGURE 7.9**: Layout of an example of a fragment separator. There is an intermediate image at the mirror symmetry plane in the middle, and the system is achromatic.

where $A$ is the atomic number, $Z$ is the charge, and $k$ and $\gamma$ are constants. Within this model, the matrix elements can be obtained as

$$(\delta|x)_d = \frac{-1}{\gamma R_0(1 - d_0/R_0)} \left(\frac{\partial d}{\partial x}\right)_{x=0},$$

$$(\delta|a)_d = 0,$$

$$(\delta|\delta)_d = \frac{1}{1 - d_0/R_0},$$

where $d_0$ and $R_0$ are values for the reference particle.

To achieve achromaticity, the system has to satisfy

$$(x|a) = 0 \quad \text{and} \quad (x|\delta) = 0.$$

By denoting the parts before and after the degrader with subscript 1 and 2, respectively, the achromatic conditions are obtained and they are

$$(x|x)_2(x|a)_1 + (x|a)_2(a|a)_1 + (x|\delta)_2\{(\delta|x)_d(x|a)_1 + (\delta|a)_d(a|a)_1\} = 0,$$
$$(x|x)_2(x|\delta)_1 + (x|a)_2(a|\delta)_1 + (x|\delta)_2\{(\delta|x)_d(x|\delta)_1 + (\delta|a)_d(a|\delta)_1 + (\delta|\delta)_d\} = 0.$$

When the previous model applies, $(\delta|a)$ vanishes. Together with the requirement that both parts are focusing, that is $(x|a)_1 = (x|a)_2 = 0$, the conditions can be reduced to

$$(x|a)_1 = (x|a)_2 = 0,$$
$$D_1 M_2 + D_2\{D_1(\delta|x)_d + (\delta|\delta)_d\} = 0, \tag{7.2}$$

where $D = (x|\delta)$ and $M = (x|x)$.

An example which uses an achromatic degrader for isotope separation is shown in Fig. 7.9. When operated on the achromatic mode, the system is mirror symmetric about the degrader and forms a dispersive image at that point. The matrices of the two parts are

$$
\hat{M}_1 = \begin{pmatrix} M_1 & 0 & D_1 \\ (a|x)_1 & 1/M_1 & 0 \\ 0 & 0 & 1 \end{pmatrix}, \quad \hat{M}_2 = \begin{pmatrix} 1/M_1 & 0 & -D_1/M_1 \\ (a|x)_1 & M_1 & -(a|x)_1 D_1 \\ 0 & 0 & 1 \end{pmatrix}.
$$

From eq. (7.2), the achromatic condition for the fragment separator is

$$
D_1(\delta|x)_d + (\delta|\delta)_d = 1.
$$

Therefore the shape of the degrader can be decided, which is a wedge with the slope

$$
\left( \frac{\partial d}{\partial x} \right) = \frac{\gamma d_0}{D_1}.
$$

It is straightforward to verify that the system with the degrader is indeed achromatic

$$
\hat{M} = \hat{M}_2 \cdot \hat{M}_d \cdot \hat{M}_1 = \begin{pmatrix} (\delta|\delta)_d & 0 & 0 \\ M_1(a|x)_1(2 - (\delta|x)_d D_1) & 1 & 0 \\ M_1(\delta|x)_d & 0 & 1 \end{pmatrix}.
$$

---

## 7.4 *Electron Microscopes and Their Correction

The field of electron optics is one of the oldest branches of beam physics, which is a direct descendant of light optics. Recently, it is also one of the most active branches due to the advancement of aberration correction in electron microscopes. In the past decade, the **TEAM** project (Transmission Electron Aberration-corrected Microscope) has been developing the next generation of electron microscopes. Initial experiments using the latest aberration-corrected scanning transmission electron microscope **(STEM)** demonstrated the scientific potential of aberration-corrected electron microscopes.

Since their invention in the early 1930s, electron microscopes have been used in various areas ranging from scientific research to industrial production, and various different types of microscopes were developed for the specific needs of those applications. The main variants are the transmission electron microscope **(TEM)**, the scanning transmission electron microscope **(STEM)**, the photoemission electron microscope **(PEEM)**, the low energy electron microscope **(LEEM)** and the scanning electron microscope **(SEM)**.

Among them, TEM and STEM are used mainly to study the bulk properties of materials with electron energy ranges from 100 keV to 1 MeV. PEEM, LEEM and SEM are used to study surface properties of materials with electron energy below 30 keV. In a PEEM, secondary electrons generated by

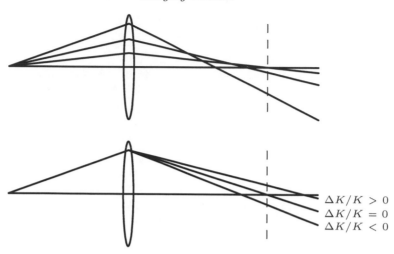

$\Delta K/K > 0$
$\Delta K/K = 0$
$\Delta K/K < 0$

**FIGURE 7.10**: Spherical (top) and chromatic (bottom) aberrations. A Gaussian image exists at the dashed line. Higher and lower energy electrons are represented by $\Delta K/K > 0$ and $\Delta K/K < 0$.

photons are imaged. In a LEEM, electrons reflected from the sample surface are imaged. In a SEM, an electron probe the size of a few angstroms is formed on the sample and secondary electrons are collected.

Except for LEEM, which needs a magnetic separator to separate the incoming and reflected electron beams, most microscopes without aberration correction consist of so-called **round lenses** only. There are two types of round lenses used in electron microscopes: the electrostatic and the magnetic lenses. **Electrostatic lenses** are used in PEEMs, LEEMs and some SEMs, whereas **magnetic lenses** are used in TEM and STEM where the higher energies of the electrons make the use of electrostatic lenses impractical.

The rotational symmetry of these lenses ensures that only a small number of aberrations remain to degrade the linear or, so-called, **Gaussian** image properties. The first kind are the **spherical aberrations**, which lead to a blurring of the image due to the opening angle of the electron beam at the object. In the electrostatic case, the first relevant terms are $(x, aaa)$ and $(y, bbb)$, which are equal due to the rotational symmetry, and usually denoted by $C_S$. There are also terms of the form $(x, a^5)$ and $(y, b^5)$ which are denoted by $C_5$, the significance of which is discussed below. Second and fourth order terms and cross terms like $(x, aab)$, etc., vanish because of the mirror symmetry.

In the magnetic case the situation is a bit more complicated since magnetic round lenses can rotate the image in the $x - y$ plane. However, considering the motion in the rotated coordinate system, it can be seen that the matter reduces to quite the same situation as in the electrostatic case. The top picture of Fig. 7.10 illustrates the effect of the spherical aberration.

The second kind of aberrations are the **chromatic aberrations**, which arise as a combination of the opening angle and the energy spread of the beam. The lowest order chromatic aberration for a round lens is 2, and in the electrostatic case has the form $(x, a\delta)$, which because of symmetry also equals $(y, b\delta)$. These chromatic aberrations are usually denoted by $C_C$. In the magnetic case after transformation into the appropriate rotated coordinate system, the situation is again the same. The effect of the chromatic aberration is illustrated in the bottom picture of Fig. 7.10.

Even in the early days of electron microscopes, the possibility of correcting the remaining aberrations had been contemplated. Yet the initial result of theoretical investigation was not very encouraging. Scherzer [62] showed that, for a round lens without reflection, the spherical and the chromatic aberrations do not change sign, the same as the focusing force of such a lens (see Section 4.4.1).

Specifically, electrons with a larger angle are focused stronger and electrons with higher energy are focused weaker. As a result, aberration correction requires violation of the above assumptions, through using either multipole elements, electron mirrors or time varying fields. Early attempts on aberration correction, between the late 1940s and the early 1990s, failed mainly due to insurmountable technical difficulties. Hence the development of electron microscopes up to the early 1990s follows mainly the line of aberration reduction through optimization of the lens design and improvement of stability. The initial success of aberration correction came when the technology was ready in the mid-1990s [59, 39].

### 7.4.1   Aberration Correction in SEM, STEM and TEM

The first successful aberration correction was reported in 1995, where the spherical aberration $C_S$ and chromatic aberration $C_C$ were corrected in a low voltage **SEM** (scanning electron microscope). The corrector consists of four multipole elements (see Fig. 7.11), which was originally proposed in the early 1960s. The two outer elements are electrostatic multipoles and the two inner ones are superimposed electrostatic and magnetic multipoles. The corrector consists of two identical **quadrupole doublets**, where the two quadrupoles are physically the same, excited at the same current but with opposite polarity.

Furthermore, it is arranged such that the so-called cosine-like ray of the horizontal plane, which in conventional transfer map terminology corresponds to the $(x, x)$ matrix element, goes through the center of the left inner element, while that of the vertical plane goes through the center of the right inner element. This entails that $(x|a\delta)$ and $(y|b\delta)$ can be corrected independently from each other.

More importantly, rays in the vertical plane coincide with those in the horizontal plane going backwards. This layout minimizes the breaking of rotational symmetry due to the introduction of multipoles. The most noticeable

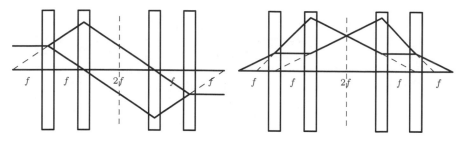

**FIGURE 7.11:** Cosine-like rays (left) and sine-like rays (right) of the quadruplet corrector.

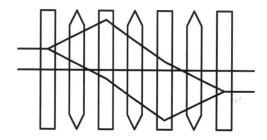

**FIGURE 7.12:** A quadrupole-octupole $C_S$ corrector. The rectangles represent quadrupoles and the hexagons represent octupoles.

consequence is that the terms $(x|a\delta)$ and $(y|b\delta)$ are equal, restoring the rotational symmetry of the chromatic aberration. The superimposed electrostatic and magnetic quadrupoles form first order **Wien filters** that can correct chromatic aberration.

In other words, the different energy dependencies of the electrostatic and the magnetic forces allow adjusting the chromatic aberration while maintaining overall linear focusing. In addition, the rotational symmetry of the spherical aberration is partially restored. Specifically, for a rotational symmetric system, we have $(x|a^3) = (x|ab^2) = (y|a^2b) = (y|b^3)$. For the present corrector, the relations among the four terms are $(x|a^3) = (y|b^3)$ and $(x|ab^2) = (y|a^2b)$. These relations show that two families of octupoles are needed to correct the spherical aberration using a corrector with the same symmetry.

For this corrector, the octupole components of the inner multipoles correct the terms $(x|a^3)$ and $(y|b^3)$, and those of the outer ones correct $(x|ab^2)$ and $(y|a^2b)$. With $C_S$ and $C_C$ corrected, the resolution of a 1 keV SEM could reach below 2 nm.

Meanwhile, another scheme was developed and used to successfully correct third order spherical aberration in a 100 keV **STEM** where a certain combination of quadrupoles and octupoles is used. It uses a similar layout for the quadrupoles as the $C_S$ and $C_C$ corrector above, which is shown in Fig. 7.12.

The linear optics consists of two identical quadrupole doublets with equal spacing between the quadrupoles and equal strength of all quadrupoles. The two outer octupoles correct the terms $(x|a^3)$ and $(y|b^3)$, and the middle one corrects $(x|ab^2)$ and $(y|a^2b)$. Due to the large difference in transverse position of the horizontal and vertical rays in the outer octupoles, the two knobs are mostly orthogonal. A resolution of 0.78 Å has been achieved using such a corrector.

While the introduction of the $C_S$ corrector into an electron microscope corrects the third order spherical aberration, it also generates much **larger fifth order spherical aberrations** $(C_5)$ through the combination of the objective lens and the octupoles and that among the octupoles, which becomes the limiting factor as the resolution reaches toward 0.5 Å. The equation below illustrates the origin of $C_5$ through combination.

$$
\begin{pmatrix} x_f \\ a_f \end{pmatrix} = \begin{pmatrix} x \\ a + k_{o2}x^3 \end{pmatrix} \circ \begin{pmatrix} (x|x)\,x + (x|a)\,a \\ (a|x)\,x + (a|a)\,a \end{pmatrix} \circ \begin{pmatrix} x_i \\ a_i + k_{o1}x_i^3 \end{pmatrix}
$$

$$
= \begin{pmatrix} (x|x)\,x_i + (x|a)\,a_i \\ (a|x)\,x_i + (a|a)\,a_i \end{pmatrix}
$$

$$
+ \begin{pmatrix} (x|a)\,k_{o1}x_i^3 \\ (a|a)\,k_{o1}x_i^3 + k_{o2}\left((x|x)\,x_i + (x|a)\,a_i + (x|a)\,k_{o1}x_i^3\right)^3 \end{pmatrix}
$$

$$
\approx \begin{pmatrix} (x|x)\,x_i + (x|a)\,a_i \\ (a|x)\,x_i + (a|a)\,a_i \end{pmatrix}
$$

$$
+ \begin{pmatrix} (x|a)\,k_{o1}x_i^3 \\ \left((a|a)\,k_{o1} + k_{o2}\,(x|x)^3\right)x_i^3 + 3\,(x|x)^2\,(x|a)\,k_{o1}k_{o2}x_i^5 \end{pmatrix}.
$$

Since $C_5$ is proportional to $(x|a)$, it vanishes when $(x|a)$ vanishes, i.e., when the first element (right) is imaged onto the second one (left). It can be seen from Fig. 7.12 that this condition is not met for this corrector. More recent designs of $C_S$ correctors have taken this into account and correct $C_5$ as well. By adjusting the image location, the value of $C_5$ can be varied and canceled.

In order to correct $C_S$ in a transmission electron microscope (**TEM**), extra attention has to be paid to maintaining a large usable object area, the so-called **field of view**. This usually requires that at least 2000 image points are well resolved in one dimension. It turns out that the simple quadrupole-octupole corrector shown in Figs. 7.11 and 7.12 does not meet this requirement. The main reason is that the cosine-like ray of the objective lens, i.e., the sine-like ray of the corrector, goes through the octupoles at large amplitude. As a result, it is deflected by the octupoles, generating large aberrations that limit the field of view.

The first successful corrector $C_S$ on TEM consists of two round lenses and two sextupoles, which is shown in Fig. 7.13. The round lenses form a so-called $-I$ transport between the centers of the sextupoles, i.e., the linear transfer matrix is a negative identity. It turns out that this cancels the second

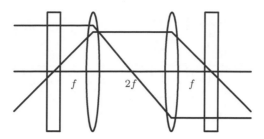

**FIGURE 7.13**: A sextupole $C_S$ corrector. The ellipses represent round lenses and the rectangles represent sextupoles.

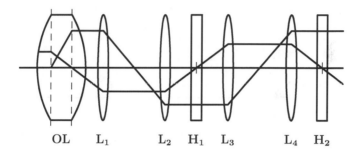

**FIGURE 7.14**: Sine-like and cosine-like rays of a $C_S$ corrected TEM from objective lens to the end of the corrector section. OL: objective lens, $L_1$ to $L_4$ : round lenses, $H_1$ and $H_2$ : sextupoles.

order aberrations generated by the sextupoles as well as $C_5$ from combination. Furthermore, the third order spherical aberration can be corrected due to the fact that $C_S$ from the sextupoles, which is proportional to $k_s^2$, is rotationally symmetric and of the opposite sign of that of the round lenses.

Fig. 7.14 shows such a corrector in a TEM, together with the objective lens and the so-called transfer lenses. Note that the cosine-like ray of the objective lens goes through the centers of the sextupoles, hence it is unaffected by the corrector, helping to maintain the field of view. A slightly modified version of such a corrector has also been used to correct $C_S$ in **STEM**. Recently, a STEM named TEAM 0.5 (Transmission Electron Aberration-corrected Microscope) has achieved the resolution of 0.5 Å at 300 keV using such a corrector.

With the success of correcting the spherical aberration in TEM, scientists and engineers in this field have set out to build a TEM that is both $C_S$ and $C_C$ corrected. Successful as it is, the sextupole corrector is not capable of correcting $C_C$ and it is not obvious how to modify the sextupole corrector to include $C_C$ correction. As a result, attention has been focused on the option of a quadrupole-octupole corrector. After many attempts, Rose [59] developed a design which satisfied the requirement and was later adopted by the TEAM

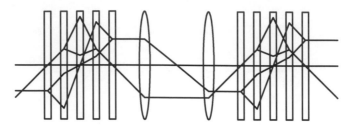

**FIGURE 7.15**:   Sine-like and cosine-like rays of the TEAM corrector. El-lipses: round transfer lenses, rectangles: multipoles. The focal length of the middle elements is half of that of the far outer ones. The ratio of the sine-like rays at the middle elements is 5.

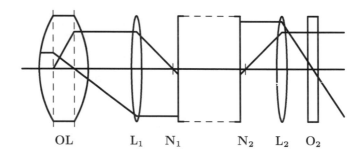

$$OL \qquad\qquad L_1 \quad N_1 \qquad\qquad N_2 \quad L_2 \quad O_2$$

**FIGURE 7.16**:   Sine-like and cosine-like rays of the TEAM microscope from objective lens (OL) to the end of the corrector section. The omitted part in the middle is the multipole corrector shown in Fig. 7.15. $L_1$ and $L_2$ : adapter lenses, $O_2$ : octupole used to cancel $(x|ab^2)$ and $(y|a^2b)$.

project and built.

As shown in Fig. 7.15, the corrector consists of two multipole quintuplets, each replacing one sextupole in the sextupole corrector (Fig. 7.13). The middle element of each quintuplet is a superimposed electrostatic and magnetic multipole which is responsible for correcting the spherical and the chromatic aberrations. Each quintuplet is mirror symmetric about its center and each half is again mirror symmetric about its own center. Each half of the quintuplet is point to parallel and parallel to point. Each quintuplet is a $-I$ transport. The result is the cancellation of a large number of aberrations. Since one of the strengths of quadrupole families is a free parameter, it is chosen such that the relative difference in the horizontal and vertical beam width at the center of the quintuplet is large. As in the case of the sextupole corrector for TEM (Fig. 7.14), the cosine-like ray of the objective lens is not affected by the aberration corrector (see Fig. 7.16). The second family of octupole can be placed either at the center of the corrector or, as shown in Fig. 7.16, after the corrector.

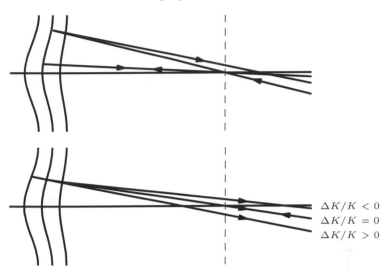

**FIGURE 7.17**:   Spherical (top) and chromatic (bottom) aberrations of an electron mirror, showing the possibility of reversing the sign of that of a regular round lens. Vertical dashed line: Gaussian image. Higher and lower energy electrons are represented by $\Delta K/K > 0$ and $\Delta K/K < 0$, respectively.

Such a corrector posed unprecedented challenges on technology in terms of tolerance on alignment errors and power supply stability. The required tolerance on alignment error is around 14 $\mu$m between adjacent elements, which is difficult but achievable. For the superimposed multipole elements which are responsible for aberration correction, the noise level of the current and voltage supplies have to be below $1.5 \cdot 10^{-8}$ ($\Delta I/|I|$) and $4 \cdot 10^{-8}$ ($\Delta U/|U|$), respectively. This level of stability was unheard of even a few years ago. Yet recently, it has been possible to achieve $\Delta I/|I| = 8.1 \cdot 10^{-9}$ and $\Delta U/|U| = 3.6 \cdot 10^{-9}$, fulfilling the design criteria. A first test of the corrector showed that the resolution of a TEM with this corrector reached 1 Å.

### 7.4.2   Aberration Correction in PEEM and LEEM

Due to the low energy of the electron beam ($< 30$ keV) in these devices, so-called electrostatic lenses that use electric fields to focus the beam are feasible. Although the multipole corrector used in low voltage SEM (scanning electron microscope) successfully corrected the spherical and the chromatic aberrations, it is not suited for **PEEM** (photoemission electron microscope) or **LEEM** (low energy electron microscope) which requires large field of view.

A sophisticated multipole corrector similar to the TEAM (Transmission Electron Aberration-corrected Microscope) corrector may be sufficient, but as it turns out there is a much simpler alternative, namely the electrostatic

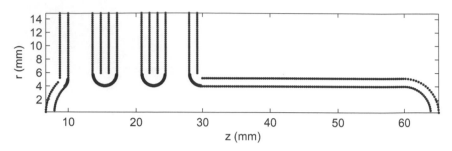

**FIGURE 7.18**:  Geometry of the tetrode mirror in PEEM3 at Lawrence Berkeley National Laboratory, California, USA.

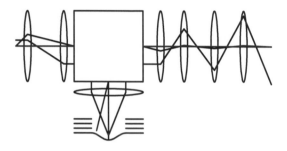

**FIGURE 7.19**:  Layout of PEEM3. Square: beam separator, ellipses: electrostatic round lenses, the mirror is on the bottom.

mirror. The reflection in the mirror makes it possible for a mirror to generate spherical and chromatic aberrations with the opposite sign of those from the regular round lenses.

The top picture in Fig. 7.17 shows that an electron with large initial angle is reflected at a location where the slope of the field line is smaller than the initial angle and can be focused less.  The bottom picture shows that an electron with higher energy penetrates deeper into the mirror, is reflected at a location where the slope of the field line is larger than that for an electron with design energy and, as a result, can be focused more.

Therefore, an electron mirror with a dent on the reflection electrode comparable to the electron beam size can form the desired field distribution for aberration correction. Fig. 7.18 shows an example in PEEM3 at Lawrence Berkeley National Laboratory, California, USA, which is an adaptation of the SMART design, and the layout of PEEM3 is shown in Fig. 7.19. SMART is a project of SpectroMicroscopy for All Relevant Techniques in Germany. The dots behind the surface denote charge rings used for numerical simulation. The first electrode from the right physically ends roughly at $z = 33$ mm. There are four electrodes used to provide tuning of the focal length, the spherical, and the chromatic aberrations.

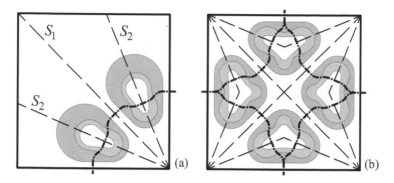

**FIGURE 7.20**: The beam separator of the first aberration-corrected PEEM. The path of the reference electron is shown by the dash-dotted curve. (From H. Rose, *Geometrical Charged-Particle Optics*, Springer-Verlag, Berlin, 2nd ed., 2012, © Springer-Verlag Berlin Heidelberg 2009, 2012 [60]. Chapter 9, Correction of Aberrations, Fig. 9.12, "Cross section of (a) the fourth quarter of the beam separator showing the double symmetry of the fields and the curved optic axis and of (b) the entire separator. The shaded areas represent the regions of the dipole field perpendicular to the pole plates. The sign and the strength of the dipole field differ for regions with different shading; the dash-dotted curve represents the optic axis." With kind permission from Springer Science and Business Media.)

Although the electron mirror itself maintains the rotational symmetry, a magnetic beam separator is needed to guide the electron beam to the detector downstream of the mirror, thus breaking the rotational symmetry of a conventional PEEM. Consequently, the most challenging part of an aberration-corrected PEEM/LEEM is the **beam separator** whose own aberrations have to be small compared to the existing ones.

The first aberration-corrected PEEM was built at Technische Universität Darmstadt, Germany, in the 1990s and was installed at BESSY II, the Berliner Elektronenspeicherring-Gesellschaft für Synchrotronstrahlung in 2001. Recently it achieved a resolution of 3 nm.

Its layout is similar to that shown in Fig. 7.19 up to the exit of the beam separator since the former has an energy filter downstream of the beam separator. The mirror column forms a so-called $-I$ transport, which ensures that the cosine-like ray turns back on the axis and is unaffected, maintaining a large field of view. The beam separator shown in Fig. 7.20 is a square magnet with 90° bending and three axes of mirror symmetry ($\theta = 27.5°$, 45° and 62.5° for each pass). The resulting optical system is an achromat with $+I$ transport, i.e., its transfer matrix is an identity matrix, and it is free of all second order geometrical aberrations.

The drawback of this separator is the difficulty in building this device to

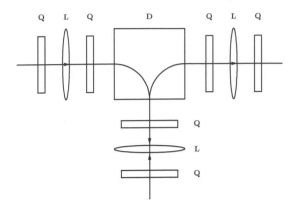

**FIGURE 7.21**:  The PEEM3 beam separator. The square, the ellipses and the rectangles represent the magnet, the electrostatic round lenses and the electrostatic quadrupoles, respectively.

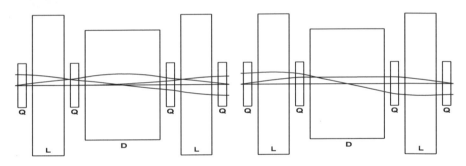

**FIGURE 7.22**:  Sine-like, cosine-like and dispersive rays of the PEEM3 beam separator in the $x$-$z$ plane (left), and in the $y$-$z$ plane (right).

the tight **machining tolerance** and in tuning it during operation due to the complexity and rigidness of the design. The fact that focusing is produced primarily by the edges entails that the slope of the grooves and the details of the field near the electron path are critical to the quality of the image.

The selection of high permeability material to fit the field distribution to the analytical model leads to magnetic material which is very soft and thus difficult to machine. As a result, the second project of an aberration-corrected PEEM, built at Lawrence Berkeley National Laboratory, turned to a simpler separator design shown in Fig. 7.21, and the rays are shown in Fig. 7.22. Since the magnet is a simple 90° sector bend, round lenses, with the help of electrostatic quadrupoles, provide the focusing. There is only one axis of mirror ($\theta = 45°$) for each pass. The system is a so-called $-I$ transport for each pass with no zero dispersion at the end. An achromat is formed after two passes.

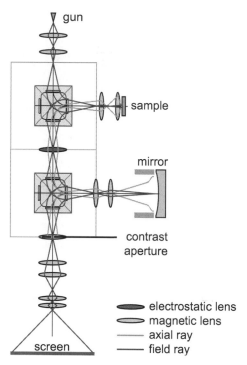

**FIGURE 7.23**:  Aberration-corrected and energy-filtered LEEM at IBM. Images of the sample are formed on the diagonal lines of the prisms, which are shown by squares. (Reprinted from *Ultramicroscopy*, v. 110, R. M. Tromp, et. al., A new aberration-corrected, energy-filtered LEEM / PEEM instrument. I. Principles and design, p. 852–861, Copyright (2010), with permission from Elsevier [68].)

Between 2007 and 2011, a third aberration-corrected PEEM/LEEM was designed and built at IBM [68]). The layout is shown in Fig. 7.23. Magnetic prisms are represented by squares, and electrostatic and magnetic round lenses are represented by ellipses. Fig. 7.23 also shows cosine-like rays (drawn in dark gray, denoted "field ray") and sine-like rays (drawn in pale gray, denoted "axial ray") from the sample. The beam separator consists of two 90° prism arrays and an electrostatic round lens. It restores the double mirror symmetry of the first separator but greatly simplifies the design and manufacture by using only commercially available components. The prism behaves like a round lens to the first order and is mirror symmetric, which entails that the dispersive ray forms a virtual image at the center. As a result, one round lens between the two prisms is sufficient to make the separator an achromat and transfer the image from the center of the first prism to that of the second one.

# Chapter 8

## The Periodic Transport

In the case of the periodic transport over long distances, the desire is not so much to give a special shape to the beam as the beam exits, but, even much more simply, to just **contain the beam**. This is of key importance in all devices in which the beam repeatedly passes through the same (or a very similar) structure. We may wonder whether this again translates into the requirement that a certain matrix element vanish, but as we shall see, this is not quite the case.

Actually it is rather straightforward to formulate a necessary condition on the linear matrix: it is not allowed to have any eigenvalue of magnitude greater than unity. If the eigenvalue is real, the argument is simple: if this were the case, any particle that has its coordinates lined up with the corresponding real eigenvector will after one period end up on the same line, but all its coordinates would have increased by a factor equal to the eigenvalue.

If on the other hand the eigenvalue is complex, there is another eigenvalue that is conjugate and hence has the same magnitude. Similar to the eigenvalues, also the eigenvectors are conjugates of each other. Now simply consider the sum of the two eigenvectors, which is real; sending this sum through the matrix multiplies the first eigenvector by the first eigenvalue, and the second one by the conjugate, resulting in a sum that is again real and increased in size by the magnitudes of the eigenvalues.

In both cases, **coordinates grow exponentially** in time, and so eigenvalues that are even only a tiny amount above unity in magnitude are detrimental. Of course the nonlinear effects also influence the motion and break the purely exponential pattern, but all experience shows that it is not possible to correct linear instability with nonlinear means; in practice, things usually work quite to the contrary.

## 8.1 The Transversal Motion

### 8.1.1 The Eigenvalues

Because of emittance preservation due to Liouville's theorem, the fact that eigenvalues greater than unity are prohibited means that, in fact, all eigen-

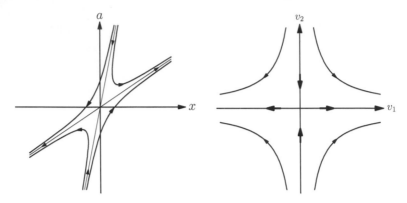

**FIGURE 8.1**: Motion in phase space (left) and in the eigenspace (right) when $|\operatorname{tr} M| > 2$.

values have to have **unit magnitude**. Of these, the cases $+1$ and $-1$ are to be excluded too, since even the slightest imperfection in the machine may otherwise lead to instability. Altogether, in a periodic system, the eigenvalues must all be **complex** and of unit magnitude.

It is particularly interesting to study the special case of a matrix with midplane symmetry. In this case, the $x$ and $y$ motion decouple and can be described by individual matrices. We obtain for the eigenvalues for the $2 \times 2$ $x$ sub-matrix, noting that the $y$ sub-matrix is treated similarly:

$$0 = \left|\hat{M} - \lambda\hat{I}\right| = \begin{vmatrix} (x|x) - \lambda & (x|a) \\ (a|x) & (a|a) - \lambda \end{vmatrix}$$
$$= \underbrace{(x|x)\,(a|a) - (x|a)\,(a|x)}_{1} - \lambda\,[(x|x) + (a|a)] + \lambda^2,$$

and so

$$\lambda_{1,2} = \frac{[(x|x) + (a|a)] \pm \sqrt{[(x|x) + (a|a)]^2 - 4}}{2}$$
$$= \frac{\operatorname{tr}\hat{M}}{2} \pm \sqrt{\left(\frac{\operatorname{tr}\hat{M}}{2}\right)^2 - 1}.$$

Hence to have complex eigenvalues requires the very simple condition

$$-2 < \operatorname{tr}\hat{M} < 2.$$

A quick check of the four cases shows that this **excludes** the point–to–point case and the parallel–to–parallel case, as in both of these, the trace just equals two or exceeds two. The parallel–to–point or point–to–parallel case each have

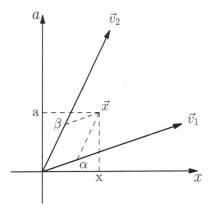

**FIGURE 8.2**: Relation between the phase space variables and the eigenvectors.

one element on the diagonal vanish, so they are permissible if the remaining diagonal matrix element is less than two in magnitude.

We also verify that for $\operatorname{tr} \hat{M} \neq 2$, the eigenvalues form a **reciprocal pair**, i.e., $\lambda_1 \lambda_2 = 1$. Let us quickly revisit the case $|\operatorname{tr} \hat{M}| > 2$, for which the eigenvalues are real, and hence one of them is greater than unity, and as we had already concluded, the motion is **unstable**. Choosing a new basis, the so-called **normal form** basis, along the real eigenvectors $\vec{v}_1$ and $\vec{v}_2$, we have that the repetitive motion asymptotically approaches the eigenvector $\vec{v}_1$ with eigenvalue greater than unity and becomes larger and larger (see the right picture in Fig. 8.1).

A detailed analysis shows that the motion indeed follows a **hyperbola**; note that $\lambda_1 \lambda_2 = 1$, and that $|\lambda_1| > 1 > |\lambda_2|$. Suppose we have a general vector expressed in the basis $(\vec{v}_1, \vec{v}_2)$ as shown in Fig. 8.2 whose coordinates are now $\alpha$ and $\beta$ (not to be confused with the Twiss parameter), and thus

$$\vec{x} \equiv \begin{pmatrix} x \\ a \end{pmatrix} = \alpha \vec{v}_1 + \beta \vec{v}_2.$$

Applying the transfer matrix, we have

$$\hat{M}\vec{x} = \alpha \hat{M}\vec{v}_1 + \beta \hat{M}\vec{v}_2 = \alpha \lambda_1 \vec{v}_1 + \beta \lambda_2 \vec{v}_2.$$

In normal form coordinates, the action of the transfer map is thus given by

$$\begin{pmatrix} \alpha \lambda_1 \\ \beta \lambda_2 \end{pmatrix} = \begin{pmatrix} \lambda_1 & 0 \\ 0 & \lambda_2 \end{pmatrix} \begin{pmatrix} \alpha \\ \beta \end{pmatrix},$$

but since $\lambda_2 = 1/\lambda_1$, the product of the coordinates stays constant, characteristic of the motion along a hyperbola. In Cartesian coordinates, the motion looks more complicated as the hyperbolic structure is deformed (see the left

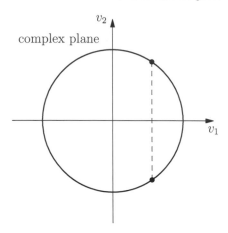

**FIGURE 8.3:**   Relation between the eigenvalues when $|\operatorname{tr} M| < 2$.

picture in Fig. 8.1). For practical purposes, this case is unstable and hence useless.

Let us now consider the case $|\operatorname{tr} \hat{M}| < 2$ in more detail. We have the complex eigenvalues that satisfy

$$\lambda_2 = \bar{\lambda}_1 \quad \text{and} \quad \lambda_2 = \lambda_1^{-1}.$$

So in the complex plane, $\lambda_1$ and $\lambda_2$ lie on a circle and form conjugate pairs, as shown in Fig. 8.3.

The eigenvalues can hence be written as

$$\lambda_{1,2} = e^{\pm i\mu},$$

where $\mu$ is called the **tune** of the system, which is

$$\mu = \arccos\left(\frac{\lambda_1 + \lambda_2}{2}\right) = \arccos\left(\frac{\operatorname{tr}\hat{M}}{2}\right).$$

The eigenvectors $\vec{v}_{1,2}$ belonging to $\lambda_{1,2}$ also form conjugate pairs, since

$$\hat{M}\overline{\vec{v}_2} = \overline{\hat{M}\vec{v}_2} = \overline{\lambda_2\vec{v}_2} = \lambda_1\overline{\vec{v}_2}.$$

Define now two new basis vectors $\vec{v}_+ = \operatorname{Re}(\vec{v}_1)$, $\vec{v}_- = \operatorname{Im}(\vec{v}_1)$ as the real and imaginary parts of the eigenvalues; they define what is called the normal form basis for stable motion. So we have

$$\vec{v}_1 = \vec{v}_+ + i\vec{v}_-, \quad \vec{v}_2 = \vec{v}_+ - i\vec{v}_-.$$

We now observe

$$\hat{M}\vec{v}_1 = \lambda_1\vec{v}_1 = e^{i\mu}\left(\vec{v}_+ + i\vec{v}_-\right) = \cos\mu \cdot \vec{v}_+ - \sin\mu \cdot \vec{v}_- + i\left(\sin\mu \cdot \vec{v}_+ + \cos\mu \cdot \vec{v}_-\right),$$

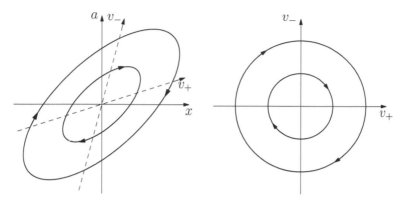

**FIGURE 8.4**: Motion in phase space (left) and in the eigenspace (right) when $|\operatorname{tr} M| < 2$.

and similarly

$$\hat{M}\vec{v}_2 = \lambda_2 \vec{v}_2 = e^{-i\mu}(\vec{v}_+ - i\vec{v}_-) = \cos\mu \cdot \vec{v}_+ - \sin\mu \cdot \vec{v}_- - i(\sin\mu \cdot \vec{v}_+ + \cos\mu \cdot \vec{v}_-).$$

Now assume we have a general vector expressed in the basis vectors $\vec{v}_\pm$ with coefficients $\alpha$ and $\beta$, i.e., $\vec{x} = \alpha\vec{v}_+ + \beta\vec{v}_-$. Then we have

$$\hat{M}\vec{x} = \alpha\hat{M}\vec{v}_+ + \beta\hat{M}\vec{v}_-$$

$$= \alpha\hat{M}\frac{\vec{v}_1 + \vec{v}_2}{2} + \beta\hat{M}\frac{\vec{v}_1 - \vec{v}_2}{2i} = \alpha\frac{\hat{M}\vec{v}_1 + \hat{M}\vec{v}_2}{2} + \beta\frac{\hat{M}\vec{v}_1 - \hat{M}\vec{v}_2}{2i}$$

$$= \alpha(\cos\mu \cdot \vec{v}_+ - \sin\mu \cdot \vec{v}_-) + \beta(\sin\mu \cdot \vec{v}_+ + \cos\mu \cdot \vec{v}_-)$$

$$= (\alpha\cos\mu + \beta\sin\mu)\vec{v}_+ + (-\alpha\sin\mu + \beta\cos\mu)\vec{v}_-.$$

So altogether, in normal form coordinates, we have

$$\hat{M}\begin{pmatrix} \alpha \\ \beta \end{pmatrix} = \begin{pmatrix} \alpha\cos\mu + \beta\sin\mu \\ -\alpha\sin\mu + \beta\cos\mu \end{pmatrix} = \begin{pmatrix} \cos\mu & \sin\mu \\ -\sin\mu & \cos\mu \end{pmatrix}\begin{pmatrix} \alpha \\ \beta \end{pmatrix},$$

and thus the transformation $\hat{M}$ simply performs a rotation as shown in the right picture in Fig. 8.4.

The angle of the rotation in normal form coordinates is simply equal to the tune $\mu$; and it is completely obvious that the motion is stable.

To obtain the motion in the original Cartesian coordinates, we have to subject the circles to a linear transformation, which turns them into ellipses; so the motion looks as in the left picture in Fig. 8.4. The angle by which particles move in the original $x$, $a$ coordinates is not necessarily $\mu$ anymore; but we can conclude that indeed if we look at the average angle advance over many turns, then this average converges to the tune $\mu$, as at least the number of full revolutions that were experienced must agree in both coordinate systems.

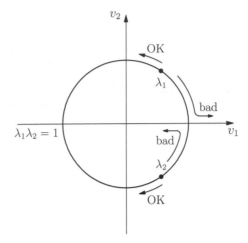

**FIGURE 8.5:** Possible movement of the eigenvalues under small perturbation near $|\operatorname{tr} M| = 2$.

It is also very illuminating to consider what happens if the system is subjected to some small errors, which in reality of course always appear. If the eigenvalues were far enough from unity, even under small errors we still have $\lambda_1 = \bar{\lambda}_2$ and $\lambda_1 = \lambda_2^{-1}$, and while the tune $\mu$ may have changed a little, the qualitative behavior of stability is totally unaffected. So as long as we maintain that $|\operatorname{tr} \hat{M}/2| < 1$ is maintained, stability prevails. If on the other hand the perturbation is so large that this is violated, the perturbation can lead to the loss of stability as shown in Fig. 8.5.

For the sake of completeness, let us consider further the case of $|\operatorname{tr} \hat{M}| = 2$. In this case, $\lambda_{1,2} = 1$, and

$$\hat{M} = \pm \hat{I}.$$

This motion is stable; but under the slightest perturbation there is danger of becoming unstable, and hence this case is practically useless.

### 8.1.2 The Invariant Ellipse

For many practical purposes it is particularly important to know in detail the parameters of the ellipse that is invariant under stable linear motion. For this purpose, let $\lambda_{1,2} = e^{\pm i\mu}$, and choose the sign of the tune $\mu$ such that $\operatorname{sign}(\mu) = \operatorname{sign}((x|a))$. We then define three parameters $\alpha_i$, $\beta_i$ and $\gamma_i$ as

$$\alpha_i = \frac{(x|x) - (a|a)}{2 \sin \mu_i}, \qquad \beta_i = \frac{(x|a)}{\sin \mu_i}, \qquad \gamma_i = -\frac{(a|x)}{\sin \mu_i}. \tag{8.1}$$

As we shall prove now, these three parameters describe the invariant ellipse via

$$\begin{pmatrix} x \\ a \end{pmatrix}^T \cdot \begin{pmatrix} \gamma_i & \alpha_i \\ \alpha_i & \beta_i \end{pmatrix} \cdot \begin{pmatrix} x \\ a \end{pmatrix} = 1,$$

where the matrix describing the ellipse is called $\hat{T}$. To prove that $\hat{T}$ is actually invariant, we first express the transfer matrix in terms of the parameters. To this end, we observe that since

$$\lambda_{1,2} = \frac{\operatorname{tr} \hat{M}}{2} \pm \sqrt{\left( \frac{\operatorname{tr} \hat{M}}{2} \right)^2 - 1},$$

we have that

$$(x|x) + (a|a) = \operatorname{tr} \hat{M} = \lambda_1 + \lambda_2 = e^{i\mu} + e^{-i\mu} = 2 \cos \mu.$$

From the definition of $\alpha_i$, we have $(x|x) - (a|a) = 2 \sin \mu_i \cdot \alpha_i$, and hence

$$(x|x) = \cos \mu_i + \alpha_i \sin \mu_i, \quad (a|a) = \cos \mu_i - \alpha_i \sin \mu_i.$$

On the other hand, from the definitions of $\beta_i$ and $\gamma_i$, we have

$$(x|a) = \beta_i \sin \mu_i, \quad (a|x) = -\gamma_i \sin \mu_i,$$

and so altogether

$$\hat{M} = \begin{pmatrix} \cos \mu_i + \alpha_i \sin \mu_i & \beta_i \sin \mu_i \\ -\gamma_i \sin \mu_i & \cos \mu_i - \alpha_i \sin \mu_i \end{pmatrix}.$$

Letting

$$\hat{I} = \begin{pmatrix} 1 & 0 \\ 0 & 1 \end{pmatrix}, \quad \hat{K} = \begin{pmatrix} \alpha_i & \beta_i \\ -\gamma_i & -\alpha_i \end{pmatrix},$$

we have

$$\hat{M} = \hat{I} \cos \mu_i + \hat{K} \sin \mu_i.$$

Computing the inverse map of $\hat{M}$, we find

$$\hat{M}^{-1} = \hat{I} \cos \mu_i - \hat{K} \sin \mu_i,$$

where we used $|\hat{M}| = 1$, and as a consequence $\beta_i \gamma_i - \alpha_i^2 = 1$, which we infer as follows:

$$1 = |\hat{M}| = (\cos \mu_i + \alpha_i \sin \mu_i)(\cos \mu_i - \alpha_i \sin \mu_i) + \beta_i \gamma_i \sin^2 \mu_i$$
$$= \cos^2 \mu_i + \left( \beta_i \gamma_i - \alpha_i^2 \right) \sin^2 \mu_i = 1 + \left( -1 + \beta_i \gamma_i - \alpha_i^2 \right) \sin^2 \mu_i,$$

but since $\mu_i$ was not allowed to be zero or $\pi$ because of our requirement of stability, we must have $\beta_i \gamma_i - \alpha_i^2 = 1$.

We now are ready to study whether indeed the ellipse defined above is invariant under $\hat{M}$. This is the case if whenever a particle satisfies the ellipse equation

$$\begin{pmatrix} x \\ a \end{pmatrix}^T \cdot \begin{pmatrix} \gamma_i & \alpha_i \\ \alpha_i & \beta_i \end{pmatrix} \cdot \begin{pmatrix} x \\ a \end{pmatrix} = 1,$$

their image under $\hat{M}$, which is given by

$$\hat{M} \cdot \begin{pmatrix} x \\ a \end{pmatrix},$$

also satisfies the ellipse equation. This means that also

$$\left[ \hat{M} \cdot \begin{pmatrix} x \\ a \end{pmatrix} \right]^T \cdot \begin{pmatrix} \gamma_i & \alpha_i \\ \alpha_i & \beta_i \end{pmatrix} \cdot \left[ \hat{M} \cdot \begin{pmatrix} x \\ a \end{pmatrix} \right] = 1.$$

This is the case if and only if

$$\hat{M}^T \cdot \hat{T} \cdot \hat{M} = \hat{T},$$

since every ellipse is described by a unique symmetric matrix and $\hat{M}^T \cdot \hat{T} \cdot \hat{M}$ is indeed symmetric. In order to execute the matrix multiplications necessary, we study various matrix products; let

$$\hat{J} \stackrel{def}{=} \begin{pmatrix} 0 & 1 \\ -1 & 0 \end{pmatrix}.$$

We then have

$$\hat{T}\hat{K} = \begin{pmatrix} \gamma_i & \alpha_i \\ \alpha_i & \beta_i \end{pmatrix} \begin{pmatrix} \alpha_i & \beta_i \\ -\gamma_i & -\alpha_i \end{pmatrix} = \begin{pmatrix} 0 & 1 \\ -1 & 0 \end{pmatrix} = \hat{J},$$

$$\hat{K}^T\hat{T} = \hat{K}^T\hat{T}^T = \left( \hat{T}\hat{K} \right)^T = \hat{J}^T = -\hat{J},$$

$$\hat{K}^T\hat{J} = \begin{pmatrix} \alpha_i & -\gamma_i \\ \beta_i & -\alpha_i \end{pmatrix} \begin{pmatrix} 0 & 1 \\ -1 & 0 \end{pmatrix} = \begin{pmatrix} \gamma_i & \alpha_i \\ \alpha_i & \beta_i \end{pmatrix} = \hat{T}.$$

Now we are ready to compute the product $\hat{M}^T \cdot \hat{T} \cdot \hat{M}$. We obtain

$$\hat{M}^T \cdot \hat{T} \cdot \hat{M} = \left( \hat{I} \cos \mu_i + \hat{K}^T \sin \mu_i \right) \hat{T} \left( \hat{I} \cos \mu_i + \hat{K} \sin \mu_i \right)$$

$$= \left( \hat{I} \cos \mu_i + \hat{K}^T \sin \mu_i \right) \left( \hat{T} \cos \mu_i + \hat{J} \sin \mu_i \right)$$

$$= \hat{T} \cos^2 \mu_i + \hat{J} \sin \mu_i \cos \mu_i - \hat{J} \sin \mu_i \cos \mu_i + \hat{T} \sin^2 \mu_i = \hat{T},$$

which is indeed what we needed to prove. To conclude we remark that there is not only one invariant ellipse, but even every ellipse that can be generated by stretching or shrinking from the original one is invariant. So altogether, we have a nested set of invariant ellipses, and particles will always stay contained

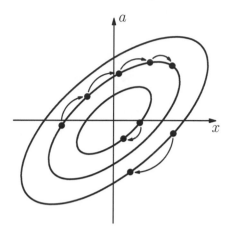

**FIGURE 8.6**:   Stable motion in phase space.

on the invariant ellipse on which they are originally lying, as shown in Fig. 8.6.

The last important question remaining in this section is to put into perspective the parameters of the **beam** $\alpha$, $\beta$, $\gamma$ and the parameters $\alpha_i$, $\beta_i$, $\gamma_i$ describing the invariant ellipse of one turn **accelerator**. Are these Greek letters equal, are they related, or do they have nothing to do with each other? This is actually a question that often throws off even die-hard accelerator physicists, and it is very much worthwhile to understand it in depth.

Regarding their origin, these two sets of parameters are actually **totally independent**. In fact, one describes some property of an accelerator, and the other describes a property of a beam; and of course we can feed any type of beam into a given accelerator.

However, if the goal is to fill the accelerator in the most efficient way, as it turns out this is accomplished if the beam's Twiss parameters **agree** with those of the accelerator. In this case, after one revolution the phase space will occupy exactly the same area (although the individual particles in it are at different positions), as shown in Fig. 8.7.

On the other hand, if one injects a beam with an ellipse that does not agree with the invariant ellipse of the accelerator, then the repetitive behavior of the beam ellipse shown solid in Fig. 8.8 is determined by the shaded invariant ellipse it touches.

As we go around the repetitive system repeatedly, the beam ellipse stays within the invariant ellipse and touches it, but, depending on the tune, will have a different orientation. In fact, if the tune is not rational — something desirable for stability reasons — over time even all different orientations will occur. If we now want to operate the accelerator, we have to make sure we can handle everything inside the invariant ellipse, leading to considerable waste

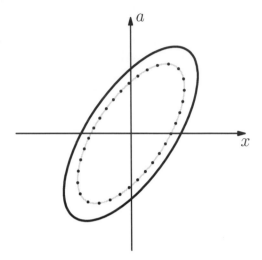

**FIGURE 8.7**: Illustration of the case where the beam ellipse (dashed) and the invariant ellipses (solid) are matched. After each revolution, the beam ellipse is exactly reproduced.

of area.

So it is best to operate a repetitive system in such a way that the beam ellipse is **matched** to the accelerator's invariant ellipse, and to avoid mismatching, the so-called **beating**.

## 8.2 Dispersive Effects

### 8.2.1 The Periodic Solution

Let $\hat{M}$ be the transfer matrix of a periodic cell,

$$\hat{M} = \begin{pmatrix} (x|x) & (x|a) & (x|\delta) \\ (a|x) & (a|a) & (a|\delta) \\ 0 & 0 & 1 \end{pmatrix}.$$

The periodic solution characterized by $D$, $D'$ satisfies

$$\begin{pmatrix} D \\ D' \\ 1 \end{pmatrix} = \begin{pmatrix} (x|x) & (x|a) & (x|\delta) \\ (a|x) & (a|a) & (a|\delta) \\ 0 & 0 & 1 \end{pmatrix} \begin{pmatrix} D \\ D' \\ 1 \end{pmatrix}. \tag{8.2}$$

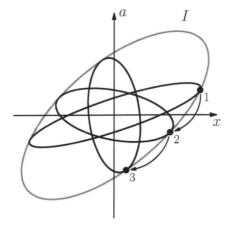

**FIGURE 8.8:** Behavior of a mismatched beam.

Thus

$$\begin{pmatrix} (x|\delta) \\ (a|\delta) \end{pmatrix} = \begin{pmatrix} 1-(x|x) & -(x|a) \\ -(a|x) & 1-(a|a) \end{pmatrix} \begin{pmatrix} D \\ D' \end{pmatrix}, \qquad (8.3)$$

$$\begin{pmatrix} D \\ D' \end{pmatrix} = \begin{pmatrix} 1-(x|x) & -(x|a) \\ -(a|x) & 1-(a|a) \end{pmatrix}^{-1} \begin{pmatrix} (x|\delta) \\ (a|\delta) \end{pmatrix},$$

when

$$\det \begin{pmatrix} 1-(x|x) & -(x|a) \\ -(a|x) & 1-(a|a) \end{pmatrix} = 2-(x|x)-(a|a) \neq 0.$$

Note that when tr $\hat{M} < 2$, which is satisfied if only stable motion is considered, $D$, $D'$ are uniquely determined.

$$\begin{pmatrix} D \\ D' \end{pmatrix} = \frac{1}{2-(x|x)-(a|a)} \begin{pmatrix} 1-(a|a) & (x|a) \\ (a|x) & 1-(x|x) \end{pmatrix} \begin{pmatrix} (x|\delta) \\ (a|\delta) \end{pmatrix}$$

$$= \frac{1}{2-(x|x)-(a|a)} \begin{pmatrix} (1-(a|a))(x|\delta)+(x|a)(a|\delta) \\ (1-(x|x))(a|\delta)+(a|x)(x|\delta) \end{pmatrix}.$$

From the point of view of the computation, a slightly different form of the same result helps to develop an algorithm that can be applied to arbitrary order with the help of the Differential Algebraic (DA) technique. Eq. (8.2) can be rewritten as

$$\begin{pmatrix} 1 & 0 & 0 \\ 0 & 1 & 0 \\ 0 & 0 & 0 \end{pmatrix} \begin{pmatrix} D \\ D' \\ 1 \end{pmatrix} + \begin{pmatrix} 0 \\ 0 \\ 1 \end{pmatrix} = \begin{pmatrix} (x|x) & (x|a) & (x|\delta) \\ (a|x) & (a|a) & (a|\delta) \\ 0 & 0 & 1 \end{pmatrix} \begin{pmatrix} D \\ D' \\ 1 \end{pmatrix},$$

and hence

$$\begin{pmatrix} D \\ D' \\ 1 \end{pmatrix} = \begin{pmatrix} (x|x) - 1 & (x|a) & (x|\delta) \\ (a|x) & (a|a) - 1 & (a|\delta) \\ 0 & 0 & 1 \end{pmatrix}^{-1} \begin{pmatrix} 0 \\ 0 \\ 1 \end{pmatrix}.$$

Defining

$$\hat{I}_H = \begin{pmatrix} 1 & 0 & 0 \\ 0 & 1 & 0 \\ 0 & 0 & 1 \end{pmatrix},$$

we obtain the more compact form

$$\begin{pmatrix} D \\ D' \\ 1 \end{pmatrix} = \left( \hat{M} - \hat{I}_H \right)^{-1} \begin{pmatrix} 0 \\ 0 \\ 1 \end{pmatrix}.$$

It is straightforward to generalize the above equation to the case of a nonlinear map. Here the equation to be solved is

$$\begin{pmatrix} D(\delta) \\ D'(\delta) \\ \delta \end{pmatrix} = \mathcal{M} \circ \begin{pmatrix} D(\delta) \\ D'(\delta) \\ \delta \end{pmatrix},$$

where $D(\delta)$ and $D'(\delta)$ are polynomials of $\delta$ without the constant part. Similar to the case for the linear map, we obtain

$$\begin{pmatrix} D(\delta) \\ D'(\delta) \\ \delta \end{pmatrix} = (\mathcal{M} - \mathcal{I}_H)^{-1} \circ \begin{pmatrix} D(\delta) \\ D'(\delta) \\ \delta \end{pmatrix},$$

where

$$\mathcal{I}_H = \begin{pmatrix} 1 & 0 & 0 \\ 0 & 1 & 0 \\ 0 & 0 & 1 \end{pmatrix} \begin{pmatrix} x \\ a \\ \delta \end{pmatrix}.$$

As described in Section 5.4.2, the map $(\mathcal{M} - \mathcal{I}_H)^{-1}$ can be obtained using the Differential Algebraic (DA) technique order-by-order up to any given order. Thus the periodic solution of dispersion up to arbitrary order can be obtained without loss of accuracy.

## 8.2.2 Chromaticity

Now let us turn our attention to the betatron tune of off-momentum particles. We know that higher momentum particles are bent less, hence focal length of quadrupoles is longer. As a result, we expect that total phase advance decreases as momentum increases. Recall that

$$k = \frac{q}{p} \frac{\partial B_y}{\partial x},$$

and when $p = p_0 + dp$, we have

$$k = \frac{q}{p}\frac{\partial B_y}{\partial x} = \frac{q}{p_0}\frac{\partial B_y}{\partial x}\frac{1}{1 + dp/p_0} =_1 k_0\left(1 - \delta\right),$$

$$k_0 = \frac{q}{p_0}\frac{\partial B_y}{\partial x}, \delta = \frac{dp}{p_0}.$$

For a small distance $ds$, we have

$$d\hat{M} = \begin{pmatrix} 1 & 0 \\ -kds & 1 \end{pmatrix}.$$

And by defining

$$d\hat{M}_0 = \begin{pmatrix} 1 & 0 \\ -k_0 ds & 1 \end{pmatrix},$$

we have

$$\hat{M} = d\hat{M}d\hat{M}_0^{-1}\begin{pmatrix} \cos(\mu_0) + \alpha\sin(\mu_0) & \beta\sin(\mu_0) \\ -\gamma\sin(\mu_0) & \cos(\mu_0) - \alpha\sin(\mu_0) \end{pmatrix}$$

$$= \begin{pmatrix} 1 & 0 \\ -(k - k_0)ds & 1 \end{pmatrix}\begin{pmatrix} \cos(\mu_0) + \alpha\sin(\mu_0) & \beta\sin(\mu_0) \\ -\gamma\sin(\mu_0) & \cos(\mu_0) - \alpha\sin(\mu_0) \end{pmatrix}$$

$$= \begin{pmatrix} 1 & 0 \\ \delta k_0 ds & 1 \end{pmatrix}\begin{pmatrix} \cos(\mu_0) + \alpha\sin(\mu_0) & \beta\sin(\mu_0) \\ -\gamma\sin(\mu_0) & \cos(\mu_0) - \alpha\sin(\mu_0) \end{pmatrix}.$$

Using the normalization transformation

$$\hat{A} = \begin{pmatrix} 1/\sqrt{\beta} & 0 \\ \alpha/\sqrt{\beta} & \sqrt{\beta} \end{pmatrix} \quad \text{and} \quad \hat{A}^{-1} = \begin{pmatrix} \sqrt{\beta} & 0 \\ -\alpha/\sqrt{\beta} & 1/\sqrt{\beta} \end{pmatrix},$$

we obtain the transfer matrix in the normalized space as

$$\widetilde{M} = \hat{A}\hat{M}\hat{A}^{-1}$$

$$= \hat{A}\begin{pmatrix} 1 & 0 \\ \delta k_0 ds & 1 \end{pmatrix}\hat{A}^{-1} \cdot \hat{A}\begin{pmatrix} \cos(\mu_0) + \alpha\sin(\mu_0) & \beta\sin(\mu_0) \\ -\gamma\sin(\mu_0) & \cos(\mu_0) - \alpha\sin(\mu_0) \end{pmatrix}\hat{A}^{-1}$$

$$= \begin{pmatrix} 1 & 0 \\ \delta\beta k_0 ds & 1 \end{pmatrix}\begin{pmatrix} \cos(\mu_0) & \sin(\mu_0) \\ -\sin(\mu_0) & \cos(\mu_0) \end{pmatrix}$$

$$= \begin{pmatrix} \cos(\mu_0) & \sin(\mu_0) \\ \delta\beta k_0 ds\cos(\mu_0) - \sin(\mu_0) & \cos(\mu_0) + \delta\beta k_0 ds\sin(\mu_0) \end{pmatrix},$$

and

$$\cos(\mu) = \frac{1}{2}\,\mathrm{tr}\left(\widetilde{M}\right) = \cos(\mu_0) + \frac{1}{2}\delta\beta k_0 ds\sin(\mu_0),$$

$$\mu = \mu_0 + d\mu \implies \cos(\mu) = \cos(\mu_0) - d\mu\sin(\mu_0) \implies d\mu = -\frac{1}{2}\delta\beta k_0 ds$$

$$\implies d\nu = \frac{d\mu}{2\pi} = -\frac{1}{4\pi}\delta\beta k_0 ds \implies \xi_N = \frac{d\nu}{dp/p_0} = -\frac{1}{4\pi}\oint\beta(s)k_0(s)ds.$$

The last step keeps only the leading order effect. The quantity $\xi_N$ is called **natural chromaticity**.

It will likely lead to beam loss since certain off-energy particles lie on resonant tunes. To remedy this problem, it is common to use sextupoles, because they do not affect tunes of on-momentum particles but at the same time provide quadratic nonlinearity that can be used to compensate the natural chromaticity. The magnetic field of a sextupole is

$$B_y = b_2 \left( x^2 - y^2 \right), \quad B_x = 2b_2 xy,$$

where $b_2 = -3M_{3,3}$. Defining $k_s = -b_2/\chi_{m0}$ and denoting

$$\hat{M} = \begin{pmatrix} \cos \mu_x + \alpha_x \sin \mu_x & \beta_x \sin \mu_x & 0 & 0 \\ -\gamma_x \sin \mu_x & \cos \mu_x - \alpha_x \sin \mu_x & 0 & 0 \\ 0 & 0 & \cos \mu_y + \alpha_y \sin \mu_y & \beta_y \sin \mu_y \\ 0 & 0 & -\gamma_y \sin \mu_y & \cos \mu_y - \alpha_y \sin \mu_y \end{pmatrix},$$

we obtain

$$\begin{pmatrix} x_1 \\ a_1 \\ y_1 \\ b_1 \end{pmatrix} = \begin{pmatrix} x \\ a + k_s ds \left( x^2 - y^2 \right) \\ y \\ b - 2k_s dsxy \end{pmatrix} \circ \left[ \hat{M} \begin{pmatrix} x_0 \\ a_0 \\ y_0 \\ b_0 \end{pmatrix} \right]$$

$$= \begin{pmatrix} x \\ a + k_s ds \left( x^2 - y^2 \right) \\ y \\ b - 2k_s dsxy \end{pmatrix} \circ \begin{pmatrix} (\cos \mu_x + \alpha_x \sin \mu_x) \, x_0 + (\beta_x \sin \mu_x) \, a_0 \\ -(\gamma_x \sin \mu_x) \, x_0 + (\cos \mu_x - \alpha_x \sin \mu_x) \, a_0 \\ (\cos \mu_y + \alpha_y \sin \mu_y) \, y_0 + (\beta_y \sin \mu_y) \, b_0 \\ -(\gamma_y \sin \mu_y) \, y_0 + (\cos \mu_y - \alpha_y \sin \mu_y) \, b_0 \end{pmatrix}.$$

Again, to make the physical picture clearer, let us apply the normalization transformation in four-dimensional space, which is

$$\begin{pmatrix} \tilde{x} \\ \tilde{a} \\ \tilde{y} \\ \tilde{b} \end{pmatrix} = \hat{A} \begin{pmatrix} x \\ a \\ y \\ b \end{pmatrix} = \begin{pmatrix} 1/\sqrt{\beta_x} & 0 & 0 & 0 \\ \alpha_x/\sqrt{\beta_x} & \sqrt{\beta_x} & 0 & 0 \\ 0 & 0 & 1/\sqrt{\beta_y} & 0 \\ 0 & 0 & \alpha_y/\sqrt{\beta_y} & \sqrt{\beta_y} \end{pmatrix} \begin{pmatrix} x \\ a \\ y \\ b \end{pmatrix}.$$

The inverse is

$$\hat{A}^{-1} = \begin{pmatrix} \sqrt{\beta_x} & 0 & 0 & 0 \\ -\alpha_x/\sqrt{\beta_x} & 1/\sqrt{\beta_x} & 0 & 0 \\ 0 & 0 & \sqrt{\beta_y} & 0 \\ 0 & 0 & -\alpha_y/\sqrt{\beta_y} & 1/\sqrt{\beta_y} \end{pmatrix}.$$

Denoting the linear normalization transformation (as opposed to the transformation matrix), its inverse and the linear transfer map as

$$A = \hat{A} \begin{pmatrix} x \\ a \\ y \\ b \end{pmatrix}, \qquad A^{-1} = \hat{A}^{-1} \begin{pmatrix} x \\ a \\ y \\ b \end{pmatrix}, \qquad M = \hat{M} \begin{pmatrix} x \\ a \\ y \\ b \end{pmatrix},$$

we obtain

$$\begin{pmatrix} \tilde{x}_1 \\ \tilde{a}_1 \\ \tilde{y}_1 \\ \tilde{b}_1 \end{pmatrix} = \left[ A \circ \begin{pmatrix} x \\ a + k_s ds \left( x^2 - y^2 \right) \\ y \\ b - 2k_s ds xy \end{pmatrix} \circ A^{-1} \right] \circ \left[ A \circ M \circ A^{-1} \right]$$

$$= \begin{pmatrix} x \\ a + k_s ds \sqrt{\beta_x} \left( \beta_x x^2 - \beta_y y^2 \right) \\ y \\ b - 2k_s ds \sqrt{\beta_x} \beta_y xy \end{pmatrix} \circ R(\mu_x, \mu_y),$$

where

$$R(\mu_x, \mu_y) = \begin{pmatrix} \hat{R}\left(\mu_x\right) & 0 \\ 0 & \hat{R}\left(\mu_y\right) \end{pmatrix} \begin{pmatrix} x \\ a \\ y \\ b \end{pmatrix}, \qquad \hat{R}\left(\mu_x\right) = \begin{pmatrix} \cos \mu & \sin \mu \\ -\sin \mu & \cos \mu \end{pmatrix}.$$

Since the kick of a sextupole is second order in the coordinates, the off-momentum particle has to go through it off the magnetic center in order to affect the linear motion and hence the tune. In other words, the dispersion has to be nonzero at the location of the sextupole to correct chromaticity. With the presence of dispersion, the coordinates are

$$x \to x - D_x \delta, \qquad a \to a - D_x' \delta,$$

and the normalized coordinates are

$$\tilde{x} \to \tilde{x} - \frac{D_x \delta}{\sqrt{\beta_x}}, \qquad \tilde{a} \to \tilde{a} - \sqrt{\beta_x} D_x' \delta.$$

The one turn map in the new normalized coordinates is

$$
\begin{pmatrix} \tilde{x}_1 \\ \tilde{a}_1 \\ \tilde{y}_1 \\ \tilde{b}_1 \end{pmatrix} = \begin{pmatrix} x + \beta_x^{-\frac{1}{2}} D_x \delta \\ a + \beta_x^{\frac{1}{2}} D_x' \delta \\ y \\ b \end{pmatrix} \circ \begin{pmatrix} x \\ a + k_s ds \beta_x^{\frac{1}{2}} \left( \beta_x x^2 - \beta_y y^2 \right) \\ y \\ b - 2k_s ds \beta_x^{\frac{1}{2}} \beta_y xy \end{pmatrix} \circ \begin{pmatrix} x - \beta_x^{-\frac{1}{2}} D_x \delta \\ a - \beta_x^{\frac{1}{2}} D_x' \delta \\ y \\ b \end{pmatrix}
$$

$$
\circ \begin{pmatrix} x + \beta_x^{-\frac{1}{2}} D_x \delta \\ a + \beta_x^{\frac{1}{2}} D_x' \delta \\ y \\ b \end{pmatrix} \circ R(\mu_x, \mu_y) \circ \begin{pmatrix} x - \beta_x^{-\frac{1}{2}} D_x \delta \\ a - \beta_x^{\frac{1}{2}} D_x' \delta \\ y \\ b \end{pmatrix} \circ \begin{pmatrix} \tilde{x}_0 \\ \tilde{a}_0 \\ \tilde{y}_0 \\ \tilde{b}_0 \end{pmatrix}
$$

$$
= \begin{pmatrix} x \\ a + k_s ds \sqrt{\beta_x} \left[ \beta_x \left( x - D_x \delta / \sqrt{\beta_x} \right)^2 - \beta_y y^2 \right] \\ y \\ b - 2k_s ds \sqrt{\beta_x} \beta_y \left( x - D_x \delta / \sqrt{\beta_x} \right) y \end{pmatrix} \circ R(\mu_x, \mu_y) \circ \begin{pmatrix} \tilde{x}_0 \\ \tilde{a}_0 \\ \tilde{y}_0 \\ \tilde{b}_0 \end{pmatrix}.
$$

The reason that the constant part vanishes after the rotation is that $D_x$ and $D_x'$ are periodic solutions of the ring. Keeping only the linear part, we obtain the transfer matrix

$$
\widetilde{M} = 2\delta k_s ds \begin{pmatrix} 1 & 0 & 0 & 0 \\ -\beta_x D_x & 1 & 0 & 0 \\ 0 & 0 & 1 & 0 \\ 0 & 0 & \beta_y D_x & 1 \end{pmatrix} \cdot \hat{R}(\mu_x, \mu_y),
$$

and the chromaticities due to the sextupole

$$
\xi_{x,s} = \frac{1}{4\pi} \oint \beta_x(s) D_x(s) k_s(s) ds,
$$

$$
\xi_{y,s} = -\frac{1}{4\pi} \oint \beta_y(s) D_x(s) k_s(s) ds.
$$

In summary the total chromaticities are

$$
\xi_x = -\frac{1}{4\pi} \oint \beta_x(s) \left[ k_x(s) - D_x(s) k_s(s) \right] ds, \tag{8.4}
$$

$$
\xi_y = -\frac{1}{4\pi} \oint \beta_y(s) \left[ k_y(s) + D_x(s) k_s(s) \right] ds, \tag{8.5}
$$

where $k_x(s)$ and $k_y(s)$ are quadrupole strength in the $x$ and $y$ planes along the ring. Usually two families of sextupoles are used to correct chromaticities in both planes. In order to make the two knobs more efficient and orthogonal, one family is placed at locations so that $\beta_x$ is large and $\beta_y$ is small and the other family at locations so that the opposite is true.

Eqs. (8.4) and (8.5) are very useful for the design process to determine where the chromaticities are generated and where the best locations are to place the sextupoles for correction. Yet computing them together with the higher order terms become almost trivial with the Differential Algebraic (DA) technique. Recall that the tunes are given by

$$\nu_{x,y} = \frac{1}{2\pi} \arccos\left(\frac{\operatorname{tr} \hat{M}_{x,y}}{2}\right).$$

The $\delta$ dependent tunes are simply

$$\nu_{x,y}(\delta) = \frac{1}{2\pi} \arccos\left(\frac{\operatorname{tr} \hat{M}_{x,y}(\delta)}{2}\right),$$

which contain the tunes and the chromaticities to arbitrary order. For example, $\nu_{x,y}$ are the constant part and $\xi_{x,y}$ are the linear coefficients.

---

## 8.3   A Glimpse at Nonlinear Effects

Linear motion around a fixed point is completely classified by the two cases we discussed previously, namely the stable or unstable case. This situation is **fundamentally different in the nonlinear case**; it is in fact much more complicated and interesting, and represents one example of the modern research field dealing with just such questions.

While this is not at all the place to try to develop a complete understanding of the nonlinear effects that may appear, let us spend some time to stake the territory and make some general observations. First we may expect that as long as the motion is **close enough** to the fixed point, it is **dominated by linear effects**, and depending on whether we have stability or not, we see either stable elliptic motion or unstable hyperbolic motion. While we may expect that linearly unstable motion will in most cases also stay unstable if we consider the nonlinear effects, **linear stable motion will not usually stay nonlinearly stable**. In fact, if the amplitudes of the motion become large, the effects of nonlinearity will become noticeable over-proportionally, and eventually they will become dominating, in most cases leading to instability for large amplitudes.

One can then try to heuristically separate the phase space into a region that appears stable for a reasonable number of turns, and a region that appears unstable. According to the previous arguments, in most cases the stable region will be near the fixed point, and the unstable region will be away from the fixed point. The region of transition between the apparently stable and apparently unstable parts is usually called the **dynamic aperture**, and it often looks like a deformed ellipse.

Let us now study a little what conditions seem to favor stable or unstable motion, respectively. If we divide the phase space regions into parts in which the nonlinear effects have a tendency to pull particles away from the origin and those that tend to push the particles toward the origin, then we may expect that we want to avoid situations where the particles spend too much time in the "pull away" regions, and it is better if we sample the phase space uniformly, and thus average out the effects as much as possible.

A nearly uniform sampling of the phase space happens if the linear **tune is not a rational** multiple of $2\pi$. On the other hand, if the tune is of the form $\mu_i = 2\pi p/q$, after $q$ turns the particle will come back to where it was before and hence can see the same effect, a situation which we call **resonance**; so it is at least not a good idea to choose $q$ too small, as repetition after large numbers of turns is not as critical. The effect of resonances in a circular accelerator is of great importance to its performance. Chapter 11 is dedicated to studying this topic in detail.

We may also wonder to what extent it is possible to perform a transformation to normal form coordinates in a similar manner as in the linear case. As it turns out, most systems cannot be brought to a normal form in which the motion is exactly circular; the existence of such a transformation is tantamount to the system being **integrable**, i.e., having one integral of motion per phase space dimension. Truly integrable systems, however, are very rare. It turns out, however, there is a powerful order-by-order iterative procedure to turn a system into nonlinear normal form up to any given order [5]. A simple example of this procedure is given in Section 11.4.

# Chapter 9

## Lattice Modules

In the design of actual devices for the manipulation of beams, it is important to employ field arrangements that achieve the basic features of **steering the beam** as a whole to its desired location, as well as **keeping the beam close together** over possibly extended distances, which is achieved through various **focusing** mechanisms. Thus in both single pass lines and rings, there exist different sections that perform different functions which require different types of lattice modules. Modern accelerators and beam transport lines focus the beam transversely using **alternate-gradient** focusing, also called **strong focusing**, which evolves from the so-called **weak focusing** used in betatrons and early weak focusing synchrotrons. In the weak focusing machines, inhomogeneous dipoles were used to bend and transversely confine the beam simultaneously. This is possible due to the fact that an inhomogeneous dipole with $0 < n < 1$ focuses the beam in both $x$ and $y$ planes.

From eq. (4.6), we have, for $0 < n < 1$,

$$\hat{M}_x = \begin{pmatrix} \cos(\sqrt{1-n}\phi) & (\rho/\sqrt{1-n})\sin(\sqrt{1-n}\phi) \\ -(\sqrt{1-n}/\rho)\sin(\sqrt{1-n}\phi) & \cos(\sqrt{1-n}\phi) \end{pmatrix},$$

$$\hat{M}_y = \begin{pmatrix} \cos(\sqrt{n}\phi) & (\rho/\sqrt{n})\sin(\sqrt{n}\phi) \\ -(\sqrt{n}/\rho)\sin(\sqrt{n}\phi) & \cos(\sqrt{n}\phi) \end{pmatrix}.$$

Yet weak focusing was eventually replaced by strong focusing because weak focusing was too weak to confine the high energy beam. Here is an example of the Tevatron at Fermi National Accelerator Laboratory (Fermilab, FNAL), Illinois, USA.

$$B_0 \simeq 4 \text{ T}, \quad E = 10^3 \text{ GeV}, \, P = 10^3 \text{ GeV}/c.$$

$$\rho = \frac{P}{B_0 q} = \frac{Pc}{qB_0 c} \simeq \frac{E}{qB_0 c} \simeq \frac{10^3 \times 10^9}{4 \cdot 3 \times 10^8} \simeq 800 \text{ m.}$$

For $n = 1/2$, we have,

$$\left| \frac{\partial B_y}{\partial x} \right| = n\frac{B_0}{\rho} \simeq \frac{1}{2}\frac{4}{800} = 2.5 \times 10^{-3} \text{ T/m,}$$

$$\Delta B_y|_{x=5cm} = 2.5 \times 10^{-3} \cdot 5 \times 10^{-2} = 1.25 \times 10^{-4} \text{ T.}$$

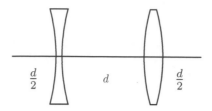

**FIGURE 9.1:**   Sketch of a FODO cell without bending magnets.

Yet for quadrupoles, $B|_{x=a} \sim$ 1–3 T. Thus we conclude that weak focusing at the 1 TeV of the Tevatron is four orders of magnitude weaker than strong focusing.

## 9.1   The FODO Cell

The most common form of strong focusing module is the so-called FODO cell, where two quadrupoles of opposite polarity (focusing (F) and defocusing (D)) are separated by drifts or homogeneous dipole magnets (O). Due to it simplicity, we can easily derive the transfer matrix of such a FODO cell. To simplify matters even further, we choose the center of the defocusing quadrupole as the start and the end of the cell and use the thin lens model of the quadrupole. First, let us consider a FODO cell without bending magnets, as shown in Fig. 9.1. The total transfer matrix of the horizontal plane is

$$
\hat{M}_x = \begin{pmatrix} 1 & 0 \\ 1/2f_2 & 1 \end{pmatrix}\begin{pmatrix} 1 & d \\ 0 & 1 \end{pmatrix}\begin{pmatrix} 1 & 0 \\ -1/f_1 & 1 \end{pmatrix}\begin{pmatrix} 1 & d \\ 0 & 1 \end{pmatrix}\begin{pmatrix} 1 & 0 \\ 1/2f_2 & 1 \end{pmatrix}
$$

$$
= \left[\begin{pmatrix} 1 & 0 \\ 1/2f_2 & 1 \end{pmatrix}\begin{pmatrix} 1 & d \\ 0 & 1 \end{pmatrix}\begin{pmatrix} 1 & 0 \\ -1/2f_1 & 1 \end{pmatrix}\right] \cdot \left[\begin{pmatrix} 1 & 0 \\ -1/2f_1 & 1 \end{pmatrix}\begin{pmatrix} 1 & d \\ 0 & 1 \end{pmatrix}\begin{pmatrix} 1 & 0 \\ 1/2f_2 & 1 \end{pmatrix}\right]
$$

$$
= \begin{pmatrix} 1-d/2f_1 & d \\ 1/2f_2-1/2f_1-d/4f_1f_2 & 1+d/2f_2 \end{pmatrix}\begin{pmatrix} 1+d/2f_2 & d \\ 1/2f_2-1/2f_1-d/4f_1f_2 & 1-d/2f_1 \end{pmatrix}
$$

$$
= \begin{pmatrix} 1-d/f_1+d/f_2-d^2/2f_1f_2 & 2d\left(1-d/2f_1\right) \\ 1/f_2-1/f_1-d/f_1f_2+d/2f_2^2-d^2/4f_1f_2^2 & 1-d/f_1+d/f_2-d^2/2f_1f_2 \end{pmatrix}.
$$

Hence we obtain the **transverse tunes**, which are

$$
\cos\left(\mu_x\right) = 1 - \frac{d}{f_1} + \frac{d}{f_2} - \frac{d^2}{2f_1f_2}, \quad \cos\left(\mu_y\right) = 1 + \frac{d}{f_1} - \frac{d}{f_2} - \frac{d^2}{2f_1f_2}.
$$

Note the second equation above is obtained through interchanging $f_1$ and $f_2$. The shaded area in Fig. 9.2 shows the range of $f_1$ and $f_2$ where motion in

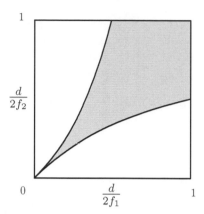

**FIGURE 9.2**: The "necktie" diagram showing the stability region of a FODO cell.

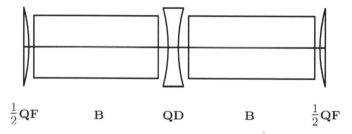

$\frac{1}{2}\mathbf{QF}$      **B**      **QD**      **B**      $\frac{1}{2}\mathbf{QF}$

**FIGURE 9.3**: Sketch of a FODO cell with bending magnets.

both planes are stable, i.e., $|1 - d/f_1 + d/f_2 - d^2/2f_1f_2| \leq 1$, $|1 + d/f_1 - d/f_2 - d^2/2f_1f_2| \leq 1$. Because of the shape of the region of stability, this and related similar figures are often referred to as a **necktie diagram**.

When $f = f_1 = f_2$, we have

$$\cos(\mu_x) = \cos(\mu_y) = 1 - \frac{d^2}{2f^2},$$

and the condition for stable motion is

$$\left|1 - \frac{d^2}{2f^2}\right| \leq 1 \Longrightarrow -1 \leq 1 - \frac{d^2}{2f^2} \leq 1 \Longrightarrow 0 \leq \frac{d^2}{2f^2} \leq 2 \Longrightarrow f \geq \frac{d}{2}.$$

From the geometrical point of view, when $f < d/2$, the particle that is parallel to the optical axis at the center of the defocusing quadrupole at the start of the cell would cross the axis before the center of the defocusing quadrupole at the end and bend further away from the axis. The transfer matrix for $f = d/2$ is $-\hat{I}$, so the maximum phase advance of a cell is $\pi$.

Next, let us consider a FODO cell with **bending magnets**, as shown in Fig. 9.3. Here we assume that the drifts between magnets are negligible. Furthermore, the bending angle $\theta$ is small. Since $l$ is nearly a constant, the transfer matrix of the dipole is

$$\hat{M} = \begin{pmatrix} \cos\theta & \rho\sin\theta & \rho\,(1-\cos\theta) \\ -(1/\rho)\sin\theta & \cos\theta & \sin\theta \\ 0 & 0 & 1 \end{pmatrix} \simeq \begin{pmatrix} 1 & l & l\theta/2 \\ 0 & 1 & \theta \\ 0 & 0 & 1 \end{pmatrix},$$

where $l = \rho\theta$ is the arc length of the dipole and $\theta/\rho = l/\rho^2 \ll f$  $\{l < 2f, \rho \gg f\}$.

Note that the factor $(1+\eta_0)/(2+\eta_0)$ does not appear due to the fact that $dp/p_0$ is used. At an energy that $\eta_0 \gg 1$, $(1+\eta_0)/(2+\eta_0) \cong 1$, and the difference between $dp/p_0$ and $dK/K_0$ becomes negligible. The matrix of the cell is

$$\hat{M}_x = \begin{pmatrix} 1 & 0 & 0 \\ -1/2f & 1 & 0 \\ 0 & l & 1 \end{pmatrix}\begin{pmatrix} 1 & l & l\theta/2 \\ 0 & 1 & \theta \\ 0 & 0 & 1 \end{pmatrix}\begin{pmatrix} 1 & 0 & 0 \\ 1/f & 1 & 0 \\ 0 & l & 1 \end{pmatrix}\begin{pmatrix} 1 & l & l\theta/2 \\ 0 & 1 & \theta \\ 0 & 0 & 1 \end{pmatrix}\begin{pmatrix} 1 & 0 & 0 \\ -1/2f & 1 & 0 \\ 0 & l & 1 \end{pmatrix}$$

$$= \begin{pmatrix} 1 - l^2/2f^2 & 2l\,(1+l/2f) & 2l\theta\,(1+l/4f) \\ -l/2f^2 + l^2/4f^3 & 1 - l^2/2f^2 & 2\theta\,(1 - l/4f - l^2/8f^2) \\ 0 & 0 & 1 \end{pmatrix}, \qquad (9.1)$$

and hence we obtain

$$\cos(\mu_x) = 1 - \frac{l^2}{2f^2}, \quad \alpha_x = 0, \quad \beta_x = \frac{2l\,[1+\sin(\mu_x/2)]}{\sin(\mu_x)}.$$

$$\alpha_x = 0 \Longrightarrow \beta_x' = 0, \text{ focusing at the ends} \Longrightarrow \beta_x = \beta_{x\,\mathrm{max}}.$$

Note that when QF and QD (see Fig. 9.3) have the same strength, there is symmetry between the two planes. The transfer matrix of the vertical plane can be obtained through changing the sign of the focusing, i.e., $(f- > -f)$. Furthermore, the transfer matrix of the cell that starts and ends at the centers of the defocusing quadrupoles can be obtained the same way. The results are summarized below.

$$\mu_y = \mu_x, \quad \beta_{y\,\mathrm{max}} = \beta_{z\,\mathrm{max}}, \quad \beta_{y\,\mathrm{min}} = \beta_{x\,\mathrm{min}} = \frac{2l\,[1-\sin(\mu/2)]}{\sin\mu}.$$

Note that $l/f$ determines $\mu$, and $l$ determines $\beta_{\mathrm{max}}$ and $\beta_{\mathrm{min}}$ when $\mu$ is fixed.

With the upper limit of the magnetic field $B$ of the dipole, which give the upper limit of $\theta$, thus the lower limit of the number of cells, the lower limit of the size of a ring can be obtained. (The upper limit of $B$ is roughly 1.5 T for warm (normal conducting) magnets and 8 T for superconducting ones.)

Similarly, we have

$$D_{\max} = \frac{l\theta\left[1 + (1/2)\sin(\mu/2)\right]}{\sin^2(\mu/2)}, \quad D_{\min} = \frac{l\theta\left[1 - (1/2)\sin(\mu/2)\right]}{\sin^2(\mu/2)},$$
$$D_F' = 0, \quad D_D' = 0,$$

where the subscripts $F$ and $D$ denote the centers of the focusing and defocusing quadrupoles, respectively. When $\mu = 90°$, we have

$$f = \frac{l}{\sqrt{2}},$$

$$\beta_{\max} = \left(2 + \sqrt{2}\right)l, \quad \beta_{\min} = \left(2 - \sqrt{2}\right)l, \quad \frac{\beta_{\max}}{\beta_{\min}} = 3 + 2\sqrt{2} \simeq 5.8,$$

$$D_{\max} = \frac{1}{2}\left(4 + \sqrt{2}\right)l\theta, \quad D_{\min} = \frac{1}{2}\left(4 - \sqrt{2}\right)l\theta.$$

As an example, let us consider the Main Injector FODO cell at Fermilab, which has the parameters

$$l = 17.2886 \text{ m}, \quad \mu = \frac{\pi}{2},$$

$$\beta_{\max} = \left(2 + \sqrt{2}\right)l \simeq 59.0 \text{ m}, \quad \beta_{\min} = \left(2 - \sqrt{2}\right)l \simeq 10.1 \text{ m},$$

and which is shown in Fig. 9.4 [33, 21]. The magnetic field and length at the momentum of 8.9 GeV/c are

$$B = 0.102 \text{ T}, \quad l_b = 6.096 \text{ m}, \quad \chi_m = 29.69 \text{ Tm}, \quad \rho = \frac{\chi_m}{B} = 291 \text{ m},$$

$$\theta = \frac{2l_b}{\rho} = 41.9 \text{ mrad (2 magnets per half cell)},$$

$$D_{\max} = \frac{1}{2}\left(4 + \sqrt{2}\right)l\theta = 1.96 \text{ m}, \quad D_{\min} = \frac{1}{2}\left(4 - \sqrt{2}\right)l\theta = 0.94 \text{ m}.$$

It is worth noting that the values of $\beta_{\max}$ and $D_{\max}$ are very close to those obtained from the exact model, which are 58.2 m and 1.95 m, respectively. This shows that the thin lens model is rather accurate for a typical FODO cell.

Now let us look at the size of the beam. The Fermilab Main Injector is designed to accelerate proton and anti-proton beams of emittance up to $40\pi$ mm mrad. Following the Fermilab convention, the emittance is defined as $\epsilon_N = 6\epsilon_{rms}\beta\gamma$, $(\beta = v/c, \gamma = 1/\sqrt{1 - \beta^2})$, where $\epsilon_{rms}$ is the *rms* area of phase space occupied by matched beam. The quantity $\epsilon_N$ is called the **normalized emittance**. The factor $\beta\gamma$ makes $\epsilon_N$ a constant through acceleration — note that $\epsilon_N \propto \epsilon_{rms}\beta\gamma \propto \beta\gamma\sqrt{\langle x^2 \rangle \langle a^2 \rangle - \langle xa \rangle^2} \propto \sqrt{\langle x^2 \rangle \langle p_x^2 \rangle - \langle xp_x \rangle^2}$. The factor 6 means that the size is $\sqrt{6}\sigma \simeq 2.45\sigma$, which contains $\sim 90\%$ particles. As a

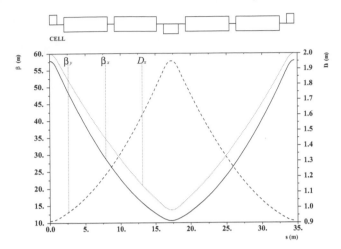

**FIGURE 9.4**:   Lattice functions of a FODO cell at the Fermilab Main Injector.

result, we obtain

$$x_{\max} = y_{\max} = \sqrt{\frac{\beta_{\max} \epsilon_N}{\beta \gamma}}.$$

At injection, with $p_0 = 8.9$ GeV/c, we have $\gamma = 9.54$, $\beta = 0.994$, so we have $x_{\max} = y_{\max} \simeq 16$ mm. For the horizontal beam size, we have to take into account the momentum spread. We assume $\delta p/p_0 \sim 0.3\%$ and have $x_D = D_{\max}\delta p/p_0 = 6$ mm. As a result the total horizontal beam size is $x_{\max}^T = \sqrt{x_{\max}^2 + x_D^2} \simeq 17$ mm. So at injection, the full beam is about 34 mm wide. At extraction, the momentum is 150 GeV/c, $\beta \simeq 1$ and $\gamma = 160$, so we have $x_{\max} = y_{\max} = 4$ mm. Since $\delta p/p_0$ scales with $\gamma$, the momentum spread becomes 0.02% and $x_D = 0.4$ mm. Thus, we have $x_{\max}^T \simeq 4$ mm, and the full beam at extraction is 8 mm wide. This is called adiabatic damping. With acceleration, all phase space variables scale the same way. As a result, the shape of a bunch does not change. It is illuminating to compare radiation damping and adiabatic damping. In the case of radiation damping, $p_0$ remains constant while $p_x$, $p_y$ and $\delta K$ decrease. (The radiated energy is recovered by the radio frequency (RF) cavity.) During adiabatic damping, $p_x$, $p_y$ and $\delta K$ remain unchanged, while $p_0$ increases. Both dipole and quadrupole magnets have to be ramped to keep the design closed orbit and tunes constant. The stronger force results in a smaller beam.

Finally, let us look at chromaticities for FODO cells.

$$
\begin{aligned}
\xi &= -\frac{1}{4\pi}\oint \beta(s)k_0(s)ds = -\frac{1}{4\pi}N\left(\frac{\beta_{\max}}{f}-\frac{\beta_{\min}}{f}\right) \\
&= -\frac{1}{4\pi}\frac{N}{f}\left(1+\sin(\tfrac{\mu}{2})-1+\sin(\tfrac{\mu}{2})\right)\frac{2l}{\sin\mu} \\
&= -\frac{N}{4\pi}\frac{4\sin(\mu/2)}{\sin\mu}2\sin(\tfrac{\mu}{2}) = -\frac{N}{\pi}\tan(\tfrac{\mu}{2}) = -\nu\frac{\tan(\mu/2)}{\mu/2}, \quad (\nu=\frac{N\mu}{2\pi}).
\end{aligned}
$$

For the Fermilab Main Injector,

$$
\mu_x = \frac{\pi}{2}, \quad \nu_x = 26.425 \implies \xi_x = -33.6,
$$

$$
\mu_y = \frac{\pi}{2}, \quad \nu_y = 25.415 \implies \xi_y = -32.4,
$$

which is not far from the exact values ($\xi_x = -33.6$ and $\xi_y = -33.9$) showing again the usefulness of the thin lens model. Without correction, and assuming the momentum spread of $\pm 1\%$, we have

$$
\Delta\nu_x = \xi_x\frac{\Delta p}{p_0} = \pm 33.6 \times 0.01 = \pm 0.336,
$$

$$
\Delta\nu_y = \xi_y\frac{\Delta p}{p_0} = \pm 32.4 \times 0.01 = \pm 0.324.
$$

Clearly, without chromaticity correction, the momentum acceptance of the ring would be very small (probably below 0.1% due to the fact that the distance between $\nu_x$ and the half-integer is only 0.075). To correct **chromaticity**, we place two sextupoles in each FODO cell, one next to the focusing quadrupole and the other next to the defocusing quadrupole. Using the thin lens model of the sextupoles and ignoring the distance between the sextupoles and their adjacent quadrupoles, we obtain the total chromaticities from eqs. (8.4) and (8.5)

$$
\begin{aligned}
\xi_x &= -\frac{1}{4\pi}\oint \beta(s)\left[k_x(s)-D_x(s)k_s(s)\right]ds \\
&= -\frac{1}{4\pi}N\left[\beta_{\max}\left(\frac{1}{f}-D_{\max}k_{sF}\right)+\beta_{\min}\left(-\frac{1}{f}-D_{\min}k_{sD}\right)\right], \\
\xi_y &= -\frac{1}{4\pi}\oint \beta(s)\left[k_x(s)+D_x(s)k_s(s)\right]ds \\
&= -\frac{1}{4\pi}N\left[\beta_{\min}\left(-\frac{1}{f}+D_{\max}k_{sF}\right)+\beta_{\max}\left(\frac{1}{f}+D_{\min}k_{sD}\right)\right],
\end{aligned}
$$

where $k_{sF}$ and $k_{sD}$ are integrated strengths of the sextupoles next to the focusing and defocusing quadrupoles, respectively. When

$$k_{sF} = \frac{1}{fD_{\max}} = \frac{\sin(\mu/2)}{2f^2\theta\,[1+\sin(\mu/2)/2]},$$

$$k_{sF} = -\frac{1}{fD_{\min}} = -\frac{\sin(\mu/2)}{2f^2\theta\,[1-\sin(\mu/2)/2]},$$

the chromaticities are corrected. Note that the relation $l = 2f\sin(\mu/2)$ is used to obtain the above expressions.

### 9.1.1 The FODO Cell Based Achromat

Achromats are needed because dispersion free straight sections are needed in both circular accelerators and beam transport lines. Achromatic sections are also required in circular machines, where, for example, straight sections that house injection and extraction kickers are dispersion free to make the beam small. The straight section where RF cavities are located is also dispersion free. Passing a RF cavity with $x$-$\delta$ correlation produces coupling between transverse and longitudinal motion, which is usually undesirable. In the case of beam transport lines, achromatic conditions have to be met when the matching requirement is such that the line is imaging or the line is isochronous.

There are mainly two types of achromats, those that utilize repetitive symmetry and those that use mirror symmetry. Let us consider a system that consists of $n$ identical cells.

$$\hat{M} = \begin{pmatrix} (x|x) & (x|a) & (x|\delta) \\ (a|x) & (a|a) & (a|\delta) \\ 0 & 0 & 1 \end{pmatrix} = \begin{pmatrix} \hat{R} & \vec{d} \\ 0 & 1 \end{pmatrix},$$

$$\hat{M}^2 = \begin{pmatrix} \hat{R} & \vec{d} \\ 0 & 1 \end{pmatrix}\begin{pmatrix} \hat{R} & \vec{d} \\ 0 & 1 \end{pmatrix} = \begin{pmatrix} \hat{R}^2 & (\hat{R}+\hat{I})\vec{d} \\ 0 & 1 \end{pmatrix},$$

$$\cdots$$

$$\hat{M}^n = \begin{pmatrix} \hat{R}^n & \left[\sum\limits_{k=0}^{n-1}\hat{R}^k\right]\vec{d} \\ 0 & 1 \end{pmatrix} = \begin{pmatrix} \hat{R}^n & (\hat{R}^n-\hat{I})/(\hat{R}-\hat{I})\cdot\vec{d} \\ 0 & 1 \end{pmatrix}.$$

When $\hat{R}^n = \hat{I}$, i.e., $\mu = (m/n)2\pi$,

$$\hat{M}^n = \begin{pmatrix} \hat{I} & \vec{0} \\ 0 & 1 \end{pmatrix},$$

which is an achromat.

For a stable FODO cell, which is what we are interested in, we can write the matrix $\hat{R}$ explicitly, which is

$$\hat{R} = \begin{pmatrix} \cos\mu + \alpha\sin\mu & \beta\sin\mu \\ -\gamma\sin\mu & \cos\mu - \alpha\sin\mu \end{pmatrix}.$$

In the normalized space,

$$\widetilde{R} = \hat{A}\hat{R}\hat{A}^{-1} = \begin{pmatrix} \cos\mu & \sin\mu \\ -\sin\mu & \cos\mu \end{pmatrix}$$

and

$$\widetilde{M} = \begin{pmatrix} \hat{A} & 0 \\ 0 & 1 \end{pmatrix} \begin{pmatrix} \hat{R} & \vec{d} \\ 0 & 1 \end{pmatrix} \begin{pmatrix} \hat{A}^{-1} & 0 \\ 0 & 1 \end{pmatrix} = \begin{pmatrix} \widetilde{R} & \hat{A}\vec{d} \\ 0 & 1 \end{pmatrix} = \begin{pmatrix} \widetilde{R} & \widetilde{\vec{d}} \\ 0 & 1 \end{pmatrix},$$

where

$$\widetilde{\vec{d}} = \hat{A}\vec{d} = \begin{pmatrix} d/\sqrt{\beta} \\ (\alpha d + \beta d')/\sqrt{\beta} \end{pmatrix}$$

with $d = (x|\delta)$ and $d' = (a|\delta)$. As a result, we have

$$\widetilde{\hat{R}}^k = \begin{pmatrix} \cos(k\mu) & \sin(k\mu) \\ -\sin(k\mu) & \cos(k\mu) \end{pmatrix},$$

$$\sum_{k=0}^{n-1}\hat{R}^k = \sum_{k=0}^{n-1}\begin{pmatrix} \cos(k\mu) & \sin(k\mu) \\ -\sin(k\mu) & \cos(k\mu) \end{pmatrix}$$

$$= \sum_{k=0}^{n-1}\begin{pmatrix} \left(e^{ik\mu} + e^{-ik\mu}\right)/2 & \left(e^{ik\mu} - e^{-ik\mu}\right)/2i \\ -\left(e^{ik\mu} - e^{-ik\mu}\right)/2i & \left(e^{ik\mu} + e^{-ik\mu}\right)/2 \end{pmatrix}$$

$$= \begin{pmatrix} (A+B)/2 & (A-B)/2i \\ -(A-B)/2i & (A+B)/2 \end{pmatrix},$$

where

$$A = \sum_{k=0}^{n-1}e^{ik\mu} = \frac{1 - e^{in\mu}}{1 - e^{i\mu}}, \quad B = \sum_{k=0}^{n-1}e^{-ik\mu} = \frac{1 - e^{-in\mu}}{1 - e^{-i\mu}}.$$

It is obvious that when $\mu = (m/n)2\pi$, we have $A = 0$ and $B = 0$, which leads to the achromatic condition.

Achromats of this kind have been used as the arcs of storage rings, beamlines and spectrometers. Examples are the 90° arcs of the South Hall Ring at the MIT-Bates Linear Accelerator Center at Massachusetts Institute of Technology, Massachusetts, USA, the 180° arcs of the storage ring at the Duke Free Electron Laser Laboratory (DFELL) at Duke University, North Carolina, USA, and the arcs of the ILC (International Linear Collider) Beam Delivery System at SLAC National Accelerator Laboratory, California, USA. There was a time-of-flight spectrometer built at Los Alamos National Laboratory, New Mexico, USA, that consists of four identical cells with $\mu = 90°$.

It turns out that this kind of system not only cancels dispersion, but also cancels all second order geometrical aberrations. Recalling the equations of motion (3.22), second order geometrical aberrations generated in the short interval $[s, s + ds]$ can be written as

$$x_f = x_i + hdsx_ia_i, \quad a_f = a_i + \sum T_{a,k,l,m,n}x_i^k a_i^l y_i^m b_i^n,$$

$$y_f = y_i + hdsx_ib_i, \quad b_f = b_i + \sum T_{b,k,l,m,n}x_i^k a_i^l y_i^m b_i^n,$$

where the summation $\sum$ is taken over $k, l, m, n$ from 0 to 2 such that $k + l + m + n = 2$; so in the above,

$$\sum \text{ reads as } \sum_{\substack{k,l,m,n=0 \\ k+l+m+n=2}}^{2} .$$

This simplified description is used in the rest of this section unless otherwise noted. Here the linear matrix from $s$ to $s+ds$ has been removed by the inverse matrix. As a result, the second order map over the interval is lumped into a point. Applying the transformation

$$\begin{pmatrix} \tilde{x} \\ \tilde{a} \\ \tilde{y} \\ \tilde{b} \end{pmatrix} = \hat{A} \begin{pmatrix} x \\ a \\ y \\ b \end{pmatrix} = \begin{pmatrix} 1/\sqrt{\beta_x} & 0 & 0 & 0 \\ \alpha_x/\sqrt{\beta_x} & \sqrt{\beta_x} & 0 & 0 \\ 0 & 0 & 1/\sqrt{\beta_y} & 0 \\ 0 & 0 & \alpha_y/\sqrt{\beta_y} & \sqrt{\beta_y} \end{pmatrix} \begin{pmatrix} x \\ a \\ y \\ b \end{pmatrix},$$

we obtain

$$\begin{pmatrix} \tilde{x}_f \\ \tilde{a}_f \\ \tilde{y}_f \\ \tilde{b}_f \end{pmatrix} = \begin{pmatrix} x/\sqrt{\beta_x} \\ (\alpha_x x + \beta_x a)/\sqrt{\beta_x} \\ y/\sqrt{\beta_y} \\ (\alpha_y y + \beta_y b)/\sqrt{\beta_y} \end{pmatrix} \circ \begin{pmatrix} x + hdsxa \\ a + \sum T_{a,k,l,m,n}x^k a^l y^m b^n \\ y + hdsxb \\ b + \sum T_{b,k,l,m,n}x^k a^l y^m b^n \end{pmatrix} \circ \begin{pmatrix} \sqrt{\beta_x}\tilde{x}_i \\ (-\alpha_x\tilde{x}_i + \tilde{a}_i)/\sqrt{\beta_x} \\ \sqrt{\beta_y}\tilde{y}_i \\ (-\alpha_y\tilde{y}_i + \tilde{b}_i)/\sqrt{\beta_y} \end{pmatrix}$$

$$= \begin{pmatrix} \tilde{x}_i + \left(hds/\sqrt{\beta_x}\right)\tilde{x}_i\left(-\alpha_x\tilde{x}_i + \tilde{a}_i\right) \\ \tilde{a}_i + \sum \tilde{T}_{a,k,l,m,n}\tilde{x}_i^k \tilde{a}_i^l \tilde{y}_i^m \tilde{b}_i^n \\ \tilde{y}_i + \left(\sqrt{\beta_x}hds/\beta_y\right)\tilde{x}_i\left(-\alpha_y\tilde{y}_i + \tilde{b}_i\right) \\ \tilde{b}_i + \sum \tilde{T}_{b,k,l,m,n}\tilde{x}_i^k \tilde{a}_i^l \tilde{y}_i^m \tilde{b}_i^n \end{pmatrix},$$

where $\tilde{T}_{a,k,l,m,n}$ and $\tilde{T}_{b,k,l,m,n}$ are linear combinations $T_{a,k,l,m,n}$ and $T_{b,k,l,m,n}$, each of which is multiplied by some powers of $\beta_x$, $\beta_y$, $\alpha_x$ and/or $\alpha_y$. Now let us consider a system that consists of $n$ identical cells with the phase advances $\mu_x = \mu_y = 2\pi/n$.

Defining

$$R\begin{pmatrix} \mu_x \\ \mu_y \end{pmatrix} = \begin{pmatrix} \tilde{R}(\mu_x) & 0 \\ 0 & \tilde{R}(\mu_x) \end{pmatrix} \begin{pmatrix} x \\ a \\ y \\ b \end{pmatrix},$$

the sum of all the $n$ kicks over the whole system is

$$
\begin{pmatrix} \widetilde{x}_f \\ \widetilde{a}_f \\ \widetilde{y}_f \\ \widetilde{b}_f \end{pmatrix} = \sum_{m=0}^{n-1} R \begin{pmatrix} (n-m)\,\mu_x - \phi_x \\ (n-m)\,\mu_y - \phi_y \end{pmatrix} \circ \begin{pmatrix} \left(hds/\sqrt{\beta_x}\right) x\left(-\alpha_x x + a\right) \\ \sum \widetilde{T}_{a,k,l,m,n} x^k a^l y^m b^n \\ \left(\sqrt{\beta_x} hds/\beta_y\right) x\left(-\alpha_y y + b\right) \\ \sum \widetilde{T}_{b,k,l,m,n} x^k a^l y^m b^n \end{pmatrix}
$$

$$
\circ\, R \begin{pmatrix} m\mu_x + \phi_x \\ m\mu_y + \phi_y \end{pmatrix}
$$

$$
= \sum_{m=0}^{n-1} R \begin{pmatrix} -2m\pi/n - \phi_x \\ -2m\pi/n - \phi_y \end{pmatrix} \circ \begin{pmatrix} \left(hds/\sqrt{\beta_x}\right) x\left(-\alpha_x x + a\right) \\ \sum \widetilde{T}_{a,k,l,m,n} x^k a^l y^m b^n \\ \left(\sqrt{\beta_x} hds/\beta_y\right) x\left(-\alpha_y y + b\right) \\ \sum \widetilde{T}_{b,k,l,m,n} x^k a^l y^m b^n \end{pmatrix}
$$

$$
\circ\, R \begin{pmatrix} 2m\pi/n + \phi_x \\ 2m\pi/n + \phi_y \end{pmatrix}.
$$

The last step uses the fact that $n\mu_x = n\mu_y = 2\pi$. Every term in the above expression can be written in the form of

$$
\sum_{m=0}^{n-1} \cos^l\left(\frac{2m\pi}{n} + \phi_j\right) \sin^{3-l}\left(\frac{2m\pi}{n} + \phi_k\right),
$$

where $l = 0, 1, 2, 3$ and $j, k = \{x, y\}$. Using the relations

$$
\cos\phi = \frac{e^{i\phi} + e^{-i\phi}}{2}, \qquad \sin\phi = \frac{e^{i\phi} - e^{-i\phi}}{2i},
$$

we have

$$
\sum_{m=0}^{n-1} \cos^l\left(\frac{2m\pi}{n} + \phi_j\right) \sin^{3-l}\left(\frac{2m\pi}{n} + \phi_k\right)
$$

$$
= \sum_{m=0}^{n-1} \left[\frac{e^{i(2m\pi/n+\phi_j)} + e^{-i(2m\pi/n+\phi_j)}}{2}\right]^l \left[\frac{e^{i(2m\pi/n+\phi_j)} - e^{-i(2m\pi/n+\phi_j)}}{2i}\right]^{3-l}.
$$

Dropping the common parts in each sum, which are functions of $\phi_x$ and $\phi_y$, there are only four kinds of sums

$$
\sum_{m=0}^{n-1} e^{i6m\pi/n} = \frac{1 - e^{in6\pi/n}}{1 - e^{i6\pi/n}} = 0, \qquad \sum_{m=0}^{n-1} e^{-i6m\pi/n} = \frac{1 - e^{-in6\pi/n}}{1 - e^{-i6\pi/n}} = 0,
$$

$$
\sum_{m=0}^{n-1} e^{i2m\pi/n} = \frac{1 - e^{in2\pi/n}}{1 - e^{i2\pi/n}} = 0, \qquad \sum_{m=0}^{n-1} e^{-i2m\pi/n} = \frac{1 - e^{-in2\pi/n}}{1 - e^{-i2\pi/n}} = 0,
$$

when $n \neq 1, 3$. In conclusion, for a system that consists of $n$ identical cells ($n > 1$, $n \neq 3$) and $\mu_x = \mu_y = 2\pi/n$, all second order geometrical aberrations vanish. Note that this is true even for systems without midplane symmetry. When coupling is present, the second order kicks for $x$ and $y$ will become more complicated but remain a polynomial of the second order. The transformation to the normalized coordinates $\hat{A}$ will be coupled as well, yet the general form of the kicks in the normalized space remains unchanged. Therefore the same proof holds.

Another result of such a system is that some chromatic aberrations are canceled. Of all the remaining chromatic terms, only two are independent. With two families of sextupoles, all second order chromatic aberrations can be corrected. Thus we obtain a system that is free of all aberrations up to the second order, which is called a second order achromat. From the symplectic condition, we know that, up to the second order, the path length depends on $\delta$ only.

Next, we are going to prove that only two independent families of chromatic terms are left. Going back to the equations of motion, the second order chromatic terms are

$$x_f = x_i + T_{x,a\delta} a_i \delta, \quad a_f = a_i + T_{a,x\delta} x_i \delta + T_{a,\delta^2} \delta^2,$$
$$y_f = y_i + T_{y,b\delta} b_i\, \delta, \quad b_f = b_i + T_{b,y\delta}\, y_i \delta.$$

In the normalized space, the map becomes

$$\begin{pmatrix} \tilde{x}_f \\ \tilde{a}_f \\ \tilde{y}_f \\ \tilde{b}_f \end{pmatrix} = \begin{pmatrix} x/\sqrt{\beta_x} \\ (\alpha_x x + \beta_x a)/\sqrt{\beta_x} \\ y/\sqrt{\beta_y} \\ (\alpha_y y + \beta_y b)/\sqrt{\beta_y} \end{pmatrix} \circ \begin{pmatrix} x + T_{x,a\delta} a\delta \\ a + T_{a,x\delta} x\delta + T_{a,\delta^2}\delta^2 \\ y + T_{y,b\delta} b\delta \\ b + T_{b,y\delta} y\delta \end{pmatrix} \circ \begin{pmatrix} \sqrt{\beta_x}\tilde{x}_i \\ (-\alpha_x \tilde{x}_i + \tilde{a}_i)/\sqrt{\beta_x} \\ \sqrt{\beta_y}\tilde{y}_i \\ \left(-\alpha_y \tilde{y}_i + \tilde{b}_i\right)/\sqrt{\beta_y} \end{pmatrix}$$

$$= \begin{pmatrix} x/\sqrt{\beta_x} \\ (\alpha_x x + \beta_x a)/\sqrt{\beta_x} \\ y/\sqrt{\beta_y} \\ (\alpha_y y + \beta_y b)/\sqrt{\beta_y} \end{pmatrix} \circ \begin{pmatrix} \sqrt{\beta_x}\tilde{x}_i + T_{x,a\delta}\left[(-\alpha_x\tilde{x}_i + \tilde{a}_i)/\sqrt{\beta_x}\right]\delta \\ (-\alpha_x\tilde{x}_i + \tilde{a}_i)/\sqrt{\beta_x} + T_{a,x\delta}\sqrt{\beta_x}\tilde{x}_i\delta + T_{a,\delta^2}\delta^2 \\ \sqrt{\beta_y}\tilde{y}_i + T_{y,b\delta}\left[\left(-\alpha_y\tilde{y}_i + \tilde{b}_i\right)/\sqrt{\beta_y}\right]\delta \\ \left(-\alpha_y\tilde{y}_i + \tilde{b}_i\right)/\sqrt{\beta_y} + T_{b,y\delta}\sqrt{\beta_y}\tilde{y}_i\delta \end{pmatrix}$$

$$= \begin{pmatrix} \tilde{x}_i + T_{x,a\delta}\left[(-\alpha_x\tilde{x}_i + \tilde{a}_i)/\beta_x\right]\delta \\ \tilde{a}_i + \left[(\beta_x^2 T_{a,x\delta} - \alpha_x^2 T_{x,a\delta})/\beta_x\right]\tilde{x}_i\delta + (\alpha_x/\beta_x)T_{x,a\delta}\tilde{a}_i\delta + \sqrt{\beta_x}T_{a,\delta^2}\delta^2 \\ \tilde{y}_i + T_{y,b\delta}\left[\left(-\alpha_y\tilde{y}_i + \tilde{b}_i\right)/\beta_y\right]\delta \\ \tilde{b}_i + \left[(\beta_y^2 T_{b,y\delta} - \alpha_y^2 T_{y,b\delta})/\beta_y\right]\tilde{y}_i\delta + (\alpha_y/\beta_y)T_{y,b\delta}\tilde{b}_i\delta \end{pmatrix}.$$

The sum over the system is

$$
\begin{pmatrix} \widetilde{x}_f \\ \widetilde{a}_f \\ \widetilde{y}_f \\ \widetilde{b}_f \end{pmatrix} = \sum_{m=0}^{n-1} R \begin{pmatrix} -2m\pi/n - \phi_x \\ -2m\pi/n - \phi_y \end{pmatrix}
$$

$$
\circ \begin{pmatrix} T_{x,a\delta}\left[(-\alpha_x x + a)/\beta_x\right]\delta \\ \left[(\beta_x^2 T_{a,x\delta} - \alpha_x^2 T_{x,a\delta})/\beta_x\right] x\delta + (\alpha_x/\beta_x)T_{x,a\delta}a\delta + \sqrt{\beta_x}\,T_{a,\delta^2}\delta^2 \\ T_{y,b\delta}\left[(-\alpha_y y + b)/\beta_y\right]\delta \\ \left[(\beta_y^2 T_{b,y\delta} - \alpha_y^2 T_{y,b\delta})/\beta_y\right] y\delta + (\alpha_y/\beta_y)\,T_{y,b\delta}b\delta \end{pmatrix}
$$

$$
\circ R \begin{pmatrix} 2m\pi/n + \phi_x \\ 2m\pi/n + \phi_y \end{pmatrix}.
$$

Since the $x$ and $y$ planes are decoupled, we can separate them. The $x$ plane is

$$
\begin{pmatrix} \widetilde{x}_f \\ \widetilde{a}_f \end{pmatrix} = \sum_{m=0}^{n-1} \begin{pmatrix} \cos(2m\pi/n + \phi_x) & -\sin(2m\pi/n + \phi_x) \\ \sin(2m\pi/n + \phi_x) & \cos(2m\pi/n + \phi_x) \end{pmatrix}
$$

$$
\cdot \begin{pmatrix} -(\alpha_x/\beta_x)\,T_{x,a\delta}\delta & (1/\beta_x)\,T_{x,a\delta}\delta \\ \left[(\beta_x^2 T_{a,x\delta} - \alpha_x^2 T_{x,a\delta})/\beta_x\right]\delta & (\alpha_x/\beta_x)\,T_{x,a\delta}\delta \end{pmatrix}
$$

$$
\cdot \begin{pmatrix} \cos(2m\pi/n + \phi_x) & \sin(2m\pi/n + \phi_x) \\ -\sin(2m\pi/n + \phi_x) & \cos(2m\pi/n + \phi_x) \end{pmatrix} \begin{pmatrix} \widetilde{x}_i \\ \widetilde{a}_i \end{pmatrix}
$$

$$
+ \sum_{m=0}^{n-1} \begin{pmatrix} \cos(2m\pi/n + \phi_x) & -\sin(2m\pi/n + \phi_x) \\ \sin(2m\pi/n + \phi_x) & \cos(2m\pi/n + \phi_x) \end{pmatrix} \begin{pmatrix} 0 \\ \sqrt{\beta_x}T_{a,\delta^2}\delta^2 \end{pmatrix},
$$

and the $y$ plane is

$$
\begin{pmatrix} \widetilde{y}_f \\ \widetilde{b}_f \end{pmatrix} = \sum_{m=0}^{n-1} \begin{pmatrix} \cos(2m\pi/n + \phi_y) & -\sin(2m\pi/n + \phi_y) \\ \sin(2m\pi/n + \phi_y) & \cos(2m\pi/n + \phi_y) \end{pmatrix}
$$

$$
\cdot \begin{pmatrix} -(\alpha_y/\beta_y)\,T_{y,b\delta}\delta & (1/\beta_y)\,T_{y,b\delta}\delta \\ \left[(\beta_y^2 T_{b,y\delta} - \alpha_y^2 T_{y,b\delta})/\beta_y\right]\delta & (\alpha_y/\beta_y)\,T_{y,b\delta}\delta \end{pmatrix}
$$

$$
\cdot \begin{pmatrix} \cos(2m\pi/n + \phi_y) & \sin(2m\pi/n + \phi_y) \\ -\sin(2m\pi/n + \phi_y) & \cos(2m\pi/n + \phi_y) \end{pmatrix} \begin{pmatrix} \widetilde{y}_i \\ \widetilde{b}_i \end{pmatrix}.
$$

The terms $(x|\delta^2)$ and $(a|\delta^2)$ are

$$
\sum_{m=0}^{n-1} \sin\left(\frac{2m\pi}{n} + \phi_x\right) = \sum_{m=0}^{n-1} \frac{e^{i(2m\pi/n + \phi_x)} - e^{-i(2m\pi/n + \phi_x)}}{2i}
$$

$$
= \frac{1}{2i}\left(e^{i\phi_x}\frac{1 - e^{in2\pi/n}}{1 - e^{i2\pi/n}} - e^{-i\phi_x}\frac{1 - e^{-in2\pi/n}}{1 - e^{-i2\pi/n}}\right) = 0,
$$

and

$$\sum_{m=0}^{n-1} \cos\left(\frac{2m\pi}{n} + \phi_x\right) = \sum_{m=0}^{n-1} \frac{e^{i(2m\pi/n+\phi_x)} + e^{-i(2m\pi/n+\phi_x)}}{2}$$

$$= \frac{1}{2}\left(e^{i\phi_x}\frac{1 - e^{in2\pi/n}}{1 - e^{i2\pi/n}} + e^{-i\phi_x}\frac{1 - e^{-in2\pi/n}}{1 - e^{-i2\pi/n}}\right) = 0.$$

Every other term in the above expression can be written in the form of

$$\sum_{m=0}^{n-1} \cos^l\left(\frac{2m\pi}{n} + \phi_j\right) \sin^{2-l}\left(\frac{2m\pi}{n} + \phi_j\right)$$

$$= \sum_{m=0}^{n-1} \left[\frac{e^{i(2m\pi/n+\phi_j)} + e^{-i(2m\pi/n+\phi_j)}}{2}\right]^l \left[\frac{e^{i(2m\pi/n+\phi_j)} - e^{-i(2m\pi/n+\phi_j)}}{2i}\right]^{2-l},$$

where $l = 0, 1, 2, 2$ and $j = \{x, y\}$. Dropping the common parts in each sum, which are functions of $\phi_j$, there are only three kinds of sums

$$\sum_{m=0}^{n-1} e^{i4m\pi/n} = \frac{1 - e^{in4\pi/n}}{1 - e^{i4\pi/n}}, \quad \sum_{m=0}^{n-1} e^{-i4m\pi/n} = \frac{1 - e^{-in4\pi/n}}{1 - e^{-i4\pi/n}}, \quad \sum_{m=0}^{n-1} 1 = n,$$

where $\sum_{m=0}^{n-1} e^{i4m\pi/n} = 0$ and $\sum_{m=0}^{n-1} e^{-i4m\pi/n} = 0$ when $n \neq 2$. As a result, we have

$$\sum_{m=0}^{n-1} \cos^2\left(\frac{2m\pi}{n} + \phi_j\right) = \frac{n}{2}, \quad \sum_{m=0}^{n-1} \sin^2\left(\frac{2m\pi}{n} + \phi_j\right) = \frac{n}{2},$$

$$\sum_{m=0}^{n-1} \sin\left(\frac{2m\pi}{n} + \phi_j\right) \cos\left(\frac{2m\pi}{n} + \phi_j\right) = 0.$$

The remaining terms are

$$\begin{pmatrix} \tilde{x}_f \\ \tilde{a}_f \end{pmatrix} = \frac{n}{2}\begin{pmatrix} 0 & C_x \\ -C_x & 0 \end{pmatrix}\begin{pmatrix} \tilde{x}_i \\ \tilde{a}_i \end{pmatrix}, \quad \begin{pmatrix} \tilde{y}_f \\ b_f \end{pmatrix} = \frac{n}{2}\begin{pmatrix} 0 & C_y \\ -C_y & 0 \end{pmatrix}\begin{pmatrix} \tilde{y}_i \\ b_i \end{pmatrix},$$

where

$$C_x = \frac{1}{\beta_x}T_{x,a\delta}\delta - \frac{\beta_x^2 T_{a,x\delta} - \alpha_x^2 T_{x,a\delta}}{\beta_x}\delta, \quad C_y = \frac{1}{\beta_y}T_{y,b\delta}\delta - \frac{\beta_y^2 T_{b,y\delta} - \alpha_y^2 T_{y,b\delta}}{\beta_y}\delta,$$

which shows that there are only two independent terms.

It turns out that there is another way to prove this point which is probably more elegant. We first observe that, from the equations of motion (3.22), the second order map of the $n$ cell achromatic system in the normalized coordinates can be written as

$$\begin{pmatrix} \tilde{x}_f \\ \tilde{a}_f \\ \tilde{y}_f \\ \tilde{b}_f \end{pmatrix} = \begin{pmatrix} \tilde{x}_i + \tilde{T}_{x,x\delta}\tilde{x}_i\delta + \tilde{T}_{x,a\delta}\tilde{a}_i\delta + \tilde{T}_{x,\delta^2}\delta^2 \\ \tilde{a}_i + \tilde{T}_{a,x\delta}\tilde{x}_i\delta + \tilde{T}_{a,a\delta}\tilde{a}_i\delta + \tilde{T}_{a,\delta^2}\delta^2 \\ \tilde{y}_i + \tilde{T}_{y,y\delta}\tilde{y}_i\delta + \tilde{T}_{y,b\delta}\tilde{b}_i\delta \\ \tilde{b}_i + \tilde{T}_{b,y\delta}\tilde{y}_i\delta + \tilde{T}_{b,b\delta}\tilde{b}_i\delta \end{pmatrix}, \tag{9.2}$$

where all geometrical terms vanish. Note that midplane symmetry is obeyed. Since the $x$ and $y$ planes are decoupled, let us study the $x$ plane first. Let us denote the second order map of the whole system

$$\mathcal{M}(n) = \mathcal{I} + \mathcal{T}(n),$$

that of one cell is

$$\mathcal{M}(1) = \mathcal{R}(1) + \mathcal{T}(1),$$

and that of $n-1$ cells is

$$\mathcal{M}(n-1) = \mathcal{R}(n-1) + \mathcal{T}(n-1).$$

Since the $n$ cells are identical, the whole system can be viewed as either one cell in front of $n-1$ cells or vice versa. Thus the following relations hold

$$\begin{aligned} \mathcal{M}(n) &=_2 \mathcal{M}(n-1) \circ \mathcal{M}(1) =_2 [\mathcal{R}(n-1) + \mathcal{T}(n-1)] \circ [\mathcal{R}(1) + \mathcal{T}(1)] \\ &=_2 \mathcal{I} + \mathcal{R}(n-1) \circ \mathcal{T}(1) + \mathcal{T}(n-1) \circ \mathcal{R}(1), \end{aligned}$$

and

$$\begin{aligned} \mathcal{M}(n) &=_2 \mathcal{M}(1) \circ \mathcal{M}(n-1) =_2 [\mathcal{R}(1) + \mathcal{T}(1)] \circ [\mathcal{R}(n-1) + \mathcal{T}(n-1)] \\ &=_2 \mathcal{I} + \mathcal{R}(1) \circ \mathcal{T}(n-1) + \mathcal{T}(1) \circ \mathcal{R}(n-1). \end{aligned}$$

Removing the first order part, we obtain

$$\mathcal{T}(n) = \mathcal{R}(n-1) \circ \mathcal{T}(1) + \mathcal{T}(n-1) \circ \mathcal{R}(1),$$

and

$$\mathcal{T}(n) = \mathcal{R}(1) \circ \mathcal{T}(n-1) + \mathcal{T}(1) \circ \mathcal{R}(n-1).$$

Furthermore, we obtain

$$\mathcal{T}(n) \circ \mathcal{R}(1)^{-1} = \mathcal{R}(n-1) \circ \mathcal{T}(1) \circ \mathcal{R}(1)^{-1} + \mathcal{T}(n-1),$$

and

$$\mathcal{R}(1)^{-1} \circ \mathcal{T}(n) = \mathcal{T}(n-1) + \mathcal{R}(1)^{-1} \circ \mathcal{T}(1) \circ \mathcal{R}(n-1).$$

Using the relation $\mathcal{R}(n-1) = \mathcal{R}(1)^{-1}$, we reach the following relation

$$\mathcal{T}(n) \circ \mathcal{R}(1) = \mathcal{R}(1) \circ \mathcal{T}(n).$$

Plugging in the first two components of eq. (9.2), we arrive at

$$
\begin{pmatrix} \cos\mu_x & \sin\mu_x \\ -\sin\mu_x & \cos\mu_x \end{pmatrix} \begin{pmatrix} \tilde{T}_{x,\delta^2}\delta^2 \\ \tilde{T}_{a,\delta^2}\delta^2 \end{pmatrix} = \begin{pmatrix} \tilde{T}_{x,\delta^2}\delta^2 \\ \tilde{T}_{a,\delta^2}\delta^2 \end{pmatrix}
$$

and

$$
\begin{pmatrix} \tilde{T}_{x,x\delta} & \tilde{T}_{x,a\delta} \\ \tilde{T}_{a,x\delta} & \tilde{T}_{a,a\delta} \end{pmatrix} \begin{pmatrix} \cos\mu_x & \sin\mu_x \\ -\sin\mu_x & \cos\mu_x \end{pmatrix} = \begin{pmatrix} \cos\mu_x & \sin\mu_x \\ -\sin\mu_x & \cos\mu_x \end{pmatrix} \begin{pmatrix} \tilde{T}_{x,x\delta} & \tilde{T}_{x,a\delta} \\ \tilde{T}_{a,x\delta} & \tilde{T}_{a,a\delta} \end{pmatrix}.
$$

For the pure chromatic terms, we have

$$
\begin{pmatrix} 1 - \cos\mu_x & -\sin\mu_x \\ \sin\mu_x & 1 - \cos\mu_x \end{pmatrix} \begin{pmatrix} \tilde{T}_{x,\delta^2} \\ \tilde{T}_{a,\delta^2} \end{pmatrix} = \begin{pmatrix} 0 \\ 0 \end{pmatrix}.
$$

Since, for $n > 1$, we have

$$
\det \begin{pmatrix} 1 - \cos\mu_x & -\sin\mu_x \\ \sin\mu_x & 1 - \cos\mu_x \end{pmatrix} = 2\left(1 - \cos\mu_x\right) \neq 0,
$$

we conclude that

$$
\tilde{T}_{x,\delta^2} = 0, \quad \tilde{T}_{a,\delta^2} = 0, \quad \text{for} \quad n > 1.
$$

For the mixed chromatic terms, we have

$$
\begin{pmatrix} \tilde{T}_{x,x\delta}\cos\mu_x - \tilde{T}_{x,a\delta}\sin\mu_x & \tilde{T}_{x,x\delta}\sin\mu_x + \tilde{T}_{x,a\delta}\cos\mu_x \\ \tilde{T}_{a,x\delta}\cos\mu_x - \tilde{T}_{a,a\delta}\sin\mu_x & \tilde{T}_{a,x\delta}\sin\mu_x + \tilde{T}_{a,a\delta}\cos\mu_x \end{pmatrix}
$$
$$
= \begin{pmatrix} \tilde{T}_{x,x\delta}\cos\mu_x + \tilde{T}_{a,x\delta}\sin\mu_x & \tilde{T}_{x,a\delta}\cos\mu_x + \tilde{T}_{a,a\delta}\sin\mu_x \\ -\tilde{T}_{x,x\delta}\sin\mu_x + \tilde{T}_{a,x\delta}\cos\mu_x & -\tilde{T}_{x,a\delta}\sin\mu_x + \tilde{T}_{a,a\delta}\cos\mu_x \end{pmatrix}.
$$

For each component, we have

$$
-\tilde{T}_{x,a\delta}\sin\mu_x = \tilde{T}_{a,x\delta}\sin\mu_x, \quad \tilde{T}_{x,x\delta}\sin\mu_x = \tilde{T}_{a,a\delta}\sin\mu_x,
$$
$$
-\tilde{T}_{a,a\delta}\sin\mu_x = -\tilde{T}_{x,x\delta}\sin\mu_x, \quad \tilde{T}_{a,x\delta}\sin\mu_x = -\tilde{T}_{x,a\delta}\sin\mu_x.
$$

For $n \neq 2$, $\sin\mu_x \neq 0$. We have

$$
\tilde{T}_{x,x\delta} - \tilde{T}_{a,a\delta} = 0, \quad \tilde{T}_{x,a\delta} + \tilde{T}_{a,x\delta} = 0.
$$

From symplectic symmetry, we have

$$
\det \begin{pmatrix} 1 + \tilde{T}_{x,x\delta} & \tilde{T}_{x,a\delta} \\ \tilde{T}_{a,x\delta} & 1 + \tilde{T}_{a,a\delta} \end{pmatrix} = 1
$$

up to the first order of $\delta$. As a result, we have

$$\tilde{T}_{x,x\delta} + \tilde{T}_{a,a\delta} = 0,$$

which, combined with the previous result, leads to

$$\tilde{T}_{x,x\delta} = 0, \quad \tilde{T}_{a,a\delta} = 0.$$

In the original space,

$$\begin{pmatrix} T_{x,\delta^2} \\ T_{a,\delta^2} \end{pmatrix} = \hat{A}^{-1} \begin{pmatrix} \tilde{T}_{x,\delta^2} \\ \tilde{T}_{a,\delta^2} \end{pmatrix} = \begin{pmatrix} 0 \\ 0 \end{pmatrix},$$

and

$$\begin{aligned}
\begin{pmatrix} T_{x,x\delta} & T_{x,a\delta} \\ T_{a,x\delta} & T_{a,a\delta} \end{pmatrix} &= \hat{A}^{-1} \begin{pmatrix} \tilde{T}_{x,x\delta} & \tilde{T}_{x,a\delta} \\ \tilde{T}_{a,x\delta} & \tilde{T}_{a,a\delta} \end{pmatrix} \hat{A} \\
&= \begin{pmatrix} \sqrt{\beta_x} & 0 \\ -\alpha_x/\sqrt{\beta_x} & 1/\sqrt{\beta_x} \end{pmatrix} \begin{pmatrix} 0 & \tilde{T}_{x,a\delta} \\ -\tilde{T}_{x,a\delta} & 0 \end{pmatrix} \begin{pmatrix} 1/\sqrt{\beta_x} & 0 \\ \alpha_x/\sqrt{\beta_x} & \sqrt{\beta_x} \end{pmatrix} \\
&= \begin{pmatrix} \alpha_x & \beta_x \\ -\gamma_x & -\alpha_x \end{pmatrix} \tilde{T}_{x,a\delta}.
\end{aligned}$$

The Twiss parameters here are the periodic solution of the cell. In conclusion, there is only one independent chromatic aberration in the $x$ plane. The same conclusion can be reached for the $y$ plane following the same procedure. This proves, from the global point of view, that only two second order chromatic aberrations are independent.

Furthermore, it is easy to show that the remaining terms $\tilde{T}_{x,a\delta}$ and $\tilde{T}_{y,b\delta}$ are simply the chromaticities. We can write the transfer matrix in the normalized space as

$$\begin{aligned}
\hat{M}_x &= \begin{pmatrix} 1 & \tilde{T}_{x,a\delta}\delta \\ -\tilde{T}_{x,a\delta}\delta & 1 \end{pmatrix} = \begin{pmatrix} \cos(2\pi) & \sin(2\pi) + \tilde{T}_{x,a\delta}\delta \\ -\sin(2\pi) - \tilde{T}_{x,a\delta}\delta & \cos(2\pi) \end{pmatrix} \\
&=_1 \begin{pmatrix} \cos\left(2\pi + \tilde{T}_{x,a\delta}\delta\right) & \sin\left(2\pi + \tilde{T}_{x,a\delta}\delta\right) \\ -\sin\left(2\pi + \tilde{T}_{x,a\delta}\delta\right) & \cos\left(2\pi + \tilde{T}_{x,a\delta}\delta\right) \end{pmatrix},
\end{aligned}$$

and we obtain

$$\xi_x = \tilde{T}_{x,a\delta}.$$

Similarly,

$$\hat{M}_y =_1 \begin{pmatrix} \cos\left(2\pi + \tilde{T}_{y,b\delta}\delta\right) & \sin\left(2\pi + \tilde{T}_{y,b\delta}\delta\right) \\ -\sin\left(2\pi + \tilde{T}_{y,b\delta}\delta\right) & \cos\left(2\pi + \tilde{T}_{y,b\delta}\delta\right) \end{pmatrix},$$

and
$$\xi_y = \tilde{T}_{y,b\delta}.$$

The simplest of such a second order achromat consists of four FODO cells with $\mu_x = \mu_y = \pi/2$ and two families of sextupoles correcting the chromaticities.

## 9.1.2    The Dispersion Suppressor

As it become clear shortly, an achromat of $n$ identical cells is not optimal in terms of minimizing $D_{\max}$. Let us consider half of an achromat,

$$\hat{M} = \begin{pmatrix} -\hat{I} & \vec{d} \\ 0 & 1 \end{pmatrix}.$$

For an $n$ cell achromat, $D$, $D'$ at the center are

$$\begin{pmatrix} D \\ D' \\ 1 \end{pmatrix} = \begin{pmatrix} -1 & 0 & d \\ 0 & -1 & d' \\ 0 & 0 & 1 \end{pmatrix} \begin{pmatrix} 0 \\ 0 \\ 1 \end{pmatrix} = \begin{pmatrix} d \\ d' \\ 1 \end{pmatrix},$$

whereas the periodic solution is

$$\begin{pmatrix} D \\ D' \\ 1 \end{pmatrix} = \begin{pmatrix} -1 & 0 & d \\ 0 & -1 & d' \\ 0 & 0 & 1 \end{pmatrix} \begin{pmatrix} D \\ D' \\ 1 \end{pmatrix}.$$

$$\Longrightarrow D = \frac{d}{2}, \quad D' = \frac{d'}{2}.$$

Obviously the dispersion at the center of the achromat is twice that of the periodic solution of a cell. There is a module called dispersion suppressor which makes the whole section an achromat while maintaining the periodic solution in the regular cells. It takes advantage of the fact that dipoles, especially when $\theta$ is small, affect only the dispersion, not focusing. A dispersion suppressor consists of two FODO cells which are the same as the standard cells except for the bending angle. Two free parameters, or "knobs," such as the the bending angles can fulfill the two conditions needed to obtain an achromat.

Recalling eq. (9.1), the transfer matrix of a FODO cell with bending is

$$\hat{M}_x = \begin{pmatrix} 1 - l^2/2f^2 & 2l\,(1+l/2f) & 2l\theta\,(1+l/4f) \\ -l/2f^2 + l^2/4f^3 & 1 - l^2/2f^2 & 2\theta\,(1 - l/4f - l^2/8f^2) \\ 0 & 0 & 1 \end{pmatrix}$$

$$= \begin{pmatrix} \cos\mu & \beta\sin\mu & 2l\,[1 + (1/2)\sin(\mu/2)]\,\theta \\ -(1/\beta)\sin\mu & \cos\mu & 2\,[1 + (1/2)\sin(\mu/2)]\,[1 - \sin(\mu/2)]\,\theta \\ 0 & 0 & 1 \end{pmatrix}.$$

For two cells of bending $\theta_1$ and $\theta_2$ per half cell, the total matrix is

$$\hat{M}_x = \begin{pmatrix} \cos\mu & \beta\sin\mu & 2l\left[1+(1/2)\sin\left(\mu/2\right)\right]\theta_2 \\ -(1/\beta)\sin\mu & \cos\mu & 2\left[1+(1/2)\sin\left(\mu/2\right)\right]\left[1-\sin\left(\mu/2\right)\right]\theta_2 \\ 0 & 0 & 1 \end{pmatrix}$$

$$\cdot \begin{pmatrix} \cos\mu & \beta\sin\mu & 2l\left[1+(1/2)\sin\left(\mu/2\right)\right]\theta_1 \\ -(1/\beta)\sin\mu & \cos\mu & 2\left[1+(1/2)\sin\left(\mu/2\right)\right]\left[1-\sin\left(\mu/2\right)\right]\theta_1 \\ 0 & 0 & 1 \end{pmatrix}$$

$$= \begin{pmatrix} \cos\left(2\mu\right) & \beta\sin\left(2\mu\right) & d \\ -(1/\beta)\sin\left(2\mu\right) & \cos\left(2\mu\right) & d' \\ 0 & 0 & 1 \end{pmatrix},$$

where

$$d = 2l\left[1+\frac{1}{2}\sin\left(\frac{\mu}{2}\right)\right]\theta_1\cos\mu + 2\left[1+\frac{1}{2}\sin\left(\frac{\mu}{2}\right)\right]\left[1-\sin\left(\frac{\mu}{2}\right)\right]\theta_1\beta\sin\mu$$

$$+ 2l\left[1+\frac{1}{2}\sin\left(\frac{\mu}{2}\right)\right]\theta_2$$

$$= 2l\left[1+\frac{1}{2}\sin\left(\frac{\mu}{2}\right)\right]\left[(2\cos\mu+1)\,\theta_1+\theta_2\right],$$

and

$$d' = 2l\left[1+\frac{1}{2}\sin\left(\frac{\mu}{2}\right)\right]\theta_1\left(-\frac{1}{\beta}\sin\mu\right)$$

$$+ 2\left[1+\frac{1}{2}\sin\left(\frac{\mu}{2}\right)\right]\left[1-\sin\left(\frac{\mu}{2}\right)\right]\theta_1\cos\mu$$

$$+ 2\left[1+\frac{1}{2}\sin\left(\frac{\mu}{2}\right)\right]\left[1-\sin\left(\frac{\mu}{2}\right)\right]\theta_2$$

$$= -4\sin^2\left(\frac{\mu}{2}\right)\left[1+\frac{1}{2}\sin\left(\frac{\mu}{2}\right)\right]\left[1-\sin\left(\frac{\mu}{2}\right)\right]\theta_1$$

$$+ 2\left[1+\frac{1}{2}\sin\left(\frac{\mu}{2}\right)\right]\left[1-\sin\left(\frac{\mu}{2}\right)\right](\cos\mu)\,\theta_1$$

$$+ 2\left[1+\frac{1}{2}\sin\left(\frac{\mu}{2}\right)\right]\left[1-\sin\left(\frac{\mu}{2}\right)\right]\theta_2$$

$$= 2\left[1+\frac{1}{2}\sin\left(\frac{\mu}{2}\right)\right]\left[1-\sin\left(\frac{\mu}{2}\right)\right]\left[(2\cos\mu-1)\,\theta_1+\theta_2\right].$$

Note that the relation

$$\beta = \frac{2l\left[1+\sin(\mu/2)\right]}{\sin\mu}$$

was used during the derivations above which lead to the final forms of $d$ and $d'$.

To match a dispersion free region to a FODO cell, we have

$$\begin{pmatrix} \cos\left(2\mu\right) & \beta\sin\left(2\mu\right) & d \\ -\sin\left(2\mu\right)/\beta & \cos\left(2\mu\right) & d' \\ 0 & 0 & 1 \end{pmatrix} \begin{pmatrix} 0 \\ 0 \\ 1 \end{pmatrix} = \begin{pmatrix} D_{\max} \\ 0 \\ 1 \end{pmatrix},$$

which leads to

$$\left(2\cos\mu + 1\right)\theta_1 + \theta_2 = \frac{\theta}{2\sin^2\left(\mu/2\right)},$$

$$\left(2\cos\mu - 1\right)\theta_1 + \theta_2 = 0.$$

The result is

$$\theta_1 = \frac{\theta}{4\sin^2\left(\mu/2\right)}, \qquad \theta_2 = \theta - \theta_1.$$

So, we have

$$\theta_1 = \theta_2 = \frac{\theta}{2}, \qquad \text{for} \quad \mu = \frac{\pi}{2}.$$

$$\theta_1 = \theta, \quad \theta_2 = 0, \quad \text{for} \quad \mu = \frac{\pi}{3}.$$

Dispersion suppressors are widely used in high energy accelerators where the achromatic straight section constitutes only a small portion of the ring. Since the main part of the ring is made up of arcs, the cost-effective way to build such a ring is to pack dipole magnets as close as possible. A FODO cell is the best choice for this purpose. Another way to save cost is to keep the beam pipe as small as possible, which saves not only due to smaller pipes themselves, but also, more importantly, smaller magnets. Dispersion suppressors help to keep beam size small by keeping the dispersion matched. In fact, there is another parameter that plays an important role in the optimization process, which is the length of the FODO cell. Both $\beta_{\max}$ and $D_{\max}$ are proportional to the length of the cell. A shorter cell leads to smaller beam size, but tends to decrease the **packing factor**, which is the ratio of the length of total bending over the total length of the cell.

## 9.2 Symmetric Achromats

Achromats based on mirror symmetry are widely used in beamlines and accelerators, especially synchrotron light sources. One difference between a synchrotron light source and a high energy accelerator is in the number of experiments it supports. While a high energy accelerator usually supports around ten fixed target experiments and a handful of collider experiments (less than five), a synchrotron light source usually supports tens and sometimes more than a hundred experiments with the circumference of the ring only a fraction of the high energy counterpart. Furthermore, insertion devices (wigglers and undulators) have become the main source of light, as opposed to bend magnets. These requirements result in a ring divided into many sections (usually identical ones) with long straight sections in between where dispersion is either zero or small. Apparently FODO cells plus dispersion suppressors are not well suited for this kind of ring. The solution has been mirror symmetric achromatic sections with relatively long straight sections at the ends.

Before going into the details of the lattice modules, let us first look into the general properties of a mirror symmetric cell. Mirror symmetry here is referred to as the symmetry between the cell and its mirror image of the $x$-$y$ plane. In other words, a mirror symmetric cell means that the optical elements of the cell are symmetric about the center, both in terms of geometry and the excitation of the fields. For example, a quadrupole is mirror symmetric and a sector bend is mirror symmetric, too. When a cell is mirror symmetric, the map of the cell is the same as that of its mirror image. To obtain the map of the mirror image cell, we observe that a particle that enters the mirror image cell with $(x_f, -a_f, y_f, -b_f)$ exits it with $(x_i, -a_i, y_i, -b_i)$, where $(x_i, a_i, y_i, b_i)$ and $(x_f, a_f, y_f, b_f)$ are the entrance and exit coordinates of the transverse phase space of the original cell. Hence the map of the mirror image cell is

$$\mathcal{M}^I = \mathcal{R} \circ \mathcal{M} \circ \mathcal{R}^{-1},$$

where

$$\mathcal{R} = \begin{pmatrix} 1 & 0 & 0 & 0 & 0 & 0 \\ 0 & -1 & 0 & 0 & 0 & 0 \\ 0 & 0 & 1 & 0 & 0 & 0 \\ 0 & 0 & 0 & -1 & 0 & 0 \\ 0 & 0 & 0 & 0 & 1 & 0 \\ 0 & 0 & 0 & 0 & 0 & 1 \end{pmatrix} \begin{pmatrix} x \\ a \\ y \\ b \\ l \\ \delta \end{pmatrix}.$$

For linear horizontal motion, we have

$$\hat{M}_x^I = \begin{pmatrix} 1 & 0 & 0 \\ 0 & -1 & 0 \\ 0 & 0 & 1 \end{pmatrix} \begin{pmatrix} (x|x) & (x|a) & (x|\delta) \\ (a|x) & (a|a) & (a|\delta) \\ 0 & 0 & 1 \end{pmatrix}^{-1} \begin{pmatrix} 1 & 0 & 0 \\ 0 & -1 & 0 \\ 0 & 0 & 1 \end{pmatrix}$$

$$= \begin{pmatrix} (a|a) & (x|a) & -(a|a)(x|\delta) + (x|a)(a|\delta) \\ (a|x) & (x|x) & -(a|x)(x|\delta) + (x|x)(a|\delta) \\ 0 & 0 & 1 \end{pmatrix}.$$

Mirror symmetry entails that $\hat{M}_x = \hat{M}_x^I$, which leads to

$$(x|x) = (a|a)$$

and

$$(a|\delta) = \frac{1 + (x|x)}{(x|a)}(x|\delta).$$

Note that the two equations from the dispersion and the dispersion prime are not linearly independent. It is easy to verify that the transverse linear matrix of a magnetic sector dipole satisfies the above relations. For a linearly stable cell, i.e., $|(x|x)| < 1$, mirror symmetry implies that

$$\alpha_x = \frac{(x|x) - (a|a)}{2\sin\mu_x} = 0,$$

and

$$D' = \frac{(1 - (x|x))(a|\delta) + (a|x)(x|\delta)}{2 - (x|x) - (a|a)} = \frac{1 - (x|x)^2 + (x|a)(a|x)}{[2 - (x|x) - (a|a)](x|a)}(x|\delta) = 0.$$

Alternatively, we can also express the transfer matrix of a mirror symmetric cell as functions of the first half of the cell. If the matrix of the first half of a mirror symmetry cell is

$$\hat{M}_x^1 = \begin{pmatrix} (x|x)_1 & (x|a)_1 & (x|\delta)_1 \\ (a|x)_1 & (a|a)_1 & (a|\delta)_1 \\ 0 & 0 & 1 \end{pmatrix},$$

the second half is

$$\hat{M}_x^2 = \begin{pmatrix} (a|a)_1 & (x|a)_1 & -(a|a)_1(x|\delta)_1 + (x|a)_1(a|\delta)_1 \\ (a|x)_1 & (x|x)_1 & -(a|x)_1(x|\delta)_1 + (x|x)_1(a|\delta)_1 \\ 0 & 0 & 1 \end{pmatrix}.$$

The matrix of the whole cell is

$$\hat{M}_x^T = \hat{M}_x^I \hat{M}_x$$
$$= \begin{pmatrix} (a|a)_1 & (x|a)_1 & -(a|a)_1(x|\delta)_1 + (x|a)_1(a|\delta)_1 \\ (a|x)_1 & (x|x)_1 & -(a|x)_1(x|\delta)_1 + (x|x)_1(a|\delta)_1 \\ 0 & 0 & 1 \end{pmatrix}\begin{pmatrix} (x|x)_1 & (x|a)_1 & (x|\delta)_1 \\ (a|x)_1 & (a|a)_1 & (a|\delta)_1 \\ 0 & 0 & 1 \end{pmatrix}$$
$$= \begin{pmatrix} (x|x)_1(a|a)_1 + (x|a)_1(a|x)_1 & 2(x|a)_1(a|a)_1 & 2(x|a)_1(a|\delta)_1 \\ 2(x|x)_1(a|x)_1 & (x|x)_1(a|a)_1 + (x|a)_1(a|x)_1 & 2(x|x)_1(a|\delta)_1 \\ 0 & 0 & 1 \end{pmatrix}.$$

In addition to the relations obtained above, this alternative expression shows that when $(a|\delta)_1 = 0$, the cell is achromatic. Furthermore it reveals another relation, which is

$$(a|\delta) = \frac{(x|x)_1}{(x|a)_1}(x|\delta).$$

To make the meaning of the relation clearer, let us add the drift of length $L$ after the cell. We obtain

$$\begin{pmatrix} 1 & L \\ 0 & 1 \end{pmatrix} \begin{pmatrix} (a|a)_1 & (x|a)_1 \\ (a|x)_1 & (x|x)_1 \end{pmatrix} = \begin{pmatrix} (a|a)_1 + L(a|x)_1 & (x|a)_1 + L(x|x)_1 \\ (a|x)_1 & (x|x)_1 \end{pmatrix},$$

and

$$\begin{pmatrix} 1 & L \\ 0 & 1 \end{pmatrix} \begin{pmatrix} 2(x|a)_1(a|\delta)_1 \\ 2(x|x)_1(a|\delta)_1 \end{pmatrix} = \begin{pmatrix} 2\left[(x|a)_1 + L(x|x)_1\right](a|\delta)_1 \\ 2(x|x)_1(a|\delta)_1 \end{pmatrix}.$$

When $L = -(x|a)_1/(x|x)_1$, the second half of the cell forms an image and the dispersive ray crosses the axis. In other words, the dispersive ray behaves the same as the axial ray from the center of the cell. The reader can check easily that it is indeed the case for a sector bend.

### 9.2.1 The Double-Bend Achromat

Now let us study the simplest mirror symmetric achromat, which consists of two bend magnets and a quadrupole in the middle. Due to the mirror symmetry, the achromatic conditions $D = D' = 0$ at the end can be satisfied requiring $D' = 0$ at the center.

$$\begin{pmatrix} D_c \\ 0 \\ 1 \end{pmatrix} = \begin{pmatrix} 1 & 0 & 0 \\ -1/2f & 1 & 0 \\ 0 & 0 & 1 \end{pmatrix} \begin{pmatrix} 1 & L_1 & 0 \\ 0 & 1 & 0 \\ 0 & 0 & 1 \end{pmatrix} \begin{pmatrix} 1 & L & L\theta/2 \\ 0 & 1 & \theta \\ 0 & 0 & 1 \end{pmatrix} \begin{pmatrix} 0 \\ 0 \\ 1 \end{pmatrix},$$

$$\begin{pmatrix} D_c \\ 0 \\ 1 \end{pmatrix} = \begin{pmatrix} 1 & L + L_1 & (L/2 + L_1)\theta \\ -1/2f & 1 - (L + L_1)/2f & [1 - (1/2f)(L/2 + L_1)]\theta \\ 0 & 0 & 1 \end{pmatrix} \begin{pmatrix} 0 \\ 0 \\ 1 \end{pmatrix}.$$

$$\implies D_c = \left(\frac{L}{2} + L_1\right)\theta, \quad f = \frac{1}{2}\left(\frac{L}{2} + L_1\right).$$

From the discussion above, the relation $f = (L/2 + L_1)/2$ is simply the result of the mirror symmetry. Even with a large bending angle where the exact matrix of the bend has to be used, being achromatic always implies that the center of the first bend is imaged to the center of the second bend. Since $f < (L + L_1)/2$, it is not possible to build a FODO cell that is stable, so a doublet or a triplet has to be used. A simple variant of the double-bend achromat (DBA) is the triplet DBA shown in Fig. 9.5, which contains a triplet between the bending magnets, with no quadrupoles outside.

Another type of DBA consists of two bending magnets, a focusing quadrupole in between and a doublet outside of each bend as shown in Fig. 9.6. The cell is symmetric about the center of QF1. Fig. 9.7 [66, 21] shows an example of the lattice function of one example of this type of achromat.

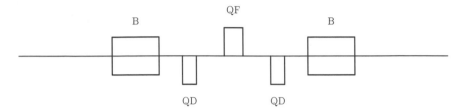

**FIGURE 9.5**: The simplest double-bend achromat (DBA).

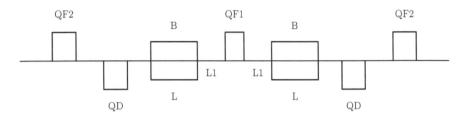

**FIGURE 9.6**: The double-bend achromat (DBA).

### 9.2.2 The Triple-Bend Achromat

The fact that the center quadrupole of the DBA images the center of the first bend to that of the second makes the DBA lattice somewhat inflexible since the horizontal phase advance between the centers of the bends is always around $\pi$. To overcome this shortcoming, **triple-bend achromat** (TBA) lattices were developed. A TBA consists of three bending magnets, at least two quadrupoles between them, and doublets (or triplets) outside, as shown in Fig. 9.8. Lattice functions for a typical example of such kind of achromat are shown in Fig. 9.9 [65, 21].

### 9.2.3 The Multiple-Bend Achromat

In the past two decades, the concept of multiple-bend achromat (MBA) has been conceived of and developed to further reduce dispersion in the bending magnets. As discussed in Section 9.2.4, this will help reduce the emittance of the electron beam and increase the brightness of the X-ray produced from synchrotron radiation. Fig. 9.10 shows the first MBA lattice, developed at the MAX IV Laboratory at Lund University, Lund, Sweden. The middle units are very similar to regular FODO cells and the end units are used as dispersion suppressors. Recent variants increase the distance between the outer most bending magnets and the middle ones to generate a dispersion bump there. As a result, the strengths of the sextupoles are reduced and the dynamic aperture is enlarged. The complexity of the lattice makes a multi-dimensional optimization tool a necessity. In that regard, the spread of various

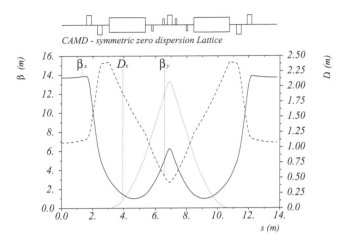

**FIGURE 9.7**: Lattice functions of a double-bend achromat (DBA), which is one of the four super-periods of the storage at Center for Advanced Microstructures and Devices (CAMD) at Louisiana State University.

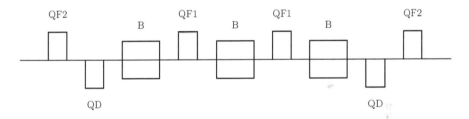

**FIGURE 9.8**: The triple-bend achromat (TBA).

numerical algorithms in the community greatly expedited the development of this concept.

## 9.2.4 The $\mathcal{H}$ Function

The design of such triple-bend achromats is mainly driven by the demand for small equilibrium emittance for synchrotron light sources. Although synchrotron radiation is not covered in this book, it is important to introduce the concept of the equilibrium emittance which results from synchrotron radiation, since it is crucial for understanding the motivation behind the development of lattice modules for the synchrotron light sources. The main difference between electron and hadron (proton, antiproton and ion) rings is synchrotron radiation. Since for a given bending radius the total radiated power is proportional to $\gamma^4$ $\left(\gamma = E/mc^2\right)$, synchrotron radiation becomes significant at a much lower energy for electrons. The energy loss is compensated with RF

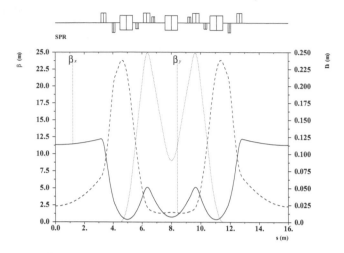

**FIGURE 9.9**:  Lattice functions of a triple-bend achromat (TBA) of the Advanced Light Source (ALS) at Lawrence Berkeley National Laboratory, California, USA.

cavities.

Since the photons are emitted into a forward pointing cone of the opening angle $1/\gamma$ which is small for electrons of GeV level energy, all three components of the momentum decrease at roughly the same rate. The RF cavity, on the other hand, increases only the longitudinal momentum. As a result, transverse momentum is damped over time.

Yet the presence of dispersion in a ring causes the emittance to grow due to synchrotron radiation. Let us consider an off-momentum electron moving along the closed orbit for the momentum in a dispersive region. After a photon is emitted, the position and slope of the electron remain unchanged but the total energy decreases. Suddenly the orbit the electron moves along is no longer the closed orbit for it and the electron starts to oscillate around the new closed orbit, resulting in emittance growth. The equilibrium emittance is reached when the damping rate equals the growth rate. It turns out that, for a ring with an identical bending field, the equilibrium emittance is proportional to $\langle \mathcal{H} \rangle_{mag}$ , where

$$\mathcal{H} = \gamma_x D^2 + 2\alpha_x DD' + \beta_x D'^2,$$

and

$$\langle \mathcal{H} \rangle_{mag} = \frac{1}{2\pi\rho} \int_{dipole} \mathcal{H} ds.$$

Note that $D$ and $D'$ are periodic solutions of the position and slope of dispersion. Furthermore, $\mathcal{H}$ is a constant outside of dipole magnets and changes inside dipole magnets. To demonstrate this point, let us consider two points

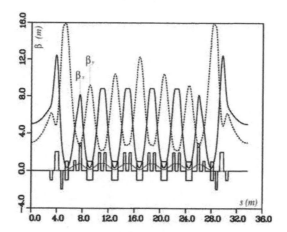

**FIGURE 9.10:** The lattice functions $\beta_x$, $\beta_y$ and dispersion multiplied by 10 in the MAX IV multiple-bend achromat (MBA) lattice at Lund University, Lund, Sweden. (© 1996 IEEE. Reprinted, with permission, from D. Einfeld, et. al., in *Proc. PAC 1995*, 1, 177, 1996 [26].)

in the ring. The linear map between them is

$$\hat{M}_{12} = \begin{pmatrix} (x|x)_{12} & (x|a)_{12} & (x|\delta)_{12} \\ (a|x)_{12} & (a|a)_{12} & (a|\delta)_{12} \\ 0 & 0 & 1 \end{pmatrix},$$

and the dispersion of those two points are related through the relation

$$\begin{pmatrix} D_2 \\ D_2' \\ 1 \end{pmatrix} = \begin{pmatrix} (x|x)_{12} & (x|a)_{12} & (x|\delta)_{12} \\ (a|x)_{12} & (a|a)_{12} & (a|\delta)_{12} \\ 0 & 0 & 1 \end{pmatrix} \begin{pmatrix} D_1 \\ D_1' \\ 1 \end{pmatrix}.$$

Hence, we have

$$\begin{aligned}
\mathcal{H}_2 =& \gamma_2 D_2^2 + 2\alpha_2 D_2 D_2' + \beta_2 D_2'^2 \\
=& (D_2, D_2') \begin{pmatrix} \gamma_2 & \alpha_2 \\ \alpha_2 & \beta_2 \end{pmatrix} \begin{pmatrix} D_2 \\ D_2' \end{pmatrix} \\
=& \left[ (D_1, D_1') \begin{pmatrix} (x|x)_{12} & (a|x)_{12} \\ (x|a)_{12} & (a|a)_{12} \end{pmatrix} + ((x|\delta)_{12}, (a|\delta)_{12}) \right] \begin{pmatrix} \gamma_2 & \alpha_2 \\ \alpha_2 & \beta_2 \end{pmatrix} \\
& \cdot \left[ \begin{pmatrix} (x|x)_{12} & (x|a)_{12} \\ (a|x)_{12} & (a|a)_{12} \end{pmatrix} \begin{pmatrix} D_1 \\ D_1' \end{pmatrix} + \begin{pmatrix} (x|\delta)_{12} \\ (a|\delta)_{12} \end{pmatrix} \right]
\end{aligned}$$

$$= (D_1, D_1') \begin{pmatrix} \gamma_1 & \alpha_1 \\ \alpha_1 & \beta_1 \end{pmatrix} \begin{pmatrix} D_1 \\ D_1' \end{pmatrix}$$

$$+ (D_1, D_1') \begin{pmatrix} (x|x)_{12} & (a|x)_{12} \\ (x|a)_{12} & (a|a)_{12} \end{pmatrix} \begin{pmatrix} \gamma_2 & \alpha_2 \\ \alpha_2 & \beta_2 \end{pmatrix} \begin{pmatrix} (x|\delta)_{12} \\ (a|\delta)_{12} \end{pmatrix}$$

$$+ ((x|\delta)_{12}, (a|\delta)_{12}) \begin{pmatrix} \gamma_2 & \alpha_2 \\ \alpha_2 & \beta_2 \end{pmatrix} \begin{pmatrix} (x|x)_{12} & (x|a)_{12} \\ (a|x)_{12} & (a|a)_{12} \end{pmatrix} \begin{pmatrix} D_1 \\ D_1' \end{pmatrix}$$

$$+ ((x|\delta)_{12}, (a|\delta)_{12}) \begin{pmatrix} \gamma_2 & \alpha_2 \\ \alpha_2 & \beta_2 \end{pmatrix} \begin{pmatrix} (x|\delta)_{12} \\ (a|\delta)_{12} \end{pmatrix}.$$

It is clear that $\mathcal{H}_2 = \mathcal{H}_1$ if $(x|\delta)_{12} = (a|\delta)_{12} = 0$, which is the case for any two points that are in the same straight section. With the expression above, we can obtain the derivative of $\mathcal{H}$ with respect to $s$ which can illustrate the matter even clearer. When the two points are close to each other, the linear map becomes

$$d\hat{M} = \begin{pmatrix} 1 & ds & 0 \\ -kds & 1 & ds/\rho \\ 0 & 0 & 1 \end{pmatrix}.$$

Carrying out the derivation one step further, we obtain

$$d\mathcal{H} = (D_1, D_1') \begin{pmatrix} 1 & -kds \\ ds & 1 \end{pmatrix} \begin{pmatrix} \gamma_2 & \alpha_2 \\ \alpha_2 & \beta_2 \end{pmatrix} \begin{pmatrix} 0 \\ ds/\rho \end{pmatrix}$$

$$+ \left(0, \frac{ds}{\rho}\right) \begin{pmatrix} \gamma_2 & \alpha_2 \\ \alpha_2 & \beta_2 \end{pmatrix} \begin{pmatrix} 1 & ds \\ -kds & 1 \end{pmatrix} \begin{pmatrix} D_1 \\ D_1' \end{pmatrix} + \left(0, \frac{ds}{\rho}\right) \begin{pmatrix} \gamma_2 & \alpha_2 \\ \alpha_2 & \beta_2 \end{pmatrix} \begin{pmatrix} 0 \\ ds/\rho \end{pmatrix}$$

$$= (D_1, D_1') \begin{pmatrix} 1 & -ds \\ kds & 1 \end{pmatrix} \begin{pmatrix} \gamma_1 & \alpha_1 \\ \alpha_1 & \beta_1 \end{pmatrix} \begin{pmatrix} 0 \\ ds/\rho \end{pmatrix}$$

$$+ \left(0, \frac{ds}{\rho}\right) \begin{pmatrix} \gamma_1 & \alpha_1 \\ \alpha_1 & \beta_1 \end{pmatrix} \begin{pmatrix} 1 & kds \\ -ds & 1 \end{pmatrix} \begin{pmatrix} D_1 \\ D_1' \end{pmatrix} + \left(0, \frac{ds}{\rho}\right) \begin{pmatrix} \gamma_2 & \alpha_2 \\ \alpha_2 & \beta_2 \end{pmatrix} \begin{pmatrix} 0 \\ ds/\rho \end{pmatrix}$$

$$=_1 2 \left(\alpha_1 D_1 + \beta_1 D_1'\right) \frac{ds}{\rho}.$$

In summary, we have

$$\mathcal{H}' = \frac{2}{\rho} \left(\alpha_1 D_1 + \beta_1 D_1'\right).$$

In order to achieve small emittance, $\mathcal{H}$ has to be small, which leads to strong quadrupoles. This in turn leads to strong sextupoles to correct chromaticities which in general would result in strong nonlinear motion and small dynamic aperture. TBA lattices can provide smaller dispersion and hence smaller emittance than DBA lattices, which result in stronger sextupoles and smaller dynamic aperture. This is one of the reasons that TBA lattices fell out of favor in the most recent synchrotron light sources. Another reason is that, when the achromatic condition is not strictly enforced, DBA lattices appear to be more flexible than TBA lattices, especially when more quadrupoles are used. For example, the DBA lattice of the Shanghai Synchrotron Radiation Facility (SSRF), Shanghai, China, contains two quadrupole doublets between the bending magnets and two triplets outside. In fact, almost all synchrotron light sources built in the past decade adopted DBA lattices.

## 9.3 Special Purpose Modules

### 9.3.1 The Low Beta Insertion

In both circular and linear colliders, the beam is focused as tightly as possible at the collision points to maximize the density of particles and hence collision rate. The simplest low beta insertion consists of **two quadrupole doublets** placed and excited symmetrically about the interaction point. The upstream doublet is roughly a parallel–to–point system, in which an initially nearly parallel beam is brought down to a small point and the downstream doublet is a point–to–parallel system. In hadron colliders where the emittance is relatively big, triplets are used to increase focusing power and reduce the width of the beam in the quadrupoles. In addition, it provides more flexibility for tuning.

Recall from eq. (6.13) that

$$\beta(s) = \beta^* + \frac{s^2}{\beta^*}.$$

Since the distance between the interaction point and the last quadrupole is on the order of 10 m and $\beta^*$ is below 1 m, the $\beta$ functions in the quadrupoles range from hundreds m to over 1 km (see Fig. 9.11). Combined with high gradient in the quadrupoles, the low beta insertion generates large chromatic and geometric aberrations. In hadron colliders, due to the relatively large emittance, the main effect of the aberrations is the additional chromaticities, which are corrected by the sextupoles in the arcs. In addition, the large beam size at the quadrupoles implies extra tight tolerances on multipole errors in those quadrupoles, which, if too large, would excite undesirable resonances causing emittance and/or beam loss. In electron-positron linear colliders, the emittance is small and the aberrations generated by the low beta insertion would cause sizable increase in beam size at the interaction point. As a result, the matching section between the low beta section and the linear accelerator (linac) is rather complicated. Telescopes are used to minimize certain aberrations and sextupoles are used to correct chromatic aberrations.

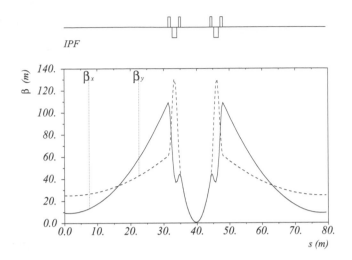

**FIGURE 9.11:** Lattice functions of a typical low beta insertion with symmetric quadrupole triplets. Here $\beta^*$ is 0.5 m.

### 9.3.2 The Chicane Bunch Compressor

A simple yet very effective and commonly used module in linac based free electron lasers (FELs) is the so-called chicane bunch compressor. It consists of four identical rectangular homogeneous bending magnets separated by drift spaces, with the middle two magnets bending in the opposite direction, and the reference orbit perpendicular to the entrance of the first and third magnets and the exit of the second and fourth magnets (see Fig. 9.12). The whole module is **mirror symmetric** about the center. Such an arrangement ensures that the bunch compressor is achromatic to all orders and that electrons with higher energy go through shorter paths. When a bunch of electrons enters the compressor with a correlation between the longitudinal position and energy, the bunch length changes at the exit of the compressor. If the slope is negative, i.e., the electrons in the head of the bunch have lower energy, the bunch is compressed.

Next, let us take a look at the basic optical properties of the chicane bunch compressor. The horizontal transfer matrix of the first bend is

$$
\hat{M}_x^1 = \begin{pmatrix} 1 & 0 & 0 \\ 1/R_0 \tan\phi & 1 & 0 \\ 0 & 0 & 1 \end{pmatrix} \begin{pmatrix} \cos\phi & R_0 \sin\phi & R_0 (1-\cos\phi) \\ -1/R_0 \sin\phi & \cos\phi & \sin\phi \\ 0 & 0 & 1 \end{pmatrix}
$$

$$
= \begin{pmatrix} \cos\phi & R_0 \sin\phi & R_0 (1-\cos\phi) \\ 0 & 1/\cos\phi & \tan\phi \\ 0 & 0 & 1 \end{pmatrix}.
$$

To obtain the transfer matrix of a bend magnet with the opposite direction

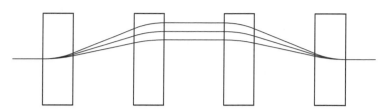

**FIGURE 9.12:** Layout of the chicane bunch compressor. The top/bottom trajectories are those of particles of lower/higher momenta than the reference trajectory in the middle.

of bending, we have to first find out the transformation between internal and external coordinate systems. Taking into account the fact that positive $x$ in the internal system (away from the center of the arc of the design orbit) is negative in the external system, the transformation is

$$\hat{S}_x = \begin{pmatrix} -1 & 0 & 0 \\ 0 & -1 & 0 \\ 0 & 0 & 1 \end{pmatrix}.$$

The transformation in the vertical plane is the identity matrix. Therefore the horizontal transfer matrix of the second bend is

$$\hat{M}_x^2 = \hat{S}_x \begin{pmatrix} \cos\phi & R_0 \sin\phi & R_0\left(1 - \cos\phi\right) \\ -1/R_0 \sin\phi & \cos\phi & \sin\phi \\ 0 & 0 & 1 \end{pmatrix} \begin{pmatrix} 1 & 0 & 0 \\ 1/R_0 \tan\phi & 1 & 0 \\ 0 & 0 & 1 \end{pmatrix} \hat{S}_x^{-1}$$

$$= \begin{pmatrix} -1 & 0 & 0 \\ 0 & -1 & 0 \\ 0 & 0 & 1 \end{pmatrix} \begin{pmatrix} 1/\cos\phi & R_0 \sin\phi & R_0\left(1 - \cos\phi\right) \\ 0 & \cos\phi & \sin\phi \\ 0 & 0 & 1 \end{pmatrix} \begin{pmatrix} -1 & 0 & 0 \\ 0 & -1 & 0 \\ 0 & 0 & 1 \end{pmatrix}$$

$$= \begin{pmatrix} 1/\cos\phi & R_0 \sin\phi & -R_0\left(1 - \cos\phi\right) \\ 0 & \cos\phi & -\sin\phi \\ 0 & 0 & 1 \end{pmatrix}.$$

The horizontal matrix of the first and the second bends separated by a drift $L_1$ is

$$\hat{M}_x^h = \begin{pmatrix} 1/\cos\phi & R_0\sin\phi & -R_0(1-\cos\phi) \\ 0 & \cos\phi & -\sin\phi \\ 0 & 0 & 1 \end{pmatrix} \begin{pmatrix} 1 & L_1 & 0 \\ 0 & 1 & 0 \\ 0 & 0 & 1 \end{pmatrix} \begin{pmatrix} \cos\phi & R_0\sin\phi & R_0(1-\cos\phi) \\ 0 & 1/\cos\phi & \tan\phi \\ 0 & 0 & 1 \end{pmatrix}$$

$$= \begin{pmatrix} 1 & 2R_0 \sin\phi + L_1/\cos^2\phi & \left[2R_0\left(1 - \cos\phi\right) + L_1 \tan\phi\right]/\cos\phi \\ 0 & 1 & 0 \\ 0 & 0 & 1 \end{pmatrix}.$$

From the mirror symmetry of the chicane, we can conclude that the module is achromatic. For $\phi \ll 1$, the dispersion between the second and the third bends is $D = L_1\phi + R_0\phi^2$. It is worth noting that the focusing in the vertical plane is insignificant. As a result the bunch compressor is transparent in transverse dynamics.

Finally, let us work out the path length difference between the reference electron and one that has a different momentum $p = (1 + \delta)\, p_0$. Due to the symmetry, the difference can be obtained analytically, which is

$$l - l_0 = 4\left(R\phi - R_0\phi_0\right) + 2L_1\left(\frac{\cos\phi_0}{\cos\phi} - 1\right),$$

where

$$R = R_0\left(1 + \delta\right),$$

and

$$\sin\phi = \frac{R_0}{R}\sin\phi_0 = \frac{\sin\phi_0}{1 + \delta}.$$

Plugging in $R$ and $\phi$, we have

$$l - l_0 = 4R_0\left[(1 + \delta)\arcsin\left(\frac{\sin\phi_0}{1 + \delta}\right) - \phi_0\right] + 2L_1\left[\frac{(1 + \delta)\cos\phi_0}{\sqrt{(1 + \delta)^2 - \sin^2\phi_0}} - 1\right].$$

In order to have a better idea about the relation between $l - l_0$ and $\delta$, we would like to learn how the low order terms behave. Before we proceed with the Taylor expansion of $l - l_0$, let us first work out that of $\arcsin(x_0 + \Delta x)$. Using $\arcsin(x\sqrt{1 - y^2} + y\sqrt{1 - x^2}) = \arcsin(x) + \arcsin(y)$ and setting $x = x_0$, we obtain

$$x_0\sqrt{1 - y^2} + y\sqrt{1 - x_0^2} = x_0 + \Delta x.$$

After straightforward algebra, we obtain

$$y = (x_0 + \Delta x)\sqrt{1 - x_0^2} - x_0\sqrt{1 - (x_0 + \Delta x)^2}.$$

As a result,

$\arcsin(x_0 + \Delta x)$

$$= \arcsin(x_0) + \arcsin\left[(x_0 + \Delta x)\sqrt{1 - x_0^2} - x_0\sqrt{1 - (x_0 + \Delta x)^2}\right]$$

$$= \arcsin(x_0) + \arcsin\left[(x_0 + \Delta x)\sqrt{1 - x_0^2} - x_0\sqrt{1 - x_0^2}\sqrt{1 - \frac{2x_0\Delta x + \Delta x^2}{1 - x_0^2}}\right]$$

$$=_2 \arcsin(x_0) + \arcsin\left\{\Delta x\sqrt{1 - x_0^2} + x_0\sqrt{1 - x_0^2}\left[\frac{2x_0\Delta x + \Delta x^2}{2(1 - x_0^2)} + \frac{4x_0^2\Delta x^2}{8(1 - x_0^2)^2}\right]\right\}$$

$$=_2 \arcsin(x_0) + \Delta x\sqrt{1 - x_0^2} + \frac{x_0}{2}\sqrt{1 - x_0^2}\left[\frac{2x_0\Delta x + \Delta x^2}{1 - x_0^2} + \frac{x_0^2\Delta x^2}{(1 - x_0^2)^2}\right]$$

$$=_2 \arcsin(x_0) + \frac{\Delta x}{\sqrt{1 - x_0^2}} + \frac{x_0\Delta x^2}{2(1 - x_0^2)^{\frac{3}{2}}}.$$

To the second order, the path length difference is

$$l - l_0 =_2 4R_0\left[(1 + \delta)\arcsin\left((1 - \delta + \delta^2)\sin\phi_0\right) - \phi_0\right]$$

$$+ 2L_1\left[\frac{(1 + \delta)\cos\phi_0}{\sqrt{\cos^2\phi_0 + 2\delta + \delta^2}} - 1\right]$$

$$=_2 4R_0\left[(1 + \delta)\arcsin\left(\sin\phi_0 - (\delta - \delta^2)\sin\phi_0\right) - \phi_0\right]$$

$$+ 2L_1\left[\frac{(1 + \delta)}{\sqrt{1 + (2\delta + \delta^2)/\cos^2\phi_0}} - 1\right]$$

$$=_2 4R_0\left\{(1 + \delta)\left[\phi_0 - (\delta - \delta^2)\tan\phi_0 + \frac{1}{2}\delta^2\tan^3\phi_0\right] - \phi_0\right\}$$

$$+ 2L_1\left[(1 + \delta)\left(1 - \frac{1}{2}\frac{2\delta + \delta^2}{\cos^2\phi_0} + \frac{3}{8}\frac{4\delta^2}{\cos^4\phi_0}\right) - 1\right]$$

$$=_2 4R_0\left[(\phi_0 - \tan\phi_0)\delta + \frac{1}{2}(\tan^3\phi_0)\delta^2\right]$$

$$- 2L_1\tan^2\phi_0\left(\delta - \frac{3}{2}\frac{\delta^2}{\cos^2\phi_0}\right).$$

For $\phi_0 \ll 1$, only the terms of the lowest order in $\phi_0$ are important and we obtain

$$l - l_0 =_2 -2L_1\phi_0^2\left(\delta - \frac{3}{2}\delta^2\right).$$

As an example, we discuss the parameters of the first bunch compressor at the Linac Coherent Light Source (LCLS) at SLAC National Accelerator Laboratory, California, USA, which operates at a beam energy of 250 MeV. It is worth noting that the bend radius $R_0$ of 2.48 m (corresponding to a

buncher cavity

just before cavity

just after cavity

further downstream

density profile

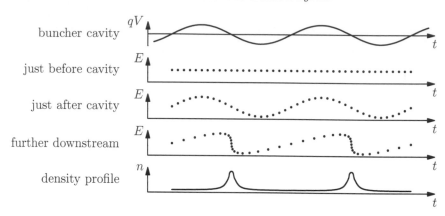

**FIGURE 9.13**:   Mechanism of an RF buncher cavity.

magnetic field $B_0$ of 3.36 kG) and $L_1$ at 2.61 m are roughly the same and that the bend angle of $\phi_0 = 4.62°$, which corresponds to a magnet length $L_b = 0.2$ m is very small. For the term $(l|\delta)$, the contribution from the magnets is $-1.74$ mm and that from the drifts is $-34.07$ mm. For the term $(l|\delta^2)$, the contribution from the magnets is 2.62 mm and that from the drifts is 51.44 mm. At this energy, the electrons are relativistic enough that the contribution from the difference in velocity is miniscule ($-27$ $\mu$m for $(l|\delta)$ and 40 $\mu$m for $(l|\delta^2)$).

### 9.3.3   Other Bunch Compressors

As shown above, chicane bunch compressors work when the higher momentum particles are in the tail of a bunch. Yet higher momentum particles are often in the head of the bunch. An important example is the DC gun, where higher momentum particles are faster and thus arrive earlier. To compress such bunches, one method is to configure electrostatic or magnetic fields in such a way that the faster particles go through longer paths.

Another method is to reverse the correlation between energy and longitudinal position before the bunch enters a chicane bunch compressor. This is achieved most commonly through an RF cavity, which because of its functionality is often called a **buncher**. It is a regular RF structure which is set up through adjusting the phase such that the mean energy of the bunch is unchanged. Meanwhile, the head of the bunch, where the high energy particles are, is decelerated, and the opposite happens to the tail of the bunch. Fig. 9.13 illustrates the mechanism of a buncher. When the particles are not highly relativistic, an RF buncher and the drift space downstream can achieve bunch compression. This is called **ballistic bunching**.

# Chapter 10

## Synchrotron Motion

Up to now we have been primarily concerned with the motion in the transverse planes. Yet, for particle accelerators, as the name implies, acceleration is the primary interest. Therefore the motion in the longitudinal phase space has to be understood. Although there are many different ways of accelerating charged particles, we restrict ourselves mostly to circular accelerators (synchrotrons, to be specific), where the acceleration is done using radio frequency (RF) cavities. The only exception is the last section, where the transverse dynamics of RF cavities is discussed, which is of great significance mainly for linacs.

The chapter is organized as follows. First, a section is devoted to a brief description of a typical RF cavity used in a ring. Second, the time-of-flight as the function of energy is derived. Next, combining the results from the previous sections, the map of the longitudinal phase space is obtained and the longitudinal motion is studied in detail. Last, the transverse effect of the RF cavities is discussed briefly.

## 10.1   RF Fundamentals

Most RF cavities used in synchrotrons are variations of the cylindrical pill-box cavity, which consists of two circular metallic plates of radius $R_c$ that are separated by the distance $l$ and that are connected with a cylindrical mantel of radius $R_c$, resulting in a geometry reminiscent of a circular pill box.

The field distribution in the interior of such a kind of a metal box can be written in a simple analytical form. Following Wangler [70], the electromagnetic field of such a cavity of length $l$ and radius $R_c$ with transverse magnetic

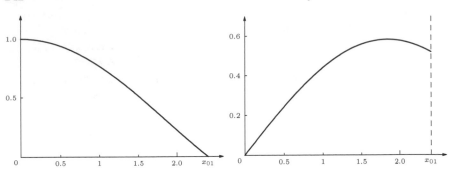

**FIGURE 10.1:** Typical RF cavity field with fundamental mode TM$_{010}$. Radial dependence of the normalized electric field $E_z(r,0)/E_0$ (left) and the normalized magnetic field $B_\theta(r, -1/4f)c/E_0$ (right) are shown as a function of $x_{01}r/R_c$.

field (mode of TM$_{mnp}$) can be written as

$$E_z = E_0 J_m (k_{mn}r) \cos (m\theta) \cos \frac{p\pi z}{l} \cos (\omega t),$$

$$E_r = -\frac{p\pi}{l} \frac{1}{k_{mn}} E_0 J'_m (k_{mn}r) \cos (m\theta) \sin \frac{p\pi z}{l} \cos (\omega t),$$

$$E_\theta = \frac{p\pi}{l} \frac{m}{k_{mn}^2 r} E_0 J_m (k_{mn}r) \sin (m\theta) \sin \frac{p\pi z}{l} \cos (\omega t),$$

$$B_z = 0,$$

$$B_r = \omega \frac{m}{k_{mn}^2 r c^2} E_0 J_m (k_{mn}r) \sin (m\theta) \cos \frac{p\pi z}{l} \sin (\omega t),$$

$$B_\theta = \omega \frac{1}{k_{mn} c^2} E_0 J'_m (k_{mn}r) \cos (m\theta) \cos \frac{p\pi z}{l} \sin (\omega t),$$

where $k_{mn} = x_{mn}/R_c$ and $\omega = c\sqrt{k_{mn}^2 + (p\pi/l)^2}$. Note that the quantity $x_{mn}$ is the $n$th zero of the **Bessel function** $J_m(x)$ (excluding the origin, $n > 0$).

Usually the fundamental mode of TM$_{010}$ is used for accelerating charged particles, whose field is

$$E_z = E_0 J_0 \left( \frac{x_{01}r}{R_c} \right) \cos (\omega t), \quad E_r = 0, \quad E_\theta = 0,$$

$$B_z = 0, \quad B_r = 0, \quad B_\theta = -\frac{E_0}{c} J_1 \left( \frac{x_{01}r}{R_c} \right) \sin (\omega t),$$

where the relation $J'_0(x) = -J_1(x)$ is used to obtain $B_\theta$ (see Fig. 10.1.) Note that $B_\theta$ is proportional to $E'_z$ with 90° phase lag, which is the result of Faraday's law and that $x_{01} = 2.405$ which, together with the design frequency, determines the size of the cavity. Specifically, for the mode of TM$_{010}$, we have

$$R_c = \frac{x_{01} c}{\omega} = \frac{x_{01} c}{2\pi f}.$$

For $f = 500$ MHz, $R_c = 0.2295$ m. In realistic cavity designs, the actual shape of the cavity is often more spherical than cylindrical. Yet the overall dimension is not very far from this crude estimate.

For $r \ll R_c$, which is usually where the beam is, the field can be approximated by the lowest order term of the Taylor expansion, which is

$$E_z =_1 E_0 \cos(\omega t), \quad B_\theta =_1 -\frac{E_0}{c} \frac{x_{01} r}{2R_c} \sin(\omega t). \tag{10.1}$$

Note that $J_0(x) = 1 + O(x^2)$ and $J_1(x) = x/2 + O(x^3)$ for $x \ll 1$. As a result, the effect of $B_\theta$ is much weaker than that of $E_z$ on the beam. Furthermore, the focusing effect of the magnetic field is usually negligible compared to main focusing elements, the quadrupole magnets in the ring.

To illustrate this, let us look at an example. Let us consider again the case of a 500 MHz cavity. Assuming that $E_0 = 20$ MV/m, which is not far from the breakdown limit of copper at this frequency, we obtain the peak gradient of the magnetic field, which is 0.35 T/m. Normal conducting quadrupoles, on the other hand, can have field gradient up to 20 T/m. Furthermore, there are usually tens to hundreds of quadrupoles with lengths between 0.2 and 1 m in a ring, whereas there are at most a handful of cavities with lengths usually below 0.5 m (around 0.3 m for a 500 MHz pillbox cavity). As a result, the integrated gradient of the cavities is on the order of up to perhaps $1/1000$ that of the quadrupoles.

Recently, the dipole mode (TM$_{110}$) has also been used to kick the beam transversely. The field of TM$_{110}$ is

$$E_z = E_0 J_1\left(\frac{x_{11} r}{R_c}\right) \cos\theta \cos(\omega t), \qquad E_r = 0, \quad E_\theta = 0,$$

$$B_z = 0, \qquad B_r = \frac{E_0}{c} \frac{R_c}{x_{11} r} J_1\left(\frac{x_{11} r}{R_c}\right) \sin\theta \sin(\omega t),$$

$$B_\theta = \frac{E_0}{c}\left[J_0\left(\frac{x_{11} r}{R_c}\right) - \frac{R_c}{x_{11} r} J_1\left(\frac{x_{11} r}{R_c}\right)\right] \cos\theta \sin(\omega t),$$

where the relation $J_1'(x) = J_0(x) - J_1(x)/x$ is used to obtain $B_\theta$. Note that $x_{11} = 3.832$. For the same cavity, the frequency of the TM$_{110}$ mode is $x_{11}/x_{01} \approx 1.6$ times that of the TM$_{010}$ mode. In order to get a clearer physical picture of the effect of the field on the beam, let us again perform Taylor expansion around the origin and keep only the leading term. The field is

$$E_z =_1 E_0 \frac{x_{11} r}{2R_c} \cos\theta \cos(\omega t),$$

$$B_r =_1 \frac{E_0}{2c} \sin\theta \sin(\omega t), \quad B_\theta =_1 \frac{E_0}{2c} \cos\theta \sin(\omega t).$$

Opposite to the $TM_{010}$ mode, the magnetic field has a much stronger effect on the beam. Furthermore, we have

$$B_y =_1 B_r \sin\theta + B_\theta \cos\theta =_1 \frac{E_0}{2c} \sin(\omega t),$$

which is an alternating current (AC) dipole and is best suited for kicking the beam transversely. Again, using the example of a 500 MHz cavity and assuming that $E_0 = 20$ MV/m, we obtain the peak field, which is 0.033 T. For such a cavity that is 0.3 m long and the energy of the electron beam being 1.9 GeV, the peak kick angle is

$$\theta_x = \frac{ev_z B_{y0}\Delta t}{p_z} = \frac{eE_0 l}{2 p_z c} \approx \frac{20 \times 10^6 \times 0.3}{2 \times 1.9 \times 10^9} = 1.6 \times 10^{-3}.$$

The approximation that equates $p_z c$ to the total energy of the electron is based on the fact that the relativistic fact $\gamma$ is around 3800 and that the divergence of the beam is usually a fraction of 1 mrad.

Now let us come back to the $TM_{010}$ mode and find out the energy gain ($\Delta K$) per pass. To simplify the matter, let us consider a particle that moves along the optical axis and the energy gain per pass is much smaller than its total kinetic energy ($K$), which entails that the change of velocity in the cavity is negligible. As a result, the energy gain is

$$\Delta K = q \int_{-\frac{l}{2}}^{\frac{l}{2}} E_z(0, z, t(z)) \, dz,$$

where $t(z) = t_0 + z/v_0$. Here we set the origin of the $z$-axis at the center of the cavity. For $TM_{010}$ mode, we have

$$\Delta K = q \int_{-\frac{l}{2}}^{\frac{l}{2}} E_0 \cos[\omega t(z)] \, dz = q \int_{-\frac{l}{2}}^{\frac{l}{2}} E_0 \cos\left[\omega\left(t_0 + \frac{z}{v_0}\right)\right] dz$$

$$= qE_0 \int_{-\frac{l}{2}}^{\frac{l}{2}} \cos\left(\phi_0 + \frac{\omega z}{v_0}\right) dz = qE_0 \int_{-\frac{l}{2}}^{\frac{l}{2}} \cos\left(\phi_0 + \frac{2\pi z}{\beta_0 \lambda}\right) dz$$

$$= qE_0 l \cos(\phi_0) \frac{\sin(\pi l/\beta_0 \lambda)}{\pi l/\beta_0 \lambda},$$

where $\lambda$ is the wavelength of the electromagnetic field and $\beta_0 = v_0/c$. The sin function in the equation above is the result of the finite length of the cavity, which is called the transit time factor ($T$). For the $TM_{010}$ mode of a pillbox cavity, the transit time factor is

$$T = \frac{\sin(\pi l/\beta_0 \lambda)}{\pi l/\beta_0 \lambda},$$

and the energy per pass is

$$\Delta K = qE_0 l T \cos(\phi_0). \tag{10.2}$$

It is clear that the transit time factor is the ratio of the energy gain of a RF cavity to that of a DC (direct current) gap of the same field. The relation between the transit time factor and the length of the cavity is shown in Fig. 10.2. It is obvious that $T \to 1$ as $l \to 0$ which means that, for constant voltage between the gap, the shorter the gap, the closer the energy gain to that of the DC gap. Yet the electric field breakdown limit of the material determines the maximum field that can be achieved, thus the energy gain is proportional to $lT$, which is in turn proportional to $\sin(\pi l/\beta_0 \lambda)$. Therefore, the maximum energy gain for the case of constant field is obtained when $l = \beta_0 \lambda/2$. For electron storage rings such as those of the synchrotron light sources, $\beta_0 \approx 1$. So we have $l = \lambda/2$, which corresponds to the fact that the time an electron takes to pass through the cavity equals half of the period of the oscillation. The transit time factor is $T = 2/\pi = 0.637$. For a 500 MHz cavity, we have $l = 0.3$ m.

In addition, other issues such as RF power efficiency also have to be taken into account. The most used parameter measuring the efficiency is called the shunt impedance, which is defined as

$$R_s = \frac{(\Delta V)^2}{P_d},$$

where

$$\Delta V = E_0 l T \cos(\phi_0),$$

which is the voltage across the accelerating gap and $P_d$ is the power dissipated in the wall. As a result, realistic normal conducting cavities are more or less spherical in shape, minimizing the total surface area, with nose cones around the beam axis to reduce the length (acceleration gap) of the cavity, maximizing the voltage across the gap. For superconducting cavities, the dissipated power is much smaller and hence the main goal of design optimization shifts to minimizing the peak field on the surface for a given on-axis field to reduce the risk of costly quench. The resulting shape is basically the bell shaped cavity, which is preferable for other practical reasons as well.

---

## 10.2   The Phase Slip Factor

Toward the end of the previous section, we studied the energy gain per pass of one particle. In this section, we will study the energy gain of a particle over many passes in a circular accelerator. Since the cavity is always designed for a given accelerator, there is at least one particle (the reference particle) that comes back to the cavity at the same phase (synchronous phase $\phi_s$) every turn. For an arbitrary particle, it may not come back to the cavity at the same phase since it may have different energy. Although an arbitrary

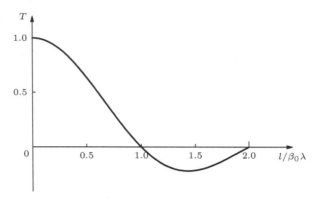

**FIGURE 10.2:** Dependence of transit time factor on length of the cavity.

particle may not move along the reference orbit, which is a closed curve, the effect of the change of arrival time due to the transverse motion tends to average out due to betatron oscillation. This will become clearer below when the large difference between the frequencies of the transverse oscillations and that of the longitudinal one is shown. As a result, we only consider a particle that moves along the close orbit of its energy. Let us consider a ring with midplane symmetry that uses pure magnetic elements for bending and transverse focusing. Hence eq. (3.17) can be simplified as

$$
\begin{aligned}
l' &= \left[ (1 + h x_\delta) \frac{1 + \eta}{1 + \eta_0} \frac{p_0}{p_s(\delta)} - 1 \right] \frac{\kappa}{v_0} \\
&= \left[ (1 + h x_\delta) \frac{1 + \eta}{1 + \eta_0} \Big/ \sqrt{\frac{\eta(2 + \eta)}{\eta_0(2 + \eta_0)} - a_\delta^2} - 1 \right] \frac{\kappa}{v_0} \\
&= \left[ (1 + h x_\delta) \left( 1 + \frac{\eta_0}{1 + \eta_0} \delta \right) \Big/ \sqrt{1 + 2 \frac{1 + \eta_0}{2 + \eta_0} \delta + \frac{\eta_0}{2 + \eta_0} \delta^2 - a_\delta^2} - 1 \right] \frac{\kappa}{v_0},
\end{aligned}
$$

where $x_\delta$ and $a_\delta$ are the horizontal position and momentum of the closed orbit of an off-momentum particle. The difference in arrival time between the reference particle is

$$
\Delta t = \frac{1}{v_0} \int_0^C \left[ (1 + h x_\delta) \left( 1 + \frac{\eta_0}{1 + \eta_0} \delta \right) \Big/ \sqrt{1 + 2 \frac{1 + \eta_0}{2 + \eta_0} \delta + \frac{\eta_0}{2 + \eta_0} \delta^2 - a_\delta^2} - 1 \right] ds.
$$

The **phase slip factor** is defined as

$$\eta^{\mathrm{ph}} \equiv -\frac{\Delta t}{t_0 \delta}$$

$$= \frac{1}{v_0 t_0 \delta} \int_0^C \left[ 1 - (1 + hx_\delta)\left(1 + \frac{\eta_0}{1+\eta_0}\delta\right) \Big/ \sqrt{1 + 2\frac{1+\eta_0}{2+\eta_0}\delta + \frac{\eta_0}{2+\eta_0}\delta^2 - a_\delta^2} \right] ds$$

$$= \eta_1^{\mathrm{ph}} + \eta_2^{\mathrm{ph}}\delta + \cdots . \tag{10.3}$$

Note that the variable $\delta$ is defined as $\Delta K / K_0$. In a pure magnetic system, momentum is the more natural variable since it scales linearly with the magnetic field. To this end, we recall eq. (3.16) and have

$$\left(\frac{p}{p_0}\right)^2 = \frac{\eta\,(2+\eta)}{\eta_0\,(2+\eta_0)}.$$

Using the relations

$$\frac{p}{p_0} = 1 + \frac{\Delta p}{p_0} \equiv 1 + \delta_p, \quad \eta = \eta_0\,(1+\delta),$$

we obtain

$$(1+\delta_p)^2 = (1+\delta)\left(1 + \frac{\eta_0}{2+\eta_0}\delta\right).$$

After a little bit of algebraic manipulations, the exact functional relation between $\delta$ and $\Delta p/p_0$ is obtained, which is

$$\delta = \frac{1+\eta_0}{\eta_0}\left[ -1 + \sqrt{1 + \frac{\eta_0\,(2+\eta_0)}{(1+\eta_0)^2}\,(2\delta_p + \delta_p^2)} \right].$$

As a result, the phase slip factor can be written as

$$\eta^{\mathrm{ph}} \equiv -\frac{\Delta t}{t_0 \delta_p}$$

$$= \frac{1}{v_0 t_0 \delta_p} \int_0^C \left[ 1 - (1 + hx_\delta)\sqrt{1 + \frac{\eta_0\,(2+\eta_0)}{(1+\eta_0)^2}\,(2\delta_p + \delta_p^2)} \Big/ \sqrt{(1+\delta_p)^2 - a_\delta^2} \right] ds$$

$$= \eta_1^{\mathrm{ph}} + \eta_2^{\mathrm{ph}}\delta_p + \cdots . \tag{10.4}$$

Taylor expanding eq. (10.4) to the leading order and taking into account the fact that

$$x_\delta = D\frac{\Delta p}{p_0},$$

we obtain

$$\eta_1^{\mathrm{ph}} = \frac{1}{(1+\eta_0)^2} - \frac{1}{C}\int_0^C \frac{D\,(s)}{\rho\,(s)}\,ds = \frac{1}{\gamma_0^2} - \frac{1}{C}\int_0^C \frac{D\,(s)}{\rho\,(s)}\,ds.$$

The first term is phase slip due to the difference in velocity and the second term is that from the difference in path length, which is called the first order momentum compaction factor $\alpha_1$. Since the velocity difference is inverse proportional to the square of $\gamma_0$, it becomes smaller as the energy increases. At the point that the two terms are equal, the phase slip factor changes sign, which is called the **transition**, which is $\gamma_{tr} = 1/\sqrt{\alpha_1}$. The significance of the transition is that the synchronous phase changes from a stable fixed point to an unstable one or vice versa. Specifically, for energy below transition, we have $\eta_1^{ph} > 0$, which entails that particles with higher energy arrive earlier. Therefore the synchronous phase is a stable fixed point when $\phi_s$ lies between $-\pi/2$ and $0$ (Fig. 10.3). For energy above transition, we have $\eta_1^{ph} < 0$, which entails that particles with higher energy arrive later. Therefore the synchronous phase is a stable fixed point when $\phi_s$ lies between $0$ and $\pi/2$ (Fig. 10.4). In practice, the phase of the RF cavity has to be changed quickly from $\phi_s$ to $-\phi_s$ in order to keep the beam confined, which is called the **transition jump**. It is not unusual that during acceleration, most beam loss occurs around transition jump. Figs. 10.3 and 10.4 also show that the maximum energy width of particles confined in the longitudinal phase space increases when the peak voltage of the RF cavity increases and/or $|\eta_1^{ph}|$ decreases.

The second order effect becomes important when the first order term is small enough. From eq. (10.3), we can easily obtain the second order phase slip factor. The only complication is that the second order term of $\delta_p$ has to be included. To the second order the relation becomes

$$\delta =_2 -\frac{v_0}{\kappa}\left(\delta_p + \frac{1}{2}\frac{1}{\gamma_0^2}\delta_p^2\right).$$

Plugging the equations

$$x_\delta = D\delta_p + D_2\delta_p^2, \quad a_\delta = D'\delta_p$$

into eq. (10.4) and expanding it to the second order, we have

$$\eta_2^{ph} = -\frac{1}{2\gamma_0^2}\left(3 - \frac{1}{\gamma_0^2}\right) - \frac{1}{C}\int_0^C \left(\frac{1}{2}\left(D'(s)\right)^2 + \frac{D_2(s)}{\rho(s)} - \frac{1}{\gamma_0^2}\frac{D(s)}{\rho(s)}\right)ds.$$

The integral form of the phase slippage factor shows clearly which quantity contributes. For example, only dispersion in the bending magnet contributes to $\eta_1^{ph}$ and the slope of the dispersion everywhere contributes to $\eta_2^{ph}$. Knowledge of this kind helps greatly during the design of a ring. The computation of the phase slip factor, on the other hand, can be done through applying the periodic solution of the dispersion function (including nonlinear terms) to the fifth variable of the one turn map, which is

$$\eta^{ph} = -\frac{\Delta t}{t_0\delta_p} = \frac{1}{C\delta_p}\mathcal{M}_l\left(x_\delta, a_\delta, \delta_p\right). \tag{10.5}$$

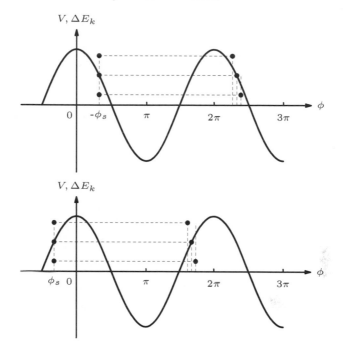

**FIGURE 10.3:** Sketch of phase stability for energy below transition, showing stable and unstable motion near the fixed points $\phi = \phi_s$ and $\phi = -\phi_s$, respectively.

Again, the fifth and the sixth variables here are defined as $-v_0 (t - t_0)$ and $\delta_p$. The one turn map $\mathcal{M}$ can be easily obtained using Differential Algebraic (DA) technique and the periodic solution of the dispersion function up to arbitrary orders can be obtained using the procedure of finding the parameter dependent fixed point (see Section 8.2.1). As an example, we study in detail the first order phase slip factor. The one turn linear matrix of the horizontal and the longitudinal phases spaces can be written as

$$
\hat{M} = \begin{pmatrix}
(x|x) & (x|a) & 0 & (x|\delta) \\
(a|x) & (a|a) & 0 & (a|\delta) \\
(l|x) & (l|a) & 1 & (l|\delta) \\
0 & 0 & 0 & 1
\end{pmatrix}.
$$

From eq. (10.5), we have

$$
\eta_1^{\mathrm{ph}} = \frac{1}{C} \left[ (l|x)D + (l|a)D' + (l|\delta) \right]. \tag{10.6}
$$

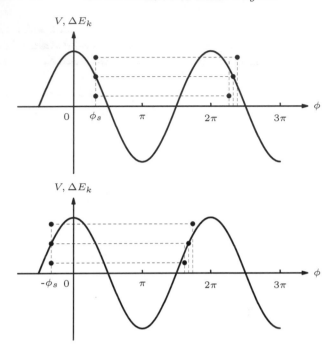

**FIGURE 10.4:** Sketch of phase stability for energy above transition, showing stable and unstable motion near the fixed points $\phi = \phi_s$ and $\phi = -\phi_s$, respectively.

Using the $(l|x)$ and the $(l|a)$ equations of (5.10) and eq. (8.3), we find that

$$
\begin{aligned}
(l|x) &= (a|x)(x|\delta) - (x|x)(a|\delta) \\
&= (a|x)\left[[1 - (x|x)]D - (x|a)D'\right] - (x|x)\left[-(a|x)D + [1 - (a|a)]D'\right] \\
&= (a|x)D + [1 - (x|x)]D',
\end{aligned}
\tag{10.7}
$$

and

$$
\begin{aligned}
(l|a) &= (a|a)(x|\delta) - (x|a)(a|\delta) \\
&= (a|a)\left[[1 - (x|x)]D - (x|a)D'\right] - (x|a)\left[-(a|x)D + [1 - (a|a)]D'\right] \\
&= -[1 - (a|a)]D - (x|a)D'.
\end{aligned}
\tag{10.8}
$$

As result, we have

$$
\begin{aligned}
\eta_1^{\mathrm{ph}} &= \frac{1}{C}\left\{(a|x)D^2 + [(a|a) - (x|x)]DD' - (x|a)D'^2 + (l|\delta)\right\} \\
&= -\frac{\sin\theta}{C}\left(\gamma_x D^2 + 2\alpha_x DD' + \beta_x D'^2\right) + \frac{(l|\delta)}{C} = -\frac{1}{C}\left[\mathcal{H}\sin\theta - (l|\delta)\right].
\end{aligned}
$$

Before finishing the section, let us find out the effect of a sextupole on the second order phase flip factor. Let us consider a sextupole with strength

$k_s = b_2/B\rho$. For a thin slice of it, the effect is $\Delta a = -k_s ds x^2$. From eq. (11.1), we obtain the change of the periodic solution of the second order dispersion function at the location of the sextupole slice, which is

$$\begin{pmatrix} \Delta D_2 \\ \Delta D_2' \end{pmatrix} = -\frac{k_s ds D^2}{2\sin(\pi\nu)} \begin{pmatrix} \beta\cos(\pi\nu) \\ \sin(\pi\nu) - \alpha\cos(\pi\nu) \end{pmatrix},$$

where $\nu = \theta/2\pi$. Plugging it into eq. (10.6) and using eqs. (10.7) and (10.8), we obtain

$$\Delta\eta_2^{ph}(s) = \frac{1}{C}[(l|x)\Delta D_2 + (l|a)\Delta D_2']$$

$$= -\frac{1}{C}\frac{k_s ds D^2}{2\sin(\pi\nu)}\{[(a|x)D + (1-(x|x))D']\beta\cos(\pi\nu)$$

$$- [(1-(a|a))D + (x|a)D'](\sin(\pi\nu) - \alpha\cos(\pi\nu))\}.$$

After straightforward algebraic and trigonometric manipulations, we arrive at a simple result, which is

$$\Delta\eta_2^{ph}(s) = \frac{1}{C}k_s D^3 ds,$$

and the total change of the second order phase slip factor is

$$\Delta\eta_2^{ph} = \frac{1}{C}\int_0^C k_s(s)D^3 ds.$$

Using the same method, we can easily obtain the result where both horizontal and vertical dispersion are present but coupling is corrected. Here the kick of the sextupole slice is

$$\begin{pmatrix} \Delta a \\ \Delta b \end{pmatrix} = -k_s ds \begin{pmatrix} x^2 - y^2 \\ -2xy \end{pmatrix},$$

and the change of the periodic solution of the second order dispersion function at the location of the sextupole slice is

$$\begin{pmatrix} \Delta D_{x2} \\ \Delta D_{x2}' \end{pmatrix} = -\frac{k_s ds\left(D_x^2 - D_y^2\right)}{2\sin(\pi\nu_x)} \begin{pmatrix} \beta_x\cos(\pi\nu_x) \\ \sin(\pi\nu_x) - \alpha_x\cos(\pi\nu_x) \end{pmatrix},$$

$$\begin{pmatrix} \Delta D_{y2} \\ \Delta D_{y2}' \end{pmatrix} = \frac{k_s ds D_x D_y}{\sin(\pi\nu_y)} \begin{pmatrix} \beta_y\cos(\pi\nu_y) \\ \sin(\pi\nu_y) - \alpha_y\cos(\pi\nu_y) \end{pmatrix}.$$

Plugging it into the 4D version of eq. (10.6), using eqs. (10.7) and (10.8) and the similar relations for $(l|y)$ and $(l|b)$, we obtain

$$\Delta \eta_2^{\mathrm{ph}}(s) = \frac{1}{C} \left[ (l|x)\Delta D_{x2} + (l|a)\Delta D'_{x2} + (l|y)\Delta D_{y2} + (l|b)\Delta D'_{y2} \right]$$

$$= -\frac{1}{C}\frac{k_s ds\left(D_x^2 - D_y^2\right)}{2\sin(\pi\nu_x)}\left\{ [(a|x)D_x + (1-(x|x))D'_x]\beta_x\cos(\pi\nu_x) \right.$$

$$\left. - [(1-(a|a))D_x + (x|a)D'_x](\sin(\pi\nu_x) - \alpha_x\cos(\pi\nu_x)) \right\}$$

$$+ \frac{1}{C}\frac{k_s ds D_x D_y}{\sin(\pi\nu_y)}\left\{ [(b|y)D_y + (1-(y|y))D'_y]\beta_y\cos(\pi\nu_y) \right.$$

$$\left. - [(1-(b|b))D_y + (y|b)D'_y](\sin(\pi\nu_y) - \alpha_y\cos(\pi\nu_y)) \right\}.$$

Taking into account the fact that the vertical part in the curly brackets is the same as that of the horizontal part, we immediately arrive at the final result, which is

$$\Delta \eta_2^{\mathrm{ph}} = \frac{1}{C}\int_0^C k_s(s)\left(D_x^3 - 3D_x D_y^2\right) ds.$$

## 10.3   Longitudinal Dynamics

Based on the previous two sections, we can construct the one turn map with the RF cavity present. From eq. (10.2), we can write the general form of energy gain

$$\Delta K(r,t) = qV_0(r)\cos[\phi(t)],$$

where $\phi(t) = \omega t$. For the TM$_{010}$ mode of a pillbox cavity, we have

$$V_0(r) = E_0 J_0\left(\frac{x_{01} r}{R_c}\right) LT,$$

where $L$, instead of $l$, is used to represent the length of the cavity to avoid confusion. Converting to the canonical coordinates, we have

$$\phi(l) = \phi_0 + \frac{\omega}{\kappa} l,$$

and

$$\delta_f(r,l) = \frac{K_0}{K_0 + \Delta K(0,0)}\delta_i + \frac{\Delta K(r,l) - \Delta K(0,0)}{K_0 + \Delta K(0,0)}$$

$$= \frac{K_0}{K_0 + qE_0 LT\cos(\phi_0)}\delta_i + \frac{qE_0 LT}{K_0 + qE_0 LT\cos(\phi_0)}$$

$$\cdot \left[ J_0\left(\frac{x_{01} r}{R_c}\right)\cos\left(\phi_0 + \frac{\omega}{\kappa} l\right) - \cos(\phi_0) \right].$$

As shown in Section 10.1, the transverse focusing is negligible. Together with the fact that the change of velocity is insignificant, the cavity is simply a drift space for the variable $x$, $a$, $y$, $b$ and $l$. As a result, we can treat the cavity as a thin slice with a kick in energy and the map of the cavity is

$$x_f = x_i, \quad a_f = \frac{p_{0i}}{p_{0f}} a_i,$$

$$y_f = y_i, \quad b_f = \frac{p_{0i}}{p_{0f}} b_i, \quad l_f = l_i,$$

$$\delta_f = \frac{K_0}{K_0 + qE_0LT\cos(\phi_0)} \delta_i + \frac{qE_0LT}{K_0 + qE_0LT\cos(\phi_0)}$$
$$\cdot \left[ J_0 \left( \frac{x_{01}\sqrt{x_i^2 + y_i^2}}{R_c} \right) \cos\left(\phi_0 + \frac{\omega}{\kappa} l_i\right) - \cos(\phi_0) \right],$$

where

$$p_{0i} = \sqrt{(K_0 + mc^2)^2 - m^2c^4},$$

$$p_{0f} = \sqrt{(K_0 + qE_0LT\cos(\phi_0) + mc^2)^2 - m^2c^4}.$$

Obviously, the relative transverse momentum decreases as the particles are accelerated and so is the phase space volume, even though that for the variables $(x, p_x, y, p_y, -\Delta t, \Delta K)$ is conserved.

It is clear that when $|\phi_0| < \pi/2$, the reference particle is accelerated and the relative energy deviation of the particles at the vicinity decreases on average. Together with the rest of the ring, we have the one turn map, which is

$$\mathcal{M}^T = \mathcal{M}^{CAV} \circ \mathcal{M}^{RING}.$$

For the most general case, $x$, $a$, $y$ and $b$ are functions of $\delta$ and $l$ is a function of $x$, $a$, $y$, $b$ and $\delta$. As a result, the cavity couples the longitudinal degree of freedom to the transverse degrees of freedom. Yet, due to the large difference in oscillation frequencies which will become clear soon, the coupling is much weaker than that between the horizontal and the vertical planes. This is particularly the case when the cavity is located in a dispersion free region, where coupling is limited to the nonlinear part of the map. In reality, it is common practice to place cavities in dispersion free regions to achieve separation of the longitudinal and the transverse motions. In the rest of this section, we always assume that there is no dispersion at the location of the cavity and ignore the chromatic terms in the nonlinear part of the transverse map ($x$, $a$, $y$ and $b$). Furthermore, we ignore the spatial dependence of the accelerating field due to the fact that $r \ll R_c$ and the difference is second order in $r$. Consequently, the longitudinal degree of freedom is decoupled from the transverse degrees of freedom and the longitudinal map is

$$l_f = l_i + \mathcal{M}_l \left( \delta_i \right),$$

$$\delta_f = \frac{K_0}{K_0 + qE_0LT\cos\left(\phi_0\right)} \delta_i + \frac{qE_0LT}{K_0 + qE_0LT\cos\left(\phi_0\right)}$$
$$\cdot \left[ \cos\left(\phi_0 + \frac{\omega}{\kappa}l_f\right) - \cos\left(\phi_0\right) \right].$$

Next, we keep only the linear terms in the map and solve for the oscillation frequency. The linear map is

$$l_f = l_i + (l|\delta)\delta_i,$$

$$\delta_f = \frac{K_0}{K_0 + qE_0LT\cos\left(\phi_0\right)} \delta_i$$
$$+ \frac{qE_0LT}{K_0 + qE_0LT\cos\left(\phi_0\right)} \frac{\omega}{\kappa} \left( l_i + (l|\delta)\delta_i \right) \sin\left(\phi_0\right).$$

Writing in matrix form, we have

$$\begin{pmatrix} l_f \\ \delta_f \end{pmatrix} = \begin{pmatrix} 1 & (l|\delta) \\ M_{\delta l} & M_{\delta\delta} \end{pmatrix} \begin{pmatrix} l_i \\ \delta_i \end{pmatrix},$$

where

$$M_{\delta l} = \frac{qE_0LT}{K_0 + qE_0LT\cos\left(\phi_0\right)} \frac{\omega}{\kappa} \sin\left(\phi_0\right),$$

$$M_{\delta\delta} = \frac{1}{K_0 + qE_0LT\cos\left(\phi_0\right)} \left[ K_0 + qE_0LT\frac{\omega}{\kappa}(l|\delta)\sin\left(\phi_0\right) \right].$$

It is easy to verify that the determinant of the matrix is

$$\frac{K_0}{K_0 + qE_0LT\cos\left(\phi_0\right)},$$

which entails that the motion is non-symplectic when $qE_0LT\cos(\phi_0) \neq 0$. Furthermore, the longitudinal emittance of the beam decreases as the beam is accelerated. It is obvious that the same is true for the transverse emittance, which is called adiabatic damping. By redefining the relative energy deviation as

$$\widetilde{\delta} = \frac{\Delta K}{K_0},$$

we obtain the new matrix

$$\begin{pmatrix} l_f \\ \widetilde{\delta}_f \end{pmatrix} = \begin{pmatrix} 1 & (l|\widetilde{\delta}) \\ M_{\widetilde{\delta}l} & M_{\widetilde{\delta}\widetilde{\delta}} \end{pmatrix} \begin{pmatrix} l_i \\ \widetilde{\delta}_i \end{pmatrix},$$

where

$$M_{\widetilde{\delta}l} = \frac{qE_0LT}{K_0}\frac{\omega}{\kappa}\sin(\phi_0),$$

$$M_{\widetilde{\delta}\widetilde{\delta}} = \frac{1}{K_0}\left[K_0 + qE_0LT\frac{\omega}{\kappa}(l|\widetilde{\delta})\sin(\phi_0)\right],$$

which is symplectic. The trace is

$$\text{tr}\,\hat{M} = 1 + \frac{1}{K_0}\left[K_0 + qE_0LT\frac{\omega}{\kappa}(l|\widetilde{\delta})\sin(\phi_0)\right]$$

$$= 2 + \frac{qE_0LT}{K_0}\frac{\omega}{\kappa}(l|\widetilde{\delta})\sin(\phi_0).$$

For $(l|\widetilde{\delta}) > 0$, $\text{tr}\,\hat{M} < 2$ if $0 < \phi_0 < \pi$; for $(l|\widetilde{\delta}) < 0$, $\text{tr}\,\hat{M} < 2$ if $-\pi < \phi_0 < 0$. In other words, for energy below transition, the synchrotron motion is stable if the synchronous phase $\phi_0 \in (-\pi, 0)$ and, for energy above transition, the synchronous motion is stable if $\phi_0 \in (0, \pi)$. If we restrict ourselves to the case of acceleration, the stable interval of the synchronous phase is $(-\pi/2, 0)$ below transition and $(0, \pi/2)$ above transition. This is the quantitative statement of the fact mentioned in the last section. Using the relation

$$\text{tr}\,\hat{M} = 2\cos(\mu_t),$$

we have

$$\mu_t = \arccos\left[1 - \frac{qE_0LT}{2K_0}(-\kappa)(l|\delta)\sin(\phi_0)\right].$$

It is worth noting that for most accelerators the relation

$$\frac{qE_0LT}{2K_0}(-\kappa)(l|\delta)\sin(\phi_0) \ll 1$$

holds. Making use of the fact that

$$\arccos x = \arcsin\sqrt{1 - x^2},$$

we obtain

$$\arccos(1 - x) = \arcsin\sqrt{1 - (1-x)^2} = \arcsin\left(\sqrt{2x - x^2}\right) =_1 \sqrt{2x}.$$

As a result, we have

$$\mu_t = \sqrt{\frac{qE_0LT}{K_0}(-\kappa)(l|\delta)\sin(\phi_0)} = \sqrt{\frac{2\pi h\,(qE_0LT)\,\eta_1^{\text{ph}}\sin(\phi_0)}{K_0}}.$$

Note that $h$ is the so-called harmonic number, which is the ratio of the RF frequency to that of the revolution frequency of the ring $\omega_0$, and $\eta_1^{\text{ph}}$ is the first order phase slippage factor defined in eq. (10.3). The quantity $\mu_t$ is called the **synchrotron tune** which is proportional to the square root of the harmonic number, the accelerating voltage and the slippage factor. Usually the slippage factor is expressed in terms of $\Delta p/p_0$, as defined in eq. (10.4), denoted here as $\eta_1^p$. Hence the relation between the two is

$$\eta_1^{\text{ph}} = \frac{\gamma_0}{1 + \gamma_0} \eta_1^p.$$

Taking into account that $\mu_t$ is the phase advance per turn, the synchrotron tune in terms of revolution per second can be written as

$$\omega_t = \frac{1}{t_0} \sqrt{\frac{2\pi h \left(qE_0 LT\right) \eta_1^p \sin\left(\phi_0\right)}{\beta_0^2 \gamma_0 mc^2}} = \omega_0 \sqrt{\frac{h \left(qE_0 LT\right) \eta_1^p \sin\left(\phi_0\right)}{2\pi \beta_0^2 \gamma_0 mc^2}},$$

which is the usual form that appears in most textbooks. For a circular accelerator with GeV level energy, the synchrotron tune $\omega_t$ is usually between 0.1% and 1% of the revolution frequency $\omega_0$. The betatron tunes, on the other hand, are usually between a few to a few hundred times of $\omega_0$. As a result, the coupling between the betatron and the synchrotron motions is usually weak. As an example, let us take a look at the storage ring of the Advanced Light Source (ALS) at Lawrence Berkeley National Laboratory (LBNL, LBL), California, USA, which is an electron machine that operates at the energy of 1.9 GeV. The main purpose of the RF cavity is to restore the energy loss due to synchrotron radiation from bending magnets and insertion devices, which is on the order of 0.5 MeV per turn per electron. The harmonic number is 328 and the slippage factor is roughly $1.4 \times 10^{-3}$. As a result, we have

$$\frac{\omega_t}{\omega_0} = \sqrt{\frac{328 \times 1.4 \times 10^{-3} \times 0.5}{2\pi \times 3718 \times 0.511}} = 4.4 \times 10^{-3}.$$

Using eqs. (8.1), we obtain

$$\alpha_t = -\frac{qE_0 LT}{2K_0 \sin \mu_t} \frac{\omega}{\kappa} (l|\delta) \sin\left(\phi_0\right) =_1 \frac{1}{2} \mu_t,$$

$$\beta_t = \frac{(l|\delta)}{\sin \mu_t} =_1 \frac{(l|\delta)}{\mu_t},$$

$$\gamma_t = -\frac{qE_0 LT}{K_0 \sin \mu_t} \frac{\omega}{\kappa} \sin\left(\phi_0\right) =_1 \frac{\mu_t}{(l|\delta)}.$$

Since $\mu_t \ll 1$, we have $\alpha_t \ll 1$. Consequently, the invariant ellipse is basically upright. For a given longitudinal emittance $\epsilon_t$, the maximum bunch length is

$$l_{\max} = -2\frac{v_0}{\kappa} \sqrt{\beta_t \epsilon_t} =_1 -2\frac{v_0}{\kappa} \sqrt{\frac{(l|\delta)\epsilon_t}{\mu_t}},$$

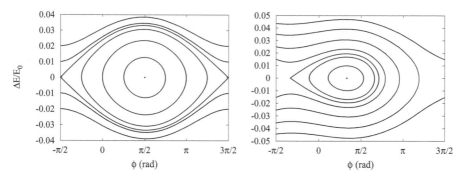

**FIGURE 10.5**: Phase space plots of longitudinal motion with $\phi_s = \pi/2$ (left) and $\phi_s = \pi/3$ (right).

and the maximum energy spread is

$$\delta_{\max} = 2\sqrt{\gamma_t \epsilon_t} =_1 2\sqrt{\frac{\mu_t \epsilon_t}{(l|\delta)}}.$$

Using the relation

$$(l|\delta) = -\kappa t_0 \eta_1^{\mathrm{ph}} = \frac{\kappa^2 C}{v_0} \eta_1^p,$$

we obtain

$$l_{\max} = 2\sqrt{C \epsilon_t} \left( \frac{\eta_1^p \beta_0^2 \gamma_0 mc^2}{2\pi h \left( q E_0 L T \right) \sin \left( \phi_0 \right)} \right)^{\frac{1}{4}},$$

and

$$\delta_{\max} = -2\frac{v_0}{\kappa} \sqrt{\frac{\epsilon_t}{C}} \left( \frac{2\pi h \left( q E_0 L T \right) \sin \left( \phi_0 \right)}{\eta_1^p \beta_0^2 \gamma_0 mc^2} \right)^{\frac{1}{4}}.$$

The main result is that $l_{\max} \propto (\eta_1^p)^{\frac{1}{4}}$ if other parameters remain unchanged. Hence, one way to reduce bunch length is to decrease the phase slippage factor. Fig. 10.5 shows the details of the dynamics of all amplitude for $\phi_s = \pi/2$ (left) and $\phi_s = \pi/3$ (right). Note that particles that are outside the stable region (called the RF bucket) lose synchronicity with the electromagnetic field in the RF cavity. For the case $\phi_s < \pi/2$, those particles would not be accelerated the same way as those inside the RF bucket and usually will be lost.

## 10.4 Transverse Dynamics of RF Cavities

Up to now, we have only treated the main effect of RF cavities, which is to accelerate (or, occasionally, decelerate) charged particles. We have shown

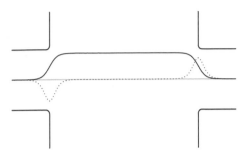

**FIGURE 10.6:** Longitudinal (solid) and transverse (dotted) field distribution along the longitudinal axis.

that, in order to accelerate the charged particles effectively, the synchronous phase has to be a stable fixed point, limiting the synchronous phase to just one quadrant (see Section 10.2). As shown below, this has significant effect on the transverse dynamics of linacs. In rings, the transverse effect of RF cavities is negligible due to the presence of magnets.

Let us start with the longitudinal distribution of the electromagnetic field of a pillbox cavity with small holes at the center of each end, which will be derived from Maxwell's equations. From eq. (10.1), we can express the longitudinal component of the electric field as

$$E_z \left( z, t \right) = E_0 \cos \left( \omega t + \varphi \right) H \left( z + L/2 \right) H \left( L/2 - z \right),$$

where $L$ is the length of the cavity and $H$ is the Heaviside step function. From $\nabla \cdot \vec{E} = 0$, we obtain

$$\frac{1}{r} \frac{\partial \left( r E_r \right)}{\partial r} + \frac{\partial E_z}{\partial z} = 0.$$

After integration, we obtain the leading order transverse component of the electric field, which is

$$E_r = -\frac{r}{2} \frac{\partial E_z}{\partial z} = -\frac{r}{2} E_0 \cos \left( \omega t + \varphi \right) \left[ \delta \left( z + L/2 \right) - \delta \left( z - L/2 \right) \right].$$

Fig. 10.6 shows the longitudinal dependence of $E_z$ and $E_r$ at an instance. From $\left( \nabla \times \vec{B} \right)_z = \left( 1/c^2 \right) \left( \partial E_z / \partial t \right)$, we obtain

$$\frac{1}{r} \frac{\partial \left( r B_\theta \right)}{\partial r} = \frac{1}{c^2} \frac{\partial E_z}{\partial t}.$$

Similarly, we obtain the leading order magnetic field, which is

$$B_\theta = \frac{r}{2c^2} \frac{\partial E_z}{\partial t} = -\frac{\omega r}{2c^2} E_0 \sin \left( \omega t + \varphi \right) H \left( z + L/2 \right) H \left( L/2 - z \right).$$

From the Lorentz force law, eq. (1.1), we obtain

$$\frac{dp_r}{dt} = q \left( E_r - v_z B_\theta \right).$$

As a result, the change in transverse momentum across the cavity is

$$\Delta p_r = q \int dt \, (E_r - v_z B_\theta) = q \int_{-L/2-\varepsilon}^{L/2+\varepsilon} dz \left( \frac{E_r}{v_z} - B_\theta \right)$$

$$= -\frac{q}{2} E_0 \left[ \frac{r_1}{v_{z1}} \cos \left( \omega t_1 + \varphi \right) - \frac{r_2}{v_{z2}} \cos \left( \omega t_2 + \varphi \right) \right]$$

$$+ \frac{q\omega}{2c^2} E_0 \int_{-L/2}^{L/2} dz r\,(z) \sin \left[ \omega t\,(z) + \varphi \right]$$

$$\simeq -\frac{q}{2c} E_0 \left[ \frac{r_1}{\beta_1} \cos \left( \omega t_1 + \varphi \right) - \frac{r_2}{\beta_2} \cos \left( \omega t_2 + \varphi \right) \right]$$

$$+ \frac{q\omega}{2c^2} E_0 \int_{-L/2}^{L/2} dz r\,(z) \sin \left[ \omega t\,(z) + \varphi \right], \tag{10.9}$$

where $r_1, v_{z1}, t_1, \beta_1$ and $r_2, v_{z2}, t_2, \beta_2$ are values of $r, v_z, t, \beta$ at $z = -L/2$ and $z = L/2$, respectively. When the charged particle is non-relativistic, i.e., $\beta \ll 1$, the contribution of the magnetic field is negligible. Hence eq. (10.9) becomes

$$\Delta p_r = -\frac{q}{2c} E_0 \left[ \frac{r_1}{\beta_1} \cos \left( \omega t_1 + \varphi \right) - \frac{r_2}{\beta_2} \cos \left( \omega t_2 + \varphi \right) \right],$$

where $t_1 = -\int_{-L/2}^{0} dz/[\beta(z)c]$ and $t_2 = \int_{0}^{L/2} dz/[\beta(z)c]$. Note that the origin of $t$ is set at the moment when the particle is located at the center of the cavity. If we make one more assumption that $|\beta_2 - \beta_1|/\beta(0) \ll 1$, which means that the energy gain (loss) through the cavity is much smaller than the total energy of the particle, we obtain that $t_1 = -L/(2\beta_0 c)$ and $t_2 = L/(2\beta_0 c)$ ($\beta_0 = \beta(0)$). Consequently, we obtain

$$\Delta p_r = -\frac{q}{2c} E_0 \left[ \frac{r_1}{\beta_1} \cos \left( \frac{\omega L}{2\beta_0} - \varphi \right) - \frac{r_2}{\beta_2} \cos \left( \frac{\omega L}{2\beta_0} + \varphi \right) \right]$$

$$= -\frac{q}{2c} E_0 \left[ \frac{r_1}{\beta_1} \cos \left( \frac{\pi L}{\beta_0 \lambda} - \varphi \right) - \frac{r_2}{\beta_2} \cos \left( \frac{\pi L}{\beta_0 \lambda} + \varphi \right) \right].$$

Let us take a look at drift tube linacs as described in Section 1.3.2. Phase stability requires that $-\pi/2 < \varphi < 0$ (see Fig. 10.3). Efficient use of energy requires that the particles are accelerated throughout the gap, i.e., $\cos \left( \varphi - \pi L/(\beta_0 \lambda) \right) > 0$ and $\cos \left( \varphi + \pi L/(\beta_0 \lambda) \right) > 0$. As a result, the particle is focused at the entrance of the gap and defocused at the exit, as shown in Fig. 1.13. Furthermore, we note that $\cos \left( \varphi - \pi L/(\beta_0 \lambda) \right) > 0$ entails that $\varphi - \pi L/(\beta_0 \lambda) > -\pi/2$, which leads to $\pi L/(\beta_0 \lambda) < \varphi + \pi/2$ and $\varphi + \pi L/(\beta_0 \lambda) < 2\varphi + \pi/2$. If $-\pi/2 < \varphi \le \pi/4$, $\varphi + \pi L/(\beta_0 \lambda) < 0$ and we obtain that $\cos \left( \varphi + \pi L/(\beta_0 \lambda) \right) > \cos \left( \varphi - \pi L/(\beta_0 \lambda) \right)$. If $-\pi/4 < \varphi < 0$, $\varphi + \pi L/(\beta_0 \lambda)$ can be positive where $\cos \left( \varphi + \pi L/(\beta_0 \lambda) \right)$ decreases as

$\varphi + \pi L/(\beta_0 \lambda)$ increases. From the fact that

$$\left| \varphi + \frac{\pi L}{\beta_0 \lambda} \right| - \left| \varphi - \frac{\pi L}{\beta_0 \lambda} \right| = \varphi + \frac{\pi L}{\beta_0 \lambda} - \left( \frac{\pi L}{\beta_0 \lambda} - \varphi \right) = 2\varphi < 0,$$

we reach the same conclusion that $\cos(\varphi + \pi L/(\beta_0 \lambda)) > \cos(\varphi - \pi L/(\beta_0 \lambda))$. Although we have $\beta_1 < \beta_2$, yet $\beta_2/\beta_1$ is usually much smaller than $\cos(\varphi + \pi L/(\beta_0 \lambda))/\cos(\varphi - \pi L/(\beta_0 \lambda))$. As a result, the net effect is defocusing. The remedy in the early dates was metal foils or grids placed on the entrance of the drift tubes (exit of the accelerating gap) to remove the defocusing force as shown in Fig. 1.13. Nowadays, quadrupole magnets are placed inside the drift tubes to provide transverse focusing.

At higher energy, the particles become relativistic and the contribution of the magnetic field has to be taken into account. Yet the matter is simplified somewhat by the fact that the difference between $\beta_1$ and $\beta_2$ can be neglected. Assuming also that $r$ remains a constant in the cavity, we obtain

$$\Delta p_r = -\frac{qr}{2\beta_0 c} E_0 \left[ \cos \left( \frac{\pi L}{\beta_0 \lambda} - \varphi \right) - \cos \left( \frac{\pi L}{\beta_0 \lambda} + \varphi \right) \right]$$

$$+ \frac{q\omega r}{2c^2} E_0 \int_{-L/2}^{L/2} dz \sin \left( \frac{2\pi z}{\beta_0 \lambda} + \varphi \right)$$

$$= -\frac{qr}{\beta_0 c} E_0 \sin \left( \frac{\pi L}{\beta_0 \lambda} \right) \sin \varphi + \frac{q\beta_0 r}{c} E_0 \sin \left( \frac{\pi L}{\beta_0 \lambda} \right) \sin \varphi$$

$$= -\frac{qr E_0}{\beta_0 \gamma_0^2 c} \sin \left( \frac{\pi L}{\beta_0 \lambda} \right) \sin \varphi$$

$$= -\frac{\pi q E_0 T L r}{\beta_0^2 \gamma_0^2 c \lambda} \sin \varphi.$$

Again, for the stable phase of $-\pi/2 < \varphi < 0$, the net effect is defocusing. It is worth noting that, for ultra-relativistic particles, this effect goes away. As $\beta_0 \to 1$, the defocusing from the electric field is canceled by the focusing from the magnetic field.

# Chapter 11

# *Resonances in Repetitive Systems

Unlike single pass systems, the dynamics of the beam in a repetitive system such as a **storage ring** is not necessarily dominated by the largest aberrations in the one turn transfer map. Due to the fact that particles go around many times, the impact of those terms that are nearly in phase with the linear motion are **amplified** and the dynamics is, to a large extent, shaped by them. The motion generated by one of those terms is called a resonance. As a result, we need a different way to evaluate the relative significance of the aberrations that is directly suited for rings.

Since **resonances** appear in various physical systems, many different methods have been developed to describe this phenomenon. Here we have adopted a method that is based on the map method that has been developed here and does not require advanced techniques such as **normal form theory** as in [5].

As shown in the previous chapters, a large class of single pass systems are imaging. Yet the entire ring cannot be imaging, since it will be linearly unstable if $|M| \neq 1$, where $M$ is the magnification. Moreover, the ring is unstable with the presence of arbitrarily small errors when $|M| = 1$.

## 11.1 Integer Resonance

In this section we will study the dynamics in a ring when one or more dipole magnets have errors in the field. Let us first consider the case that one magnet has a dipole error in the field, which is $\Delta B$. Without lost of generality, we adopt the thin lens approximation, since a thick dipole can always be cut into a number of thin slices and the contribution of the whole is the sum of each slice. The kick resulting from the error is $\Delta Bl/B\rho$. Thus the position and angle after one turn is

$$\begin{pmatrix} x_1 \\ a_1 \end{pmatrix} = \begin{pmatrix} 0 \\ \Delta Bl/B\rho \end{pmatrix} + \hat{M} \begin{pmatrix} x_0 \\ a_0 \end{pmatrix},$$

where $\hat{M}$ is the one turn map of the ideal ring and the kick happens at the end of one turn. Similarly, the position and angle after two turns is

$$
\begin{pmatrix} x_2 \\ a_2 \end{pmatrix} = \begin{pmatrix} 0 \\ \Delta Bl/B\rho \end{pmatrix} + \hat{M} \begin{pmatrix} x_1 \\ a_1 \end{pmatrix}
$$

$$
= \begin{pmatrix} 0 \\ \Delta Bl/B\rho \end{pmatrix} + \hat{M} \begin{pmatrix} 0 \\ \Delta Bl/B\rho \end{pmatrix} + \hat{M}^2 \begin{pmatrix} x_0 \\ a_0 \end{pmatrix}
$$

$$
= \left( \hat{I} + \hat{M} \right) \begin{pmatrix} 0 \\ \Delta Bl/B\rho \end{pmatrix} + \hat{M}^2 \begin{pmatrix} x_0 \\ a_0 \end{pmatrix}.
$$

Now let us assume

$$
\begin{pmatrix} x_{n-1} \\ a_{n-1} \end{pmatrix} = \left( \hat{I} + \hat{M} + \cdots + \hat{M}^{n-2} \right) \begin{pmatrix} 0 \\ \Delta Bl/B\rho \end{pmatrix} + \hat{M}^{n-1} \begin{pmatrix} x_0 \\ a_0 \end{pmatrix},
$$

and we obtain

$$
\begin{pmatrix} x_n \\ a_n \end{pmatrix} = \begin{pmatrix} 0 \\ \Delta Bl/B\rho \end{pmatrix} + \hat{M} \begin{pmatrix} x_{n-1} \\ a_{n-1} \end{pmatrix}
$$

$$
= \left( \hat{I} + \hat{M} + \cdots + \hat{M}^{n-1} \right) \begin{pmatrix} 0 \\ \Delta Bl/B\rho \end{pmatrix} + \hat{M}^n \begin{pmatrix} x_0 \\ a_0 \end{pmatrix}
$$

$$
= \left( \hat{I} - \hat{M}^n \right) \left( \hat{I} - \hat{M} \right)^{-1} \begin{pmatrix} 0 \\ \Delta Bl/B\rho \end{pmatrix} + \hat{M}^n \begin{pmatrix} x_0 \\ a_0 \end{pmatrix},
$$

where

$$
\left( \hat{I} - \hat{M}^n \right) \left( \hat{I} - \hat{M} \right)^{-1}
$$

$$
= \frac{1}{2(1 - \cos\mu)} \begin{pmatrix} 1 - \cos n\mu - \alpha \sin n\mu & -\beta \sin n\mu \\ \gamma \sin n\mu & 1 - \cos n\mu + \alpha \sin n\mu \end{pmatrix}
$$

$$
\cdot \begin{pmatrix} 1 - \cos\mu + \alpha \sin\mu & \beta \sin\mu \\ -\gamma \sin\mu & 1 - \cos\mu - \alpha \sin\mu \end{pmatrix}
$$

$$
= \frac{\sin(n\mu/2)}{\sin(\mu/2)} \begin{pmatrix} \sin(n\mu/2) - \alpha \cos(n\mu/2) & -\beta \cos(n\mu/2) \\ \gamma \cos(n\mu/2) & \sin(n\mu/2) + \alpha \cos(n\mu/2) \end{pmatrix}
$$

$$
\cdot \begin{pmatrix} \sin(\mu/2) + \alpha \cos(\mu/2) & \beta \cos(\mu/2) \\ -\gamma \cos(\mu/2) & \sin(\mu/2) - \alpha \cos(\mu/2) \end{pmatrix}
$$

$$
= \frac{\sin(n\mu/2)}{\sin(\mu/2)} \hat{M}^{(n-1)/2}.
$$

Note that

$$
\hat{M}^{(n-1)/2} = \begin{pmatrix} \cos\mu(n-1)/2 + \alpha \sin\mu(n-1)/2 & \beta \sin\mu(n-1)/2 \\ -\gamma \sin\mu(n-1)/2 & \cos\mu(n-1)/2 - \alpha \sin\mu(n-1)/2 \end{pmatrix}.
$$

When $\mu \rightarrow 2\pi k$, $\sin(n\mu/2)/\sin(\mu/2) \rightarrow n$. The result is that the motion is **divergent** in phase space and eventually the beam will be lost. In other words, the particle is called in **resonance** when $\mu = 2\pi k$ for some $k$. Since errors in dipole magnets are unavoidable, the only way to avoid this resonance is to adjust the tune away from any integer.

In theory, the phase space coordinates become arbitrarily large only when the tune is infinitely close to an integer. But in practice, the finite size of the beam pipe defines a finite interval around an integer such that the beam will be lost, which is called the **stop band**. In order to determine the stop band of a given machine, the size of the beam pipe and the dipole errors of the magnets have to be known. The source of dipole errors can be either the field errors in the dipole magnets and dipole component generated from misalignment of multipoles (quadrupoles, mainly, and sextupoles to a lesser extent). Since the dipole errors act on every particle the same way, it is sufficient to study the motion of the so-called centroid of the beam only. In a ring, this centroid, which is called the closed orbit, is the periodic solution of the one turn transfer map. With the presence of dipole errors, the one turn map is not origin preserving. Because the distorted closed orbit is usually close to the design orbit, only the linear part of the map is included in the treatment. The periodic solution of a single error at $s_0$ is obtained through the equation

$$\begin{pmatrix} x_0 \\ a_0 \end{pmatrix} = \begin{pmatrix} 0 \\ (\Delta Bl)(s_0)/B\rho \end{pmatrix} + \hat{M} \begin{pmatrix} x_0 \\ a_0 \end{pmatrix}.$$

As a result, we have

$$\begin{pmatrix} x_0 \\ a_0 \end{pmatrix} = \left( \hat{I} - \hat{M} \right)^{-1} \begin{pmatrix} 0 \\ (\Delta Bl)(s_0)/B\rho \end{pmatrix}$$

$$= \frac{(\Delta Bl)(s_0)}{B\rho} \frac{1}{2\sin(\mu/2)} \begin{pmatrix} \beta \cos(\mu/2) \\ \sin(\mu/2) - \alpha \cos(\mu/2) \end{pmatrix}. \tag{11.1}$$

From eq. (6.10), we obtain the closed orbit at an arbitrary location $s$

$$x_{co}(s) = \sqrt{\beta_s/\beta_{s0}} \left( \cos\bar{\phi} + \alpha \sin\bar{\phi} \right) x_0 + \sqrt{\beta_s\beta_{s0}} \sin\bar{\phi} \, a_0$$

$$= \frac{(\Delta Bl)_{s0}}{B\rho} \frac{1}{2\sin(\mu/2)} \left[ \sqrt{\beta_s/\beta_{s0}} \left( \cos\bar{\phi} + \alpha_{s0}\sin\bar{\phi} \right) \beta_{s0}\cos\frac{\mu}{2} \right.$$

$$\left. + \sqrt{\beta_s\beta_{s0}} \sin\bar{\phi} \left( \sin\frac{\mu}{2} - \alpha_{s0}\cos\frac{\mu}{2} \right) \right]$$

$$= \frac{(\Delta Bl)_{s0}}{B\rho} \frac{\cos\left(\pi\nu - \bar{\phi}\right)}{2\sin(\mu/2)},$$

where $\bar{\phi}$, $\beta_s$, $\beta_{s0}$, $\alpha_{s0}$ and $(\Delta Bl)_{s0}$ represent $\phi(s) - \phi(s_0)$, $\beta(s)$, $\beta(s_0)$, $\alpha(s_0)$ and $(\Delta Bl)(s_0)$, respectively. When $n$ dipole errors are present, the closed orbit is

$$x_{co}(s) = \frac{1}{2\sin(\mu/2)} \sum_{m=1}^{n} \frac{(\Delta Bl)(s_m)}{B\rho} \cos\left[\mu/2 - (\phi(s) - \phi(s_m))\right].$$

## 11.2   Half–Integer Resonance

In this section we study the effect of **quadrupole errors**. Again, let us first assume that only one quadrupole has an error in the field gradient, which is denoted as $\Delta k$. Without loss of generality, we assume that this quadrupole is located at the end of a turn and that it is thin. As a result, the linear one turn map is

$$
\begin{pmatrix} x_1 \\ a_1 \end{pmatrix} = \begin{pmatrix} 1 & 0 \\ -\Delta kl & 1 \end{pmatrix} \hat{M} \begin{pmatrix} x_0 \\ a_0 \end{pmatrix}
$$

$$
= \begin{pmatrix} 1 & 0 \\ -\Delta kl & 1 \end{pmatrix} \begin{pmatrix} \cos\mu + \alpha\sin\mu & \beta\sin\mu \\ -\gamma\sin\mu & \cos\mu - \alpha\sin\mu \end{pmatrix} \begin{pmatrix} x_0 \\ a_0 \end{pmatrix}.
$$

In order to simplify the calculation, we adopt a new coordinate system such that the unperturbed motion is a rotation. As a reminder of the general theory of transformation, let us assume that a matrix $\hat{A}$ transforms $\vec{x}$ into $\vec{y}$, i.e., $\vec{y} = \hat{A}\vec{x}$. Assuming another matrix $\hat{M}$ is the linear one turn map in the space of $\vec{x}$, which means that

$$
\vec{x}_1 = \hat{M}\vec{x}_0.
$$

Multiplying $\hat{A}$ to the left and inserting $\hat{A}^{-1}\hat{A}$ between $\hat{M}$ and $\vec{x}_0$, we obtain

$$
\hat{A}\vec{x}_1 = \hat{A}\hat{M}\hat{A}^{-1} \cdot \hat{A}\vec{x}_0,
$$

and hence

$$
\vec{y}_1 = \hat{A}\hat{M}\hat{A}^{-1}\vec{y}_0.
$$

It is straightforward to verify that

$$
\hat{A} = \begin{pmatrix} 1/\sqrt{\beta} & 0 \\ \alpha/\sqrt{\beta} & \sqrt{\beta} \end{pmatrix}
$$

implies

$$
\hat{A}\hat{M}\hat{A}^{-1} = \begin{pmatrix} 1/\sqrt{\beta} & 0 \\ \alpha/\sqrt{\beta} & \sqrt{\beta} \end{pmatrix} \begin{pmatrix} \cos\mu + \alpha\sin\mu & \beta\sin\mu \\ -\gamma\sin\mu & \cos\mu - \alpha\sin\mu \end{pmatrix} \begin{pmatrix} \sqrt{\beta} & 0 \\ -\alpha/\sqrt{\beta} & 1/\sqrt{\beta} \end{pmatrix}
$$

$$
= \begin{pmatrix} \cos\mu & \sin\mu \\ -\sin\mu & \cos\mu \end{pmatrix} \equiv \hat{R}(\mu).
$$

The relation between the coordinate systems is

$$
\begin{pmatrix} \tilde{x} \\ \tilde{a} \end{pmatrix} = \begin{pmatrix} 1/\sqrt{\beta} & 0 \\ \alpha/\sqrt{\beta} & \sqrt{\beta} \end{pmatrix} \begin{pmatrix} x \\ a \end{pmatrix}.
$$

The one turn perturbed map in the new coordinate system is

$$\begin{pmatrix} \tilde{x}_1 \\ \tilde{a}_1 \end{pmatrix} = \hat{A} \begin{pmatrix} 1 & 0 \\ -\Delta kl & 1 \end{pmatrix} \hat{M}\hat{A}^{-1} \begin{pmatrix} \tilde{x}_0 \\ \tilde{a}_0 \end{pmatrix} = \hat{A} \begin{pmatrix} 1 & 0 \\ -\Delta kl & 1 \end{pmatrix} \hat{A}^{-1} \cdot \hat{A}\hat{M}\hat{A}^{-1} \begin{pmatrix} \tilde{x}_0 \\ \tilde{a}_0 \end{pmatrix}$$

$$= \begin{pmatrix} 1 & 0 \\ -K_n & 1 \end{pmatrix} \hat{R}(\mu) \begin{pmatrix} \tilde{x}_0 \\ \tilde{a}_0 \end{pmatrix},$$

where $\Delta K$ above is defined as $\Delta K = \Delta kl\beta$. After $n$ turns, the coordinates are

$$\begin{pmatrix} \tilde{x}_n \\ \tilde{a}_n \end{pmatrix} = \left[ \begin{pmatrix} 1 & 0 \\ -\Delta K & 1 \end{pmatrix} \hat{R}(\mu) \right]^n \begin{pmatrix} \tilde{x}_0 \\ \tilde{a}_0 \end{pmatrix}.$$

In order to illustrate clearly the nature of the dynamics, we treat the problem perturbatively. To the first order, the coordinates are

$$\begin{pmatrix} \tilde{x}_n \\ \tilde{a}_n \end{pmatrix} = \left[ \left( \hat{R}(\mu) \right)^n + \begin{pmatrix} 0 & 0 \\ -\Delta K & 0 \end{pmatrix} \left( \hat{R}(\mu) \right)^n + \hat{R}(\mu) \begin{pmatrix} 0 & 0 \\ -\Delta K & 0 \end{pmatrix} \left( \hat{R}(\mu) \right)^{n-1} \right.$$

$$\left. + \cdots + \left( \hat{R}(\mu) \right)^{n-1} \begin{pmatrix} 0 & 0 \\ -\Delta K & 0 \end{pmatrix} \hat{R}(\mu) \right] \begin{pmatrix} \tilde{x}_0 \\ \tilde{a}_0 \end{pmatrix}$$

$$= \left[ \hat{R}(n\mu) - \Delta K \begin{pmatrix} 0 & 0 \\ \cos n\mu & \sin n\mu \end{pmatrix} \right.$$

$$\left. - \Delta K \sum_{m=1}^{n-1} \begin{pmatrix} \sin m\mu \cos[(n-m)\mu] & \sin m\mu \sin[(n-m)\mu] \\ \cos m\mu \cos[(n-m)\mu] & \cos m\mu \sin[(n-m)\mu] \end{pmatrix} \right] \begin{pmatrix} \tilde{x}_0 \\ \tilde{a}_0 \end{pmatrix}$$

$$= \left[ \hat{R}(n\mu) - \Delta K \begin{pmatrix} 0 & 0 \\ \cos n\mu & \sin n\mu \end{pmatrix} - \frac{\Delta K}{2}(n-1) \begin{pmatrix} \sin n\mu & -\cos n\mu \\ \cos n\mu & \sin n\mu \end{pmatrix} \right.$$

$$\left. - \frac{\Delta K}{2} \begin{pmatrix} \sum_{m=1}^{n-1} \sin[(2m-n)\mu] & \sum_{m=1}^{n-1} \cos[(2m-n)\mu] \\ \sum_{m=1}^{n-1} \cos[(2m-n)\mu] & -\sum_{m=1}^{n-1} \sin[(2m-n)\mu] \end{pmatrix} \right] \begin{pmatrix} \tilde{x}_0 \\ \tilde{a}_0 \end{pmatrix}.$$

In order to simplify the expression further, we take advantage of the trigonometrical series

$$\sum_{m=0}^{n-1} \sin(x+my) = \frac{\sin[x+y(n-1)/2]\sin(yn/2)}{\sin(y/2)},$$

$$\sum_{m=0}^{n-1} \cos(x+my) = \frac{\cos[x+y(n-1)/2]\sin(yn/2)}{\sin(y/2)}.$$

Specifically, we have

$$\sum_{m=1}^{n-1} \sin\left[(2m-n)\mu\right]$$

$$= \sum_{m=1}^{n-1} \sin\left[2\left(m-1\right)\mu - \left(n-2\right)\mu\right] = \sum_{m=0}^{n-2} \sin\left[-\left(n-2\right)\mu + 2m\mu\right]$$

$$= \frac{\sin\left[-\left(n-2\right)\mu + 2\mu\left(n-2\right)/2\right]\cdot\sin\left[(n-1)\mu\right]}{\sin\mu} = 0,$$

and

$$\sum_{m=1}^{n-1} \cos\left[(2m-n)\mu\right]$$

$$= \sum_{m=1}^{n-1} \cos\left[2\left(m-1\right)\mu - \left(n-2\right)\mu\right] = \sum_{m=0}^{n-2} \cos\left[-\left(n-2\right)\mu + 2m\mu\right]$$

$$= \frac{\cos\left[-\left(n-2\right)\mu + 2\mu\left(n-2\right)/2\right]\cdot\sin\left[(n-1)\mu\right]}{\sin\mu} = \frac{\sin\left[(n-1)\mu\right]}{\sin\mu}.$$

Hence the final result is

$$\begin{pmatrix} \tilde{x}_n \\ \tilde{a}_n \end{pmatrix} = \left[ \hat{R}\left(n\mu\right) - \Delta K \begin{pmatrix} 0 & 0 \\ \cos n\mu & \sin n\mu \end{pmatrix} \right.$$

$$\left. -\frac{\Delta K}{2}\left(n-1\right)\begin{pmatrix} \sin n\mu & -\cos n\mu \\ \cos n\mu & \sin n\mu \end{pmatrix} - \frac{\Delta K}{2}\frac{\sin\left[(n-1)\mu\right]}{\sin\mu}\begin{pmatrix} 0 & 1 \\ 1 & 0 \end{pmatrix} \right] \begin{pmatrix} \tilde{x}_0 \\ \tilde{a}_0 \end{pmatrix}.$$

When $\mu \to 2\pi k$ or $2\pi(k+1/2)$, $\sin[(n-1)\mu]/\sin\mu \to n-1$. Hence the resonance is called **half–integer resonance**.

When the tune is a certain distance away from the half–integer, the motion is still stable, but the invariant ellipse and the tune change. To obtain the perturbed invariant ellipse and tune, only the one turn map is needed. From the one turn matrix

$$\hat{M}^q = \begin{pmatrix} 1 & 0 \\ -\Delta kl & 1 \end{pmatrix} \begin{pmatrix} \cos\mu + \alpha\sin\mu & \beta\sin\mu \\ -\gamma\sin\mu & \cos\mu - \alpha\sin\mu \end{pmatrix}$$

$$= \begin{pmatrix} \cos\mu + \alpha\sin\mu & \beta\sin\mu \\ -\gamma\sin\mu - \left(\Delta K/\beta\right)\left(\cos\mu + \alpha\sin\mu\right) & -\Delta K\sin\mu + \cos\mu - \alpha\sin\mu \end{pmatrix},$$

we obtain

$$2\cos\left(\mu + \Delta\mu\right) = 2\cos\mu - \Delta K\sin\mu,$$
$$\cos\left(\mu + \Delta\mu\right) + \left(\alpha + \Delta\alpha\right)\sin\left(\mu + \Delta\mu\right) = \cos\mu + \alpha\sin\mu,$$
$$\left(\beta + \Delta\beta\right)\sin\left(\mu + \Delta\mu\right) = \beta\sin\mu.$$

As a result the changes are

$$\Delta\mu = \arccos\left(\cos\mu - \frac{\Delta K}{2}\sin\mu\right) - \mu =_1 \frac{-\frac{\Delta K}{2}\sin\mu}{-\sin\mu} =_1 \frac{\Delta K}{2},$$

$$\Delta\alpha = \frac{\cos\mu + \alpha\sin\mu - \cos(\mu+\Delta\mu)}{\sin(\mu+\Delta\mu)} - \alpha = \frac{(\alpha + \Delta K/2)\sin\mu}{\sin\mu\cos(\Delta\mu) + \cos\mu\sin(\Delta\mu)} - \alpha$$

$$=_1 \left(\alpha + \frac{\Delta K}{2}\right)\left(1 - \frac{\Delta K}{2}\cot\mu\right) - \alpha =_1 \frac{\Delta K}{2}(1 - \alpha\cot\mu),$$

$$\Delta\beta = \frac{\beta\sin\mu}{\sin(\mu+\Delta\mu)} - \beta = \frac{\beta\sin\mu}{\sin\mu\cos(\Delta\mu) + \cos\mu\sin(\Delta\mu)} - \beta$$

$$=_1 \beta\left(1 - \frac{\Delta K}{2}\cot\mu\right) - \beta =_1 -\frac{\Delta K\beta}{2}\cot\mu,$$

where we remind ourselves of $\Delta K = \Delta kl\beta$. The final relations are

$$\Delta\mu =_1 \frac{\Delta K}{2},$$

$$\Delta\alpha =_1 \frac{\Delta K}{2}\frac{\sin\mu - \alpha\cos\mu}{\sin\mu},$$

$$\frac{\Delta\beta}{\beta} =_1 -\frac{\Delta K}{2}\frac{\cos\mu}{\sin\mu}.$$

It is worth noting that the invariant ellipse becomes infinitely large when $\nu \to k/2$. Like the closed orbit, the invariant ellipse is the periodic solution.

Now let us extend the calculation to multiple errors. The one turn matrix is

$$\hat{M}^q = \hat{M}_{n0}\begin{pmatrix} 1 & 0 \\ -(\Delta kl)_n & 1 \end{pmatrix}\cdots\hat{M}_{23}\begin{pmatrix} 1 & 0 \\ -(\Delta kl)_2 & 1 \end{pmatrix}\hat{M}_{12}\begin{pmatrix} 1 & 0 \\ -(\Delta kl)_1 & 1 \end{pmatrix}\hat{M}_{01},$$

where, from eq. (6.9),

$$\hat{M}_{ij} = \begin{pmatrix} \sqrt{\beta_j} & 0 \\ -\alpha_j/\sqrt{\beta_j} & 1/\sqrt{\beta_j} \end{pmatrix}\hat{R}(\phi_{ij})\begin{pmatrix} 1/\sqrt{\beta_i} & 0 \\ \alpha_i/\sqrt{\beta_i} & \sqrt{\beta_i} \end{pmatrix}.$$

The factorization gives us a way to simplify the calculation by working in the normalized space where the matrices between the kicks are rotations. Specifically, we have

$$\hat{M}^q = \begin{pmatrix} \sqrt{\beta_0} & 0 \\ -\alpha_0/\sqrt{\beta_0} & 1/\sqrt{\beta_0} \end{pmatrix}\cdot\widetilde{M}^q\cdot\begin{pmatrix} 1/\sqrt{\beta_0} & 0 \\ \alpha_0/\sqrt{\beta_0} & \sqrt{\beta_0} \end{pmatrix},$$

and, denoting $\Delta K_m = (\Delta kl)_m\beta_m$,

$$\widetilde{M}^q = \hat{R}(\phi_{n0})\begin{pmatrix} 1 & 0 \\ -\Delta K_m & 1 \end{pmatrix}\cdot\ \cdots\ \cdot\hat{R}(\phi_{12})\begin{pmatrix} 1 & 0 \\ -\Delta K_1 & 1 \end{pmatrix}\hat{R}(\phi_{01}).$$

To the first order of the errors, the one turn matrix becomes

$$\widetilde{M}^q = \hat{R}(\mu) - \sum_{m=1}^{n} \Delta K_m \begin{pmatrix} \sin\phi_{m0}\cos\phi_{0m} & \sin\phi_{m0}\sin\phi_{0m} \\ \cos\phi_{m0}\cos\phi_{0m} & \cos\phi_{m0}\sin\phi_{0m} \end{pmatrix}$$

$$= \hat{R}(\mu) - \frac{1}{2}\sum_{m=1}^{n} \Delta K_m \begin{pmatrix} \sin\mu & -\cos\mu \\ \cos\mu & \sin\mu \end{pmatrix} - \frac{1}{2}\sum_{m=1}^{n} \Delta K_m \begin{pmatrix} \sin\widetilde{\mu}_m & \cos\widetilde{\mu}_m \\ \cos\widetilde{\mu}_m & -\sin\widetilde{\mu}_m \end{pmatrix},$$

where $\widetilde{\mu}$ denotes

$$\widetilde{\mu}_m = \mu - 2\phi_{0m}.$$

The change in tune can be obtained through the relation

$$\cos(\mu + \Delta\mu) = \frac{1}{2}\mathrm{tr}\left(\widetilde{M}^q\right) = \cos\mu - \left[\frac{1}{2}\sum_{m=1}^{n}\Delta K_m\right]\sin\mu, \qquad (11.2)$$

which is, to the first order of $\Delta k$,

$$\Delta\mu = \frac{1}{2}\sum_{m=1}^{n}\Delta K_m.$$

If the change in focusing is due to the difference in momentum, this equation gives the chromaticity. Since the tune is changed due to the presence of gradient errors, one important question is how far away it has to be from an integer or a half–integer in order to maintain stability for a given set of errors. This interval in the tune space that the motion is unstable is called the **stop band**. Assuming that the errors are small, the stop band $\Delta\mu$ is small, too. As a result, the unperturbed tune can be written as $\mu = 2\pi(p + \epsilon)$ or $\mu = 2\pi(p + 1/2 + \epsilon)$, where $\epsilon$ is small. Therefore, we have $\sin\mu =_1 \pm 2\pi\epsilon$. Plugging it into eq. (11.2), we find that the term $-[(1/2)\cdot\sum_{m=1}^{n}\Delta K_m]\sin\mu =_2 \mp\pi\epsilon\sum_{m=1}^{n}\Delta K_m$, which is a second order one. Hence we have to go one step further to include the contribution up to the second order of $\Delta k$. The one turn matrix in the normalized space is

$$\widetilde{M}^q = \hat{R}(\mu) - \frac{1}{2}\sum_{m=1}^{n}\Delta K_m \begin{pmatrix} \sin\mu & -\cos\mu \\ \cos\mu & \sin\mu \end{pmatrix} - \frac{1}{2}\sum_{m=1}^{n}\Delta K_m \begin{pmatrix} \sin\widetilde{\mu} & \cos\widetilde{\mu} \\ \cos\widetilde{\mu} & -\sin\widetilde{\mu} \end{pmatrix}$$

$$+ \sum_{l,m=1,l<m}^{n}\Delta K_l\Delta K_m\hat{R}(\phi_{m0})\begin{pmatrix} 0 & 0 \\ 1 & 0 \end{pmatrix}\hat{R}(\phi_{lm})\begin{pmatrix} 0 & 0 \\ 1 & 0 \end{pmatrix}\hat{R}(\phi_{l0})$$

$$= \hat{R}(\mu) - \frac{1}{2}\sum_{m=1}^{n}\Delta K_m \begin{pmatrix} \sin\mu & -\cos\mu \\ \cos\mu & \sin\mu \end{pmatrix} - \frac{1}{2}\sum_{m=1}^{n}\Delta K_m \begin{pmatrix} \sin\widetilde{\mu} & \cos\widetilde{\mu} \\ \cos\widetilde{\mu} & -\sin\widetilde{\mu} \end{pmatrix}$$

$$+ \sum_{l,m=1,l<m}^{n}\Delta K_l\Delta K_m \begin{pmatrix} \sin\phi_{m0}\sin\phi_{lm}\cos\phi_{l0} & \sin\phi_{m0}\sin\phi_{lm}\sin\phi_{l0} \\ \cos\phi_{m0}\sin\phi_{lm}\cos\phi_{l0} & \cos\phi_{m0}\sin\phi_{lm}\sin\phi_{l0} \end{pmatrix}.$$

The trace of the matrix is

$$\operatorname{tr}\left(\widetilde{M}^{q}\right) = 2\cos\mu - \left[\sum_{m=1}^{n} \Delta K_m\right]\sin\mu + \sum_{l,m=1,l<m}^{n} \Delta K_l \Delta K_m \sin\phi_{lm} \sin(\mu - \phi_{lm}).$$

To the second order, the trace is

$$\operatorname{tr}\left(\widetilde{M}^{q}\right) = \pm 2\left[1 - 2\pi^2\epsilon^2 - \pi\epsilon\sum_{m=1}^{n} \Delta K_m\right]$$

$$+ \sum_{l,m=1,l<m}^{n} \Delta K_l \Delta K_m \sin\phi_{lm} \sin\left(2\pi\nu - \phi_{lm}\right).$$

The last term can be simplified, which is

$$\sum_{l,m=1,l<m}^{n} \Delta K_l \Delta K_m \sin\phi_{lm} \sin\left(2\pi\nu - \phi_{lm}\right)$$

$$= \frac{1}{4}\sum_{l,m=1}^{n} \Delta K_l \Delta K_m \left[\cos\left(2\phi_{lm}\right) - 1\right]$$

$$= \frac{1}{4}\sum_{l,m=1}^{n} \Delta K_l \Delta K_m \cos\left(2\Psi_l\right)\cos\left(2\Psi_m\right)$$

$$+ \frac{1}{4}\sum_{l,m=1}^{n} \Delta K_l \Delta K_m \sin\left(2\Psi_l\right)\sin\left(2\Psi_m\right) - \frac{1}{4}\sum_{l,m=1}^{n} \Delta K_l \Delta K_m$$

$$= \frac{1}{4}\left[\sum_{m=1}^{n} \Delta K_m \cos\left(2\Psi_m\right)\right]^2 + \frac{1}{4}\left[\sum_{m=1}^{n} \Delta K_m \sin\left(2\Psi_m\right)\right]^2 - \frac{1}{4}\left[\sum_{m=1}^{n} \Delta K_m\right]^2,$$

where $\Psi_m$ is the phase at the $m$th quadrupole. For the integer stop band, we have

$$\operatorname{tr}\left(\widetilde{M}^{q}\right) = 2\left[1 - 2\pi^2\epsilon^2 - \pi\epsilon\sum_{m=1}^{n} \Delta K_m\right] + \frac{1}{4}\left[\sum_{m=1}^{n} \Delta K_m \cos\left(2\Psi_m\right)\right]^2$$

$$+ \frac{1}{4}\left[\sum_{m=1}^{n} \Delta K_m \sin\left(2\Psi_m\right)\right]^2 - \frac{1}{4}\left[\sum_{m=1}^{n} \Delta K_m\right]^2.$$

The unstable region of the tune is determined by the relation

$$2\pi^2\epsilon^2 + \pi\epsilon\sum_{m=1}^{n} \Delta K_m - \frac{1}{8}\left[\sum_{m=1}^{n} \Delta K_m \cos\left(2\Psi_m\right)\right]^2$$

$$+ \frac{1}{8}\left[\sum_{m=1}^{n} \Delta K_m \sin\left(2\Psi_m\right)\right]^2 + \frac{1}{8}\left[\sum_{m=1}^{n} \Delta K_m\right]^2 < 0.$$

Denoting

$$\Delta\mu = \sqrt{\left[\sum_{m=1}^{n}\Delta K_m \cos\left(2\Psi_m\right)\right]^2 + \left[\sum_{m=1}^{n}\Delta K_m \sin\left(2\Psi_m\right)\right]^2},$$

we see the unstable interval of the tune is

$$\left|2\pi\epsilon + \frac{1}{2}\sum_{m=1}^{n}\Delta K_m\right| < \frac{\Delta\mu}{2},$$

and $\Delta\mu$ is called the integer stop band. Similar calculation shows that the same expression also gives the half–integer stop band [16].

In order to obtain the change in the invariant ellipse, we have to go back to the original space, where

$$
\hat{M}^q = \begin{pmatrix} \sqrt{\beta_0} & 0 \\ -\alpha_0/\sqrt{\beta_0} & 1/\sqrt{\beta_0} \end{pmatrix} \cdot \widetilde{M}^q \cdot \begin{pmatrix} 1/\sqrt{\beta_0} & 0 \\ \alpha_0/\sqrt{\beta_0} & \sqrt{\beta_0} \end{pmatrix}
$$

$$
= \begin{pmatrix} \cos\mu + \alpha_0\sin\mu & \beta_0\sin\mu \\ -\gamma_0\sin\mu & \cos\mu - \alpha_0\sin\mu \end{pmatrix}
$$

$$
- \frac{1}{2}\sum_{m=1}^{n}\Delta K_m \begin{pmatrix} \sin\mu - \alpha_0\cos\mu & -\beta_0\cos\mu \\ \gamma_0\cos\mu & \sin\mu + \alpha_0\cos\mu \end{pmatrix}
$$

$$
- \frac{1}{2}\sum_{m=1}^{n}\Delta K_m \begin{pmatrix} \sin\widetilde{\mu} + \alpha_0\cos\widetilde{\mu} & \beta_0\cos\widetilde{\mu} \\ (1-\alpha_0^2)\cos\widetilde{\mu}/\beta_0 + 2\alpha_0/\beta_0 & -\sin\widetilde{\mu} - \alpha_0\cos\widetilde{\mu} \end{pmatrix}.
$$

Immediately we have

$$\cos\left(\mu + \Delta\mu\right) + \left(\alpha_0 + \Delta\alpha\right)\sin\left(\mu + \Delta\mu\right)$$

$$= \cos\mu + \alpha_0\sin\mu - \frac{1}{2}\sum_{m=1}^{n}\Delta K_m\left[\sin\mu - \alpha_0\cos\mu + \sin\widetilde{\mu} + \alpha_0\cos\widetilde{\mu}\right],$$

$$\left(\beta_0 + \Delta\beta\right)\sin\left(\mu + \Delta\mu\right)$$

$$= \beta_0\sin\mu - \frac{1}{2}\sum_{m=1}^{n}\Delta K_m\left[-\beta_0\cos\mu + \beta_0\cos\widetilde{\mu}\right].$$

To the first order of $\Delta k$, we have

$$\Delta\alpha = -\frac{1}{2\sin\mu}\sum_{m=1}^{n}\Delta K_m\left[\sin\widetilde{\mu}_m + \alpha_0\cos\widetilde{\mu}_m\right],$$

$$\frac{\Delta\beta}{\beta_0} = \frac{1}{2\sin\mu}\sum_{m=1}^{n}\Delta K_m\cos\widetilde{\mu}_m,$$

and we remind ourselves of $\widetilde{\mu}_m = \mu - 2\phi_{0m}$ and $\Delta K_m = (\Delta kl)_m\beta_m$. It is clear that when the tune is close to the half–integer resonance, the size of the beam becomes larger and eventually goes to infinity as the tune approaches the half–integer.

## 11.3 Linear Coupling Resonance

Linear coupling refers to mixing of linear motion between the horizontal and vertical planes. Coupling between transversal and longitudinal motion is present as well, but it is usually weaker. Linear coupling between transversal planes arises from solenoids, roll of quadrupoles, vertical misalignment and orbit offset at sextupole locations. In this section we study the effect of **skew quadrupoles**, which can be present intentionally or due to the roll of normal quadrupoles. Again, let us first assume that only one skew quadrupole is present, which is denoted as $k_s$. Without loss of generality, we assume that this skew quadrupole is located at the end of a turn and that it is thin. As a result, the linear one turn map is

$$
\begin{pmatrix} x_1 \\ a_1 \\ y_1 \\ b_1 \end{pmatrix} = \begin{pmatrix} 1 & 0 & 0 & 0 \\ 0 & 1 & k_s l & 0 \\ 0 & 0 & 1 & 0 \\ k_s l & 0 & 0 & 1 \end{pmatrix} \hat{M} \begin{pmatrix} x_0 \\ a_0 \\ y_0 \\ b_0 \end{pmatrix},
$$

where

$$
\hat{M} = \begin{pmatrix} \cos\mu_x + \alpha_x \sin\mu_x & \beta_x \sin\mu_x & 0 & 0 \\ -\gamma_x \sin\mu_x & \cos\mu_x - \alpha_x \sin\mu_x & 0 & 0 \\ 0 & 0 & \cos\mu_y + \alpha_y \sin\mu_y & \beta_y \sin\mu_y \\ 0 & 0 & -\gamma_y \sin\mu_y & \cos\mu_y - \alpha_y \sin\mu_y \end{pmatrix}.
$$

Applying the same coordinate transformation, we obtain the normalized coordinate system, which is

$$
\begin{pmatrix} \tilde{x} \\ \tilde{a} \\ \tilde{y} \\ \tilde{b} \end{pmatrix} = \begin{pmatrix} 1/\sqrt{\beta_x} & 0 & 0 & 0 \\ \alpha_x/\sqrt{\beta_x} & \sqrt{\beta_x} & 0 & 0 \\ 0 & 0 & 1/\sqrt{\beta_y} & 0 \\ 0 & 0 & \alpha_y/\sqrt{\beta_y} & \sqrt{\beta_y} \end{pmatrix} \begin{pmatrix} x \\ a \\ y \\ b \end{pmatrix}.
$$

As a result, we have

$$
\begin{pmatrix} \tilde{x}_1 \\ \tilde{a}_1 \\ \tilde{y}_1 \\ \tilde{b}_1 \end{pmatrix} = \begin{pmatrix} 1 & 0 & 0 & 0 \\ 0 & 1 & K & 0 \\ 0 & 0 & 1 & 0 \\ K & 0 & 0 & 1 \end{pmatrix} \begin{pmatrix} \cos\mu_x & \sin\mu_x & 0 & 0 \\ -\sin\mu_x & \cos\mu_x & 0 & 0 \\ 0 & 0 & \cos\mu_y & \sin\mu_y \\ 0 & 0 & -\sin\mu_y & \cos\mu_y \end{pmatrix} \begin{pmatrix} \tilde{x}_0 \\ \tilde{a}_0 \\ \tilde{y}_0 \\ \tilde{b}_0 \end{pmatrix},
$$

where $K$ above denotes

$$
K = k_s l \sqrt{\beta_x \beta_y}. \tag{11.3}
$$

To shorten the equations, let us use again the following symbol

$$\hat{R}\begin{pmatrix}\mu_x\\\mu_y\end{pmatrix}=\begin{pmatrix}\cos\mu_x & \sin\mu_x & 0 & 0\\-\sin\mu_x & \cos\mu_x & 0 & 0\\0 & 0 & \cos\mu_y & \sin\mu_y\\0 & 0 & -\sin\mu_y & \cos\mu_y\end{pmatrix}.$$

After $n$ turns, the coordinates are

$$\begin{pmatrix}\tilde{x}_n\\\tilde{a}_n\\\tilde{y}_n\\\tilde{b}_n\end{pmatrix}=\left[\begin{pmatrix}1 & 0 & 0 & 0\\0 & 1 & K & 0\\0 & 0 & 1 & 0\\K & 0 & 0 & 1\end{pmatrix}\hat{R}\begin{pmatrix}\mu_x\\\mu_y\end{pmatrix}\right]^n\begin{pmatrix}\tilde{x}_0\\\tilde{a}_0\\\tilde{y}_0\\\tilde{b}_0\end{pmatrix}.$$

To the first order, the coordinates are

$$\begin{pmatrix}\tilde{x}_n\\\tilde{a}_n\\\tilde{y}_n\\\tilde{b}_n\end{pmatrix}=\left[\hat{R}\begin{pmatrix}n\mu_x\\n\mu_y\end{pmatrix}+\begin{pmatrix}0 & 0 & 0 & 0\\0 & 0 & K & 0\\0 & 0 & 0 & 0\\K & 0 & 0 & 0\end{pmatrix}\hat{R}\begin{pmatrix}n\mu_x\\n\mu_y\end{pmatrix}\right.$$

$$+\hat{R}\begin{pmatrix}\mu_x\\\mu_y\end{pmatrix}\begin{pmatrix}0 & 0 & 0 & 0\\0 & 0 & K & 0\\0 & 0 & 0 & 0\\K & 0 & 0 & 0\end{pmatrix}\hat{R}\begin{pmatrix}(n-1)\mu_x\\(n-1)\mu_y\end{pmatrix}+\cdots$$

$$\left.+\hat{R}\begin{pmatrix}(n-1)\mu_x\\(n-1)\mu_y\end{pmatrix}\begin{pmatrix}0 & 0 & 0 & 0\\0 & 0 & K & 0\\0 & 0 & 0 & 0\\K & 0 & 0 & 0\end{pmatrix}\hat{R}\begin{pmatrix}\mu_x\\\mu_y\end{pmatrix}\right]\begin{pmatrix}\tilde{x}_0\\\tilde{a}_0\\\tilde{y}_0\\\tilde{b}_0\end{pmatrix}$$

$$=\left[\hat{R}\begin{pmatrix}n\mu_x\\n\mu_y\end{pmatrix}+K\begin{pmatrix}0 & 0 & 0 & 0\\0 & 0 & \cos(n\mu_y) & \sin(n\mu_y)\\0 & 0 & 0 & 0\\\cos(n\mu_x) & \sin(n\mu_x) & 0 & 0\end{pmatrix}\right.$$

$$\left.+K\sum_{m=1}^{n-1}\begin{pmatrix}0 & 0 & & \\0 & 0 & \hat{R}_{UR} & \\ & & 0 & 0\\\hat{R}_{LL} & & 0 & 0\end{pmatrix}\right]\begin{pmatrix}\tilde{x}_0\\\tilde{a}_0\\\tilde{y}_0\\\tilde{b}_0\end{pmatrix},$$

where

$$\hat{R}_{LL} = \begin{pmatrix} \cos\left[(n-m)\,\mu_x\right]\sin\left(m\mu_y\right) & \sin\left[(n-m)\,\mu_x\right]\sin\left(m\mu_y\right) \\ \cos\left[(n-m)\,\mu_x\right]\cos\left(m\mu_y\right) & \sin\left[(n-m)\,\mu_x\right]\cos\left(m\mu_y\right) \end{pmatrix},$$

$$\hat{R}_{UR} = \begin{pmatrix} \cos\left[(n-m)\,\mu_y\right]\sin\left(m\mu_x\right) & \sin\left[(n-m)\,\mu_y\right]\sin\left(m\mu_x\right) \\ \cos\left[(n-m)\,\mu_y\right]\cos\left(m\mu_x\right) & \sin\left[(n-m)\,\mu_y\right]\cos\left(m\mu_x\right) \end{pmatrix}.$$

The nonzero terms in the transfer matrix can be further simplified using trigonometric relations. Therefore we have

$$\sum_{m=1}^{n-1} \cos\left[(n-m)\mu_x\right]\sin\left(m\mu_y\right) = \sum_{m=0}^{n-1} \cos\left[(n-m)\mu_x\right]\sin\left(m\mu_y\right)$$

$$= \frac{1}{2}\sum_{m=0}^{n-1}\left\{\sin\left[n\mu_x - m\left(\mu_x - \mu_y\right)\right] - \sin\left[n\mu_x - m\left(\mu_x + \mu_y\right)\right]\right\}$$

$$= \frac{1}{2}\left\{\sin\left[n\mu_x - \frac{1}{2}\left(n-1\right)\left(\mu_x - \mu_y\right)\right]\frac{\sin\left[n\left(\mu_x - \mu_y\right)/2\right]}{\sin\left[\left(\mu_x - \mu_y\right)/2\right]}\right.$$

$$\left. - \sin\left[n\mu_x - \frac{1}{2}\left(n-1\right)\left(\mu_x + \mu_y\right)\right]\frac{\sin\left[n\left(\mu_x + \mu_y\right)/2\right]}{\sin\left[\left(\mu_x + \mu_y\right)/2\right]}\right\}$$

$$= \frac{1}{2}\left\{\sin\left[\frac{1}{2}n\left(\mu_x + \mu_y\right)\right]\cos\left[\frac{1}{2}\left(\mu_x - \mu_y\right)\right]\frac{\sin\left[n\left(\mu_x - \mu_y\right)/2\right]}{\sin\left[\left(\mu_x - \mu_y\right)/2\right]}\right.$$

$$+ \cos\left[\frac{1}{2}n\left(\mu_x + \mu_y\right)\right]\sin\left[\frac{1}{2}n\left(\mu_x - \mu_y\right)\right]$$

$$- \sin\left[\frac{1}{2}n\left(\mu_x - \mu_y\right)\right]\cos\left[\frac{1}{2}\left(\mu_x + \mu_y\right)\right]\frac{\sin\left[n\left(\mu_x + \mu_y\right)/2\right]}{\sin\left[\left(\mu_x + \mu_y\right)/2\right]}$$

$$\left. - \cos\left[\frac{1}{2}n\left(\mu_x - \mu_y\right)\right]\sin\left[\frac{1}{2}n\left(\mu_x + \mu_y\right)\right]\right\}$$

$$= \frac{1}{2}\left\{\sin\left[\frac{1}{2}n\left(\mu_x + \mu_y\right)\right]\cos\left[\frac{1}{2}\left(\mu_x - \mu_y\right)\right]\frac{\sin\left[n\left(\mu_x - \mu_y\right)/2\right]}{\sin\left[\left(\mu_x - \mu_y\right)/2\right]}\right.$$

$$\left. - \sin\left[\frac{1}{2}n\left(\mu_x - \mu_y\right)\right]\cos\left[\frac{1}{2}\left(\mu_x + \mu_y\right)\right]\frac{\sin\left[n\left(\mu_x + \mu_y\right)/2\right]}{\sin\left[\left(\mu_x + \mu_y\right)/2\right]} - \sin\left(n\mu_y\right)\right\}.$$

When $\mu_y \to \pm\mu_x + 2\pi N$ ($N$ is an integer), we obtain

$$\sum_{m=1}^{n-1} \cos\left[(n-m)\mu_x\right]\sin\left(m\mu_y\right) \to \frac{1}{2}\left(n-1\right)\sin\left(n\mu_y\right).$$

Similarly, we have

$$\sum_{m=1}^{n-1} \cos\left[(n-m)\mu_x\right] \cos\left(m\mu_y\right) = \sum_{m=0}^{n-1} \cos\left[(n-m)\mu_x\right] \cos\left(m\mu_y\right) - \cos\left(n\mu_x\right)$$

$$= \frac{1}{2}\left\{ \cos\left[\frac{1}{2}n\left(\mu_x + \mu_y\right)\right] \cos\left[\frac{1}{2}\left(\mu_x - \mu_y\right)\right] \frac{\sin\left[n\left(\mu_x - \mu_y\right)/2\right]}{\sin\left[\left(\mu_x - \mu_y\right)/2\right]} \right.$$

$$\left. + \cos\left[\frac{1}{2}n\left(\mu_x - \mu_y\right)\right] \cos\left[\frac{1}{2}\left(\mu_x + \mu_y\right)\right] \frac{\sin\left[n\left(\mu_x + \mu_y\right)/2\right]}{\sin\left[\left(\mu_x + \mu_y\right)/2\right]} \right\}$$

$$- \cos\left[\frac{1}{2}n\left(\mu_x + \mu_y\right)\right] \cos\left[\frac{1}{2}n\left(\mu_x - \mu_y\right)\right],$$

so we obtain when $\mu_y \to \pm\mu_x + 2\pi N$,

$$\sum_{m=1}^{n-1} \cos\left[(n-m)\mu_x\right] \cos\left(m\mu_y\right) \to \frac{1}{2}\left\{ (n-1)\cos\left(n\mu_y\right) + \frac{\sin\left[(n-1)\mu_y\right]}{\sin\mu_y} \right\}.$$

Furthermore, we have

$$\sum_{m=1}^{n-1} \sin\left[(n-m)\mu_x\right] \sin\left(m\mu_y\right) = \sum_{m=0}^{n-1} \sin\left[(n-m)\mu_x\right] \sin\left(m\mu_y\right)$$

$$= \frac{1}{2}\left\{ -\cos\left[\frac{1}{2}n\left(\mu_x + \mu_y\right)\right] \cos\left[\frac{1}{2}\left(\mu_x - \mu_y\right)\right] \frac{\sin\left[n\left(\mu_x - \mu_y\right)/2\right]}{\sin\left[\left(\mu_x - \mu_y\right)/2\right]} \right.$$

$$\left. + \cos\left[\frac{1}{2}n\left(\mu_x - \mu_y\right)\right] \cos\left[\frac{1}{2}\left(\mu_x + \mu_y\right)\right] \frac{\sin\left[n\left(\mu_x + \mu_y\right)/2\right]}{\sin\left[\left(\mu_x + \mu_y\right)/2\right]} \right\},$$

and we obtain when $\mu_y \to \pm\mu_x + 2\pi N$,

$$\sum_{m=1}^{n-1} \sin\left[(n-m)\mu_x\right] \sin\left(m\mu_y\right) \to \mp\frac{1}{2}\left\{ (n-1)\cos\left(n\mu_y\right) - \frac{\sin\left[(n-1)\mu_y\right]}{\sin\mu_y} \right\}.$$

Lastly, we have

$$\sum_{m=1}^{n-1} \sin\left[(n-m)\mu_x\right] \cos\left(m\mu_y\right) = \sum_{m=0}^{n-1} \sin\left[(n-m)\mu_x\right] \cos\left(m\mu_y\right) - \sin\left(n\mu_x\right)$$

$$= \frac{1}{2}\left\{ \sin\left[\frac{1}{2}n\left(\mu_x + \mu_y\right)\right] \cos\left[\frac{1}{2}\left(\mu_x - \mu_y\right)\right] \frac{\sin\left[n\left(\mu_x - \mu_y\right)/2\right]}{\sin\left[\left(\mu_x - \mu_y\right)/2\right]} \right.$$

$$\left. + \sin\left[\frac{1}{2}n\left(\mu_x - \mu_y\right)\right] \cos\left[\frac{1}{2}\left(\mu_x + \mu_y\right)\right] \frac{\sin\left[n\left(\mu_x + \mu_y\right)/2\right]}{\sin\left[\left(\mu_x + \mu_y\right)/2\right]} - \sin\left(n\mu_x\right) \right\},$$

and we obtain when $\mu_y \to \pm\mu_x + 2\pi N$,

$$\sum_{m=1}^{n-1} \sin\left[(n-m)\mu_x\right] \cos\left(m\mu_y\right) \to \pm\frac{1}{2}(n-1)\sin\left(n\mu_y\right).$$

The upper right block of the matrix can be obtained by switching $\mu_y$ and $\mu_y$ of the results above. In summary the transfer matrix at the limit of $\mu_y \to \pm\mu_x + 2\pi N$ is given as

$$\vec{z}_n = \left\{ \hat{M}_1 + K \left[ \hat{M}_2 + \frac{1}{2}\hat{M}_3 + \frac{1}{2}(n-1)\hat{M}_4 \right] \right\} \vec{z}_0,$$

where

$$\vec{z}_n = \begin{pmatrix} \widetilde{x}_n \\ \widetilde{a}_n \\ \widetilde{y}_n \\ \widetilde{b}_n \end{pmatrix}, \quad \vec{z}_0 = \begin{pmatrix} \widetilde{x}_0 \\ \widetilde{a}_0 \\ \widetilde{y}_0 \\ \widetilde{b}_0 \end{pmatrix},$$

$$\hat{M}_1 = \begin{pmatrix} \cos(n\mu_x) & \sin(n\mu_x) & 0 & 0 \\ -\sin(n\mu_x) & \cos(n\mu_x) & 0 & 0 \\ 0 & 0 & \cos(n\mu_y) & \sin(n\mu_y) \\ 0 & 0 & -\sin(n\mu_y) & \cos(n\mu_y) \end{pmatrix},$$

$$\hat{M}_2 = \begin{pmatrix} 0 & 0 & 0 & 0 \\ 0 & 0 & \cos(n\mu_y) & \sin(n\mu_y) \\ 0 & 0 & 0 & 0 \\ \cos(n\mu_x) & \sin(n\mu_x) & 0 & 0 \end{pmatrix},$$

$$\hat{M}_3 = \begin{pmatrix} 0 & 0 & 0 & \pm s_x \\ 0 & 0 & s_x & 0 \\ 0 & \pm s_y & 0 & 0 \\ s_y & 0 & 0 & 0 \end{pmatrix},$$

$$\hat{M}_4 = \begin{pmatrix} 0 & 0 & \sin(n\mu_x) & \mp\cos(n\mu_x) \\ 0 & 0 & \cos(n\mu_x) & \pm\sin(n\mu_x) \\ \sin(n\mu_y) & \mp\cos(n\mu_y) & 0 & 0 \\ \cos(n\mu_y) & \pm\sin(n\mu_y) & 0 & 0 \end{pmatrix}, \quad (11.4)$$

and the coupling terms $s_x$ and $s_y$ above are

$$s_x = \frac{\sin[(n-1)\mu_x]}{\sin\mu_x}, \quad s_y = \frac{\sin[(n-1)\mu_y]}{\sin\mu_y},$$

indicating that an arbitrarily small perturbation leads to arbitrarily large coupling between the horizontal and the vertical spaces.

Similar to the half–integer resonance, the presence of the skew quadrupole component leads to a stop band gap in which the beam becomes unstable. To illustrate this point, let us first work the formalism of the eigenvalues of a $4 \times 4$ symplectic matrix.

Before we treat linear coupling, let us look at a few general properties of symplectic matrices and their eigenvalues. Recall that the symplectic condition is

$$\hat{M}^T \hat{J} \hat{M} = \hat{J},$$

where $\hat{M}$ is a $4 \times 4$ real matrix. The eigenvalue of $\hat{M}$ can be obtained through solving

$$\det\left(\hat{M} - \lambda\hat{I}\right) = 0. \tag{11.5}$$

Since $\hat{M}$ is real,

$$\hat{M}\vec{v} = \lambda\vec{v} \implies \hat{M}^\dagger\vec{v}^\dagger = \lambda^\dagger\vec{v}^\dagger \implies \hat{M}\vec{v}^\dagger = \lambda^\dagger\vec{v}^\dagger.$$

So $\lambda^\dagger$ is also an eigenvalue of $\hat{M}$ and $\vec{v}^\dagger$ is the eigenvector. Keep in mind that

$$\det(\hat{A}\hat{B}) = \det(\hat{A})\det(\hat{B}), \quad \det(\hat{A}^T) = \det(\hat{A}).$$

With these identities, we can transform eq. (11.5).

$$\det\left(\hat{M} - \lambda\hat{I}\right) = 0 \implies \det\hat{J} \cdot \det(\hat{J}\hat{M} - \lambda\hat{I}) = 0 \implies \det(\hat{J}\hat{M} - \lambda\hat{J}) = 0$$

$$\implies \det\hat{M}^T \cdot \det(\hat{J}\hat{M} - \lambda\hat{J}) = 0 \implies \det(\hat{J} - \lambda\hat{M}^T\hat{J}) = 0$$

$$\implies \det(\hat{I} - \lambda\hat{M}^T) = 0 \implies \lambda^{2n}\det(\lambda^{-1}\hat{I} - \hat{M}^T) = 0$$

$$\implies \det(\hat{M}^T - \lambda^{-1}\hat{I}) = 0.$$

Therefore $\lambda^{-1}$ is also an eigenvalue. Together with $\lambda^\dagger$, we reach the conclusion that if $\hat{M}$ has an eigenvalue $\lambda$ then $\lambda^\dagger, \lambda^{-1}, \lambda^{\dagger-1}$ are also eigenvalues. We know that for $\hat{M}$ to be stable, $|\lambda|$ has to be smaller or equal to 1 for all eigenvalues of $\hat{M}$. As a result, all eigenvalues of $\hat{M}$ lie on the unit circle. Apparently, we have, for this case, $\lambda = \lambda^{\dagger-1}$ and $\lambda^{-1} = \lambda^\dagger$.

The next question we can ask is that supposing $\hat{M}$ is stable and every eigenvalue is reasonably far away from each other, what happens when $\hat{M}$ is perturbed? In terms of betatron motion, it means that $\mu_x$ and $\mu_y$ are reasonably far away.

To frame it in a more mathematical way, we say that there is a neighborhood around each eigenvalue so that there is only one eigenvalue in it. When $\hat{M}$ is perturbed its eigenvalues will move. They may all stay on the unit circle, or some of them may move away from it. Let us say one of them, $\lambda$, moves away from the unit circle, then $\lambda^{\dagger-1}$ will also move away from it and be in the same neighborhood that $\lambda$ is in, making the total number of eigenvalues greater than 4, which is impossible. So the conclusion is that every one of them will stay on the unit circle (see Fig. 11.1). This is consistent with our experience, which is that when we change quadrupole strength, $\mu_x$ and $\mu_y$ change, but the motion is stable.

The question that relates to linear coupling is what happens if two or more of the eigenvalues get close to each other? The answer is more complicated. It has been proven that when two colliding eigenvalues have the same sign of phase, motion remains stable after collision. Otherwise, instability may occur. In other words, difference resonance $\mu_x - \mu_y$ does not lead to instability, sum resonance $\mu_x + \mu_y$ does.

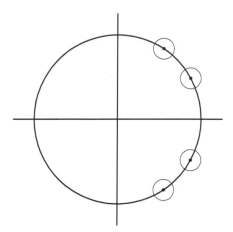

**FIGURE 11.1:** Distinct eigenvalues around the unit circle. Small circles show the neighborhood of each eigenvalue.

Let us focus on coupling between horizontal and vertical planes. The matrix $\hat{M}$ is a $4 \times 4$ symplectic matrix, which we describe as

$$\hat{M} = \begin{pmatrix} \hat{A} & \hat{B} \\ \hat{C} & \hat{D} \end{pmatrix},$$

where $\hat{A}$, $\hat{B}$, $\hat{C}$ and $\hat{D}$ are $2 \times 2$ matrices. From eq. (5.11), we have

$$\hat{M}^{-1} = \begin{pmatrix} \bar{A} & \bar{C} \\ \bar{B} & \bar{D} \end{pmatrix},$$

where $\bar{A}$ is defined as $\bar{A} = -\hat{J}\hat{A}^T\hat{J}$, as well as $\bar{B}$, $\bar{C}$ and $\bar{D}$. Let us solve for the eigenvalue $\Lambda$ of $\hat{M} + \hat{M}^{-1}$. Obviously $\Lambda = \lambda + 1/\lambda$, because

$$\hat{M}\vec{v} = \lambda\vec{v} \implies \vec{v} = \lambda\hat{M}^{-1}\vec{v} \implies \frac{1}{\lambda}\vec{v} = \hat{M}^{-1}\vec{v}$$

$$\implies \left(\hat{M} + \hat{M}^{-1}\right)\vec{v} = \left(\lambda + \frac{1}{\lambda}\right)\vec{v}.$$

$$\hat{M} + \hat{M}^{-1} = \begin{pmatrix} \hat{A} & \hat{B} \\ \hat{C} & \hat{D} \end{pmatrix} + \begin{pmatrix} \bar{A} & \bar{C} \\ \bar{B} & \bar{D} \end{pmatrix} = \begin{pmatrix} \hat{A}+\bar{A} & \hat{B}+\bar{C} \\ \hat{C}+\bar{B} & \hat{D}+\bar{D} \end{pmatrix}.$$

Note that

$$\hat{A} + \bar{A} = \begin{pmatrix} a_{11} & a_{12} \\ a_{21} & a_{22} \end{pmatrix} + \begin{pmatrix} a_{22} & -a_{12} \\ -a_{21} & a_{11} \end{pmatrix} = \operatorname{tr}\hat{A} \cdot \begin{pmatrix} 1 & 0 \\ 0 & 1 \end{pmatrix} = \operatorname{tr}\hat{A} \cdot \hat{I}.$$

Using the relation

$$\left(\hat{M} + \hat{M}^{-1}\right) = \begin{pmatrix} \operatorname{tr}\hat{A} \cdot \hat{I} & \hat{B}+\bar{C} \\ \hat{C}+\bar{B} & \operatorname{tr}\hat{D} \cdot \hat{I} \end{pmatrix},$$

we proceed and obtain

$$\det \left( \hat{M} + \hat{M}^{-1} - \Lambda \hat{I} \right) = \det \begin{pmatrix} (\operatorname{tr} \hat{A} - \Lambda) \cdot \hat{I} & \hat{B} + \bar{C} \\ \hat{C} + \bar{B} & (\operatorname{tr} \hat{D} - \Lambda) \cdot \hat{I} \end{pmatrix}$$

$$= \det \left[ (\operatorname{tr} \hat{A} - \Lambda)(\operatorname{tr} \hat{D} - \Lambda)\hat{I} - (\operatorname{tr} \hat{A} - \Lambda)\hat{I}(\hat{C} + \bar{B})((\operatorname{tr} \hat{A} - \Lambda)\hat{I})^{-1}(\hat{B} + \bar{C}) \right]$$

$$= \det \left[ (\operatorname{tr} \hat{A} - \Lambda)(\operatorname{tr} \hat{D} - \Lambda)\hat{I} - (\hat{C} + \bar{B})(\hat{B} + \bar{C}) \right].$$

To find $\Lambda$, we have to solve the equation

$$\det \left[ (\operatorname{tr} \hat{A} - \Lambda)(\operatorname{tr} \hat{D} - \Lambda) \cdot \hat{I} - (\hat{C} + \bar{B})(\hat{B} + \bar{C}) \right] = 0.$$

Now we have to find out what $(\hat{C} + \bar{B})(\hat{B} + \bar{C})$ is.

$$\hat{B} + \bar{C} = \begin{pmatrix} b_{11} & b_{12} \\ b_{21} & b_{22} \end{pmatrix} + \begin{pmatrix} c_{22} & -c_{12} \\ -c_{21} & c_{11} \end{pmatrix} = \begin{pmatrix} b_{11} + c_{22} & b_{12} - c_{12} \\ b_{21} - c_{21} & b_{22} + c_{11} \end{pmatrix} \equiv \begin{pmatrix} e & f \\ g & h \end{pmatrix},$$

$$\hat{C} + \bar{B} = \begin{pmatrix} c_{11} & c_{12} \\ c_{21} & c_{22} \end{pmatrix} + \begin{pmatrix} b_{22} & -b_{12} \\ -b_{21} & b_{11} \end{pmatrix} = \begin{pmatrix} c_{11} + b_{22} & c_{12} - b_{12} \\ c_{21} - b_{21} & c_{22} + b_{11} \end{pmatrix} \equiv \begin{pmatrix} h & -f \\ -g & e \end{pmatrix}.$$

We have just shown that

$$\hat{C} + \bar{B} = \overline{\hat{B} + \bar{C}}.$$

Then we obtain that

$$(\hat{C} + \bar{B})(\hat{B} + \bar{C}) = \det(\hat{C} + \bar{B}) \cdot \hat{I}.$$

So the equation for solving $\Lambda$ becomes

$$(\operatorname{tr} \hat{A} - \Lambda)(\operatorname{tr} \hat{D} - \Lambda) - \det(\hat{C} + \bar{B}) = 0.$$

Then

$$\Lambda^2 - (\operatorname{tr} \hat{A} + \operatorname{tr} \hat{D})\Lambda + \operatorname{tr} \hat{A} \cdot \operatorname{tr} \hat{D} - \det(\hat{C} + \bar{B}) = 0$$

$$\implies \Lambda = \frac{1}{2}(\operatorname{tr} \hat{A} + \operatorname{tr} \hat{D}) \pm \sqrt{\frac{(\operatorname{tr} \hat{A} - \operatorname{tr} \hat{D})^2}{4} + \det(\hat{C} + \bar{B})} . \qquad (11.6)$$

This is the standard result of coupled motion, where the eigenvalue of the motion in one plane depends on the motion in the other plane and vice versa. The main features of eq. (11.6) are that $\Lambda_+ \to \operatorname{tr} \hat{A}$ when $\operatorname{tr} \hat{A} \gg \operatorname{tr} \hat{D}$ and $\Lambda_+ \to \operatorname{tr} \hat{D}$ when $\operatorname{tr} \hat{A} \ll \operatorname{tr} \hat{D}$; and that $\Lambda_- \to \operatorname{tr} \hat{D}$ when $\operatorname{tr} \hat{A} \gg \operatorname{tr} \hat{D}$ and $\Lambda_- \to \operatorname{tr} \hat{A}$ when $\operatorname{tr} \hat{A} \ll \operatorname{tr} \hat{D}$.

Courant and Snyder [16] have shown that the sign of $\det(\hat{C} + \bar{B})$ determines the stability of the motion when $\operatorname{tr} \hat{A} \simeq \operatorname{tr} \hat{D}$. Specifically, when $\mu_x - \mu_y = 2\pi N$ (difference resonance), $\det(\hat{C} + \bar{B})$ is positive, $\Lambda$ is real, $\lambda$ is on the unit circle, so motion is stable. When $\mu_x + \mu_y = 2\pi N$ (sum resonance), $\det(\hat{C} + \bar{B})$ is

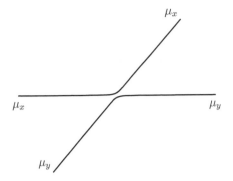

**FIGURE 11.2:** Crossing the difference resonance. Before: $\mu_x$ is constant and $\mu_y$ increases. After: $\mu_x$ and $\mu_y$ reverse roles. Throughout the process, the ratio of the knobs (quadrupoles in synchrotrons) remains unchanged.

negative, $\Lambda$ can be complex, $\lambda$ moves away from the unit circle, so motion is unstable.

Let us now look at the difference $(\mu_x - \mu_y = 2\pi N)$ and sum resonance $(\mu_x + \mu_y = 2\pi N)$ separately. In case of difference resonance, we have

$$\Lambda_x - \Lambda_y = 2\sqrt{\frac{(\operatorname{tr}\hat{A} - \operatorname{tr}\hat{D})^2}{4} + \det(\hat{C} + \bar{B})},$$

$$\Lambda_x = e^{i\mu_x} + e^{-i\mu_x} = 2\cos\mu_x$$

$$\Longrightarrow \cos\mu_x - \cos\mu_y = \sqrt{\frac{(\operatorname{tr}\hat{A} - \operatorname{tr}\hat{D})^2}{4} + \det(\hat{C} + \bar{B})}$$

$$\Longrightarrow (\cos\mu_x - \cos\mu_y)^2 = \frac{(\operatorname{tr}\hat{A} - \operatorname{tr}\hat{D})^2}{4} + \det(\hat{C} + \bar{B}).$$

As a result, there is a minimum separation between the tunes when the difference becomes small, as shown in Fig. 11.2. The minimum separation of tunes can be used to determine the amount of coupling in an accelerator.

In case of sum resonance, motion becomes unstable when

$$\frac{(\operatorname{tr}\hat{A} - \operatorname{tr}\hat{D})^2}{4} < -\det(\hat{C} + \bar{B}).$$

It is illuminating to compare the result above with that of eq. (11.4). There both sum and difference resonances show unbounded growth in coupling, yet only the sum resonance leads to instability in the 4D phase space.

To illustrate this, let us follow Courant and Snyder [16] and work out the one turn map when a skew quadrupole is present.

$$\hat{M} = \begin{pmatrix} 1 & 0 & 0 & 0 \\ 0 & 1 & K & 0 \\ 0 & 0 & 1 & 0 \\ K & 0 & 0 & 1 \end{pmatrix} \begin{pmatrix} \cos\mu_x & \sin\mu_x & 0 & 0 \\ -\sin\mu_x & \cos\mu_x & 0 & 0 \\ 0 & 0 & \cos\mu_y & \sin\mu_y \\ 0 & 0 & -\sin\mu_y & \cos\mu_y \end{pmatrix}$$

$$= \begin{pmatrix} \cos\mu_x & \sin\mu_x & 0 & 0 \\ -\sin\mu_x & \cos\mu_x & K\cos\mu_y & K\sin\mu_y \\ 0 & 0 & \cos\mu_y & \sin\mu_y \\ K\cos\mu_x & K\sin\mu_x & -\sin\mu_y & \cos\mu_y \end{pmatrix}.$$

Hence

$$\hat{B} = K\begin{pmatrix} 0 & 0 \\ \cos\mu_y & \sin\mu_y \end{pmatrix}, \qquad \hat{C} = K\begin{pmatrix} 0 & 0 \\ \cos\mu_x & \sin\mu_x \end{pmatrix},$$

and

$$\hat{C} + \bar{B} = K\left[\begin{pmatrix} 0 & 0 \\ \cos\mu_x & \sin\mu_x \end{pmatrix} + \begin{pmatrix} \sin\mu_y & 0 \\ -\cos\mu_y & 0 \end{pmatrix}\right] = K\begin{pmatrix} \sin\mu_y & 0 \\ \cos\mu_x - \cos\mu_y & \sin\mu_x \end{pmatrix}.$$

Furthermore, we have

$$\det(\hat{C} + \bar{B}) = K^2 \sin\mu_x \sin\mu_y.$$

For sum resonance, we have $\sin\mu_x = -\sin\mu_y$ and $\det(\hat{C}+\bar{B}) < 0$. The motion is unstable. When there are many skew quadrupole components in a ring, to the leading order and near the sum resonance, we have

$$\det(\hat{C} + \bar{B}) = -\sin^2\mu \sum_{m=1}^{n} K_m^2,$$

where we use the abbreviation $K_m = (k_s l)_m \sqrt{\beta_{xm}\beta_{ym}}$ as eq. (11.3). The stop band is

$$(\cos\mu_x - \cos\mu_y)^2 < \sin^2\mu \sum_{m=1}^{n} K_m^2,$$

and the full width, to the leading order, is

$$\Delta\mu = 2\sqrt{\sum_{m=1}^{n} K_m^2}.$$

For difference resonance, we have $\sin\mu_x = \sin\mu_y$ and $\det(\hat{C} + \bar{B}) > 0$. The motion is stable and the minimum tune difference is

$$\Delta\mu_{\min} = \sqrt{\sum_{m=1}^{n} K_m^2}.$$

## 11.4   Third–Integer Resonance

In this section we deal with the third–integer resonance, which is generated by the sextupoles introduced into the ring to compensate for chromaticity. Similar to the previous section, we will start with a particle with an arbitrary position and angle and demonstrate the resonant behavior when the tune is close to the resonance. Again, let us first consider the case that a thin sextupole is located at the end of the ring. The one turn map is

$$
\begin{pmatrix} x_1 \\ a_1 \end{pmatrix} = \begin{pmatrix} x \\ a + k_s x^2 \end{pmatrix} \circ \left[ \hat{M} \begin{pmatrix} x_0 \\ a_0 \end{pmatrix} \right]
$$

$$
= \begin{pmatrix} x \\ a + k_s x^2 \end{pmatrix} \circ \left[ \begin{pmatrix} \cos\mu + \alpha\sin\mu & \beta\sin\mu \\ -\gamma\sin\mu & \cos\mu - \alpha\sin\mu \end{pmatrix} \begin{pmatrix} x_0 \\ a_0 \end{pmatrix} \right].
$$

In the normalized space

$$
\begin{pmatrix} \widetilde{x} \\ \widetilde{a} \end{pmatrix} = \hat{A} \begin{pmatrix} x \\ a \end{pmatrix} = \begin{pmatrix} 1/\sqrt{\beta} & 0 \\ \alpha/\sqrt{\beta} & \sqrt{\beta} \end{pmatrix} \begin{pmatrix} x \\ a \end{pmatrix},
$$

the one turn map is

$$
\begin{pmatrix} \widetilde{x}_1 \\ \widetilde{a}_1 \end{pmatrix} = \left[ \hat{A} \begin{pmatrix} \widetilde{x} \\ \widetilde{a} \end{pmatrix} \right] \circ \begin{pmatrix} \widetilde{x} \\ \widetilde{a} + k_s \widetilde{x}^2 \end{pmatrix} \circ \left[ \hat{M} \hat{A}^{-1} \begin{pmatrix} \widetilde{x}_0 \\ \widetilde{a}_0 \end{pmatrix} \right]
$$

$$
= \left[ \hat{A} \begin{pmatrix} \widetilde{x} \\ \widetilde{a} \end{pmatrix} \right] \circ \begin{pmatrix} \widetilde{x} \\ \widetilde{a} + k_s \widetilde{x}^2 \end{pmatrix} \circ \left[ \hat{A}^{-1} \begin{pmatrix} \widetilde{x} \\ \widetilde{a} \end{pmatrix} \right] \circ \left[ \hat{A} \hat{M} \hat{A}^{-1} \begin{pmatrix} \widetilde{x}_0 \\ \widetilde{a}_0 \end{pmatrix} \right]
$$

$$
= \begin{pmatrix} \widetilde{x} \\ \widetilde{a} + k_s \beta^{\frac{3}{2}} \widetilde{x}^2 \end{pmatrix} \circ \left[ \hat{R}(\mu) \begin{pmatrix} \widetilde{x}_0 \\ \widetilde{a}_0 \end{pmatrix} \right].
$$

The position and angle after $n$ turns are

$$
\begin{pmatrix} \widetilde{x}_n \\ \widetilde{a}_n \end{pmatrix} = \begin{pmatrix} \widetilde{x} \\ \widetilde{a} + k_s \beta^{\frac{3}{2}} \widetilde{x}^2 \end{pmatrix} \circ \left[ \hat{R}(\mu) \begin{pmatrix} \widetilde{x} \\ \widetilde{a} \end{pmatrix} \right] \circ \cdots \circ \begin{pmatrix} \widetilde{x} \\ \widetilde{a} + k_s \beta^{\frac{3}{2}} \widetilde{x}^2 \end{pmatrix} \circ \left[ \hat{R}(\mu) \begin{pmatrix} \widetilde{x}_0 \\ \widetilde{a}_0 \end{pmatrix} \right].
$$

To the first order of $k_s$, we obtain

$$
\begin{pmatrix} \widetilde{x}_n \\ \widetilde{a}_n \end{pmatrix} = \left( \hat{R}(\mu) \right)^n \begin{pmatrix} \widetilde{x}_0 \\ \widetilde{a}_0 \end{pmatrix}
$$

$$
+ \sum_{m=0}^{n-1} \left[ \left( \hat{R}(\mu) \right)^m \begin{pmatrix} \widetilde{x} \\ \widetilde{a} \end{pmatrix} \right] \circ \begin{pmatrix} 0 \\ k_s \beta^{\frac{3}{2}} \widetilde{x}^2 \end{pmatrix} \circ \left[ \left( \hat{R}(\mu) \right)^{n-m} \begin{pmatrix} \widetilde{x}_0 \\ \widetilde{a}_0 \end{pmatrix} \right]
$$

$$= \begin{pmatrix} \tilde{x}_0 \cos(n\mu) + \tilde{a}_0 \sin(n\mu) \\ -\tilde{x}_0 \sin(n\mu) + \tilde{a}_0 \cos(n\mu) \end{pmatrix} + k_s \beta^{\frac{3}{2}} \begin{pmatrix} 0 \\ [\tilde{x}_0 \cos(n\mu) + \tilde{a}_0 \sin(n\mu)]^2 \end{pmatrix}$$

$$+ k_s \beta^{\frac{3}{2}} \sum_{m=1}^{n-1} \begin{pmatrix} \{\tilde{x}_0 \cos[(n-m)\mu] + \tilde{a}_0 \sin[(n-m)\mu]\}^2 \sin(m\mu) \\ \{\tilde{x}_0 \cos[(n-m)\mu] + \tilde{a}_0 \sin[(n-m)\mu]\}^2 \cos(m\mu) \end{pmatrix}$$

$$= \begin{pmatrix} \tilde{x}_0 \cos(n\mu) + \tilde{a}_0 \sin(n\mu) \\ -\tilde{x}_0 \sin(n\mu) + \tilde{a}_0 \cos(n\mu) \end{pmatrix} + k_s \beta^{\frac{3}{2}} \begin{pmatrix} 0 \\ [\tilde{x}_0 \cos(n\mu) + \tilde{a}_0 \sin(n\mu)]^2 \end{pmatrix}$$

$$+ \frac{1}{2} k_s \beta^{\frac{3}{2}} (\tilde{x}_0^2 + \tilde{a}_0^2) \sum_{m=1}^{n-1} \begin{pmatrix} \sin(m\mu) \\ \cos(m\mu) \end{pmatrix}$$

$$+ \frac{1}{4} k_s \beta^{\frac{3}{2}} (\tilde{x}_0^2 - \tilde{a}_0^2) \sum_{m=1}^{n-1} \begin{pmatrix} \sin[(2n-m)\mu] - \sin[(2n-3m)\mu] \\ \cos[(2n-m)\mu] + \cos[(2n-3m)\mu] \end{pmatrix}$$

$$+ \frac{1}{2} k_s \beta^{\frac{3}{2}} \tilde{x}_0 \tilde{a}_0 \sum_{m=1}^{n-1} \begin{pmatrix} \cos[(2n-3m)\mu] - \cos[(2n-m)\mu] \\ \sin[(2n-m)\mu] + \sin[(2n-3m)\mu] \end{pmatrix}.$$

Again, using the formula

$$\sum_{m=0}^{n-1} \cos(x + my) = \frac{\cos(x + (n-1)y/2) \sin(ny/2)}{\sin(y/2)}, \qquad (11.7)$$

we obtain

$$\sum_{m=1}^{n-1} \cos(m\mu) = \sum_{m=1}^{n-1} \cos[\mu + (m-1)\mu] = \sum_{m=0}^{n-2} \cos(\mu + m\mu)$$

$$= \frac{\cos[\mu + (n-2)\mu/2] \sin[(n-1)\mu/2]}{\sin(\mu/2)} = \frac{\cos(n\mu/2) \sin[(n-1)\mu/2]}{\sin(\mu/2)},$$

$$\sum_{m=1}^{n-1} \cos[(2n-m)\mu] = \sum_{m=1}^{n-1} \cos[(2n-1)\mu - (m-1)\mu]$$

$$= \sum_{m=0}^{n-2} \cos[(2n-1)\mu - m\mu] = \frac{\cos(3n\mu/2) \sin[(n-1)\mu/2]}{\sin(\mu/2)},$$

$$\sum_{m=1}^{n-1} \cos[(2n-3m)\mu] = \sum_{m=1}^{n-1} \cos[(2n-3)\mu - 3(m-1)\mu]$$

$$= \sum_{m=0}^{n-2} \cos[(2n-3)\mu - 3m\mu] = \frac{\cos(n\mu/2) \sin[3(n-1)\mu/2]}{\sin(3\mu/2)}.$$

From eq. (11.7), we have

$$\sum_{m=0}^{n-1} \sin\left(x+my\right) = \sum_{m=0}^{n-1} \cos\left(x+my-\frac{\pi}{2}\right)$$

$$= \frac{\cos\left[x+(n-1)\,y/2-\pi/2\right]\sin\left(ny/2\right)}{\sin\left(y/2\right)} = \frac{\sin\left[x+(n-1)\,y/2\right]\sin\left(ny/2\right)}{\sin\left(y/2\right)}.$$

As a result, we obtain

$$\sum_{m=1}^{n-1} \sin\left(i\mu\right) = \sum_{m=1}^{n-1} \sin\left[\mu+(m-1)\,\mu\right] = \sum_{m=0}^{n-2} \sin\left(\mu+m\mu\right)$$

$$= \frac{\sin\left(n\mu/2\right)\sin\left[(n-1)\,\mu/2\right]}{\sin\left(\mu/2\right)},$$

$$\sum_{m=1}^{n-1} \sin\left[(2n-m)\,\mu\right] = \sum_{m=1}^{n-1} \sin\left[(2n-1)\,\mu-(m-1)\,\mu\right]$$

$$= \sum_{m=0}^{n-2} \sin\left[(2n-1)\,\mu-m\mu\right] = \frac{\sin\left(3n\mu/2\right)\sin\left[(n-1)\,\mu/2\right]}{\sin\left(\mu/2\right)},$$

$$\sum_{m=1}^{n-1} \sin\left[(2n-3m)\,\mu\right] = \sum_{m=1}^{n-1} \sin\left[(2n-3)\,\mu-3\,(m-1)\,\mu\right]$$

$$= \sum_{m=0}^{n-2} \sin\left[(2n-3)\,\mu-3m\mu\right] = \frac{\sin\left(n\mu/2\right)\sin\left[3\,(n-1)\,\mu/2\right]}{\sin\left(3\mu/2\right)}.$$

The position and angle after $n$ turns are

$$\begin{pmatrix} \widetilde{x}_n \\ \widetilde{a}_n \end{pmatrix} = \begin{pmatrix} \widetilde{x}_0\cos\left(n\mu\right)+\widetilde{a}_0\sin\left(n\mu\right) \\ -\widetilde{x}_0\sin\left(n\mu\right)+\widetilde{a}_0\cos\left(n\mu\right) \end{pmatrix} + k_s\beta^{\frac{3}{2}} \begin{pmatrix} 0 \\ \left[\widetilde{x}_0\cos\left(n\mu\right)+\widetilde{a}_0\sin\left(n\mu\right)\right]^2 \end{pmatrix}$$

$$+ \frac{1}{2}k_s\beta^{\frac{3}{2}}\left(\widetilde{x}_0^2+\widetilde{a}_0^2\right)\frac{\sin\left[(n-1)\,\mu/2\right]}{\sin\left(\mu/2\right)}\begin{pmatrix} \sin\left(n\mu/2\right) \\ \cos\left(n\mu/2\right) \end{pmatrix}$$

$$+ \frac{1}{4}k_s\beta^{\frac{3}{2}}\left(\widetilde{x}_0^2-\widetilde{a}_0^2\right)\frac{\sin\left[(n-1)\,\mu/2\right]}{\sin\left(\mu/2\right)}\begin{pmatrix} \sin\left(3n\mu/2\right) \\ \cos\left(3n\mu/2\right) \end{pmatrix}$$

$$+ \frac{1}{4}k_s\beta^{\frac{3}{2}}\left(\widetilde{x}_0^2-\widetilde{a}_0^2\right)\frac{\sin\left[3\,(n-1)\,\mu/2\right]}{\sin\left(3\mu/2\right)}\begin{pmatrix} -\sin\left(n\mu/2\right) \\ \cos\left(n\mu/2\right) \end{pmatrix}$$

$$+ \frac{1}{2}k_s\beta^{\frac{3}{2}}\widetilde{x}_0\widetilde{a}_0\frac{\sin\left[(n-1)\,\mu/2\right]}{\sin\left(\mu/2\right)}\begin{pmatrix} -\cos\left(3n\mu/2\right) \\ \sin\left(3n\mu/2\right) \end{pmatrix}$$

$$+ \frac{1}{2}k_s\beta^{\frac{3}{2}}\widetilde{x}_0\widetilde{a}_0\frac{\sin\left[3\,(n-1)\,\mu/2\right]}{\sin\left(3\mu/2\right)}\begin{pmatrix} \cos\left(n\mu/2\right) \\ \sin\left(n\mu/2\right) \end{pmatrix},$$

which shows that the position and angle become arbitrarily large when $\mu \to 2\pi k$ or $2\pi(k \pm 1/3)$.

Next, we will study the deformation of the invariant. Since the sextupole affect only the nonlinear part of the motion, the perturbation of the invariant is of the third order and higher. One way to obtain the new invariant is to find a new coordinate system in which the motion is a circle. The new invariant can be found via the relation between the new and the old coordinates. This method is called the normal form theory. In general, there is no analytical solution to the perturbed invariant. We can only obtain it perturbatively. In order to demonstrate the essence of the method, the lowest order perturbation of the invariant will be derived. Let $\tilde{\tilde{x}}$ and $\tilde{\tilde{a}}$ be the new coordinates such that, in this coordinate system, the motion is a circle up to the second order. Since the linear motion in the coordinate system of $(\tilde{x}, \tilde{a})$ is already a circle, the general form of the relation between the new and the old coordinate systems can be written as

$$
\begin{pmatrix} \tilde{\tilde{x}} \\ \tilde{\tilde{a}} \end{pmatrix} = \mathcal{A} \circ \begin{pmatrix} \tilde{x} \\ \tilde{a} \end{pmatrix} = \begin{pmatrix} \tilde{x} \\ \tilde{a} \end{pmatrix} + \begin{pmatrix} A_{11}\tilde{x}^2 + A_{12}\tilde{x}\tilde{a} + A_{22}\tilde{a}^2 \\ B_{11}\tilde{x}^2 + B_{12}\tilde{x}\tilde{a} + B_{22}\tilde{a}^2 \end{pmatrix},
$$

where $\mathcal{A}$ denote a second order transfer map that transforms $(x, a)$. The inverse, to the second order, is

$$
\begin{pmatrix} \tilde{x} \\ \tilde{a} \end{pmatrix} = \mathcal{A}^{-1} \circ \begin{pmatrix} \tilde{\tilde{x}} \\ \tilde{\tilde{a}} \end{pmatrix} = \begin{pmatrix} \tilde{\tilde{x}} \\ \tilde{\tilde{a}} \end{pmatrix} - \begin{pmatrix} A_{11}\tilde{\tilde{x}}^2 + A_{12}\tilde{\tilde{x}}\tilde{\tilde{a}} + A_{22}\tilde{\tilde{a}}^2 \\ B_{11}\tilde{\tilde{x}}^2 + B_{12}\tilde{\tilde{x}}\tilde{\tilde{a}} + B_{22}\tilde{\tilde{a}}^2 \end{pmatrix}.
$$

From the relation

$$
\begin{pmatrix} \tilde{x}_1 \\ \tilde{a}_1 \end{pmatrix} = \begin{pmatrix} \tilde{x} \\ \tilde{a} + k_s \beta^{\frac{3}{2}} \tilde{x}^2 \end{pmatrix} \circ \left[ \hat{R}(\mu) \begin{pmatrix} \tilde{x}_0 \\ \tilde{a}_0 \end{pmatrix} \right],
$$

we obtain the one turn map in the new coordinates, which is

$$
\begin{pmatrix} \tilde{\tilde{x}}_1 \\ \tilde{\tilde{a}}_1 \end{pmatrix} = \mathcal{A} \circ \begin{pmatrix} x \\ a + k_s \beta^{\frac{3}{2}} x^2 \end{pmatrix} \circ \begin{pmatrix} x \cos \mu + a \sin \mu \\ -x \sin \mu + a \cos \mu \end{pmatrix} \circ \mathcal{A}^{-1} \circ \begin{pmatrix} \tilde{\tilde{x}}_0 \\ \tilde{\tilde{a}}_0 \end{pmatrix}. \quad (11.8)
$$

Expanding it to the second order, we have

$$
\begin{pmatrix} \tilde{\tilde{x}}_1 \\ \tilde{\tilde{a}}_1 \end{pmatrix} = \begin{pmatrix} \tilde{\tilde{x}}_0 \cos \mu + \tilde{\tilde{a}}_0 \sin \mu \\ -\tilde{\tilde{x}}_0 \sin \mu + \tilde{\tilde{a}}_0 \cos \mu \end{pmatrix} + \begin{pmatrix} z_{1x} \\ z_{1a} \end{pmatrix} + \begin{pmatrix} z_{2x} \\ z_{2a} \end{pmatrix} + \begin{pmatrix} z_{3x} \\ z_{3a} \end{pmatrix},
$$

where

$$z_{1x} = A_{11}\left(\widetilde{x}_0^2 \cos^2\mu + \widetilde{x}_0\widetilde{a}_0 \sin(2\mu) + \widetilde{a}_0^2 \sin^2\mu\right)$$

$$+ A_{12}\left(-\frac{1}{2}\widetilde{x}_0^2 \sin(2\mu) + \widetilde{x}_0\widetilde{a}_0 \cos(2\mu) + \frac{1}{2}\widetilde{a}_0^2 \sin(2\mu)\right)$$

$$+ A_{22}\left(\widetilde{x}_0^2 \sin^2\mu - \widetilde{x}_0\widetilde{a}_0 \sin(2\mu) + \widetilde{a}_0^2 \cos^2\mu\right),$$

$$z_{1a} = B_{11}\left(\widetilde{x}_0^2 \cos^2\mu + \widetilde{x}_0\widetilde{a}_0 \sin(2\mu) + \widetilde{a}_0^2 \sin^2\mu\right)$$

$$+ B_{12}\left(-\frac{1}{2}\widetilde{x}_0^2 \sin(2\mu) + \widetilde{x}_0\widetilde{a}_0 \cos(2\mu) + \frac{1}{2}\widetilde{a}_0^2 \sin(2\mu)\right)$$

$$+ B_{22}\left(\widetilde{x}_0^2 \sin^2\mu - \widetilde{x}_0\widetilde{a}_0 \sin(2\mu) + \widetilde{a}_0^2 \cos^2\mu\right),$$

$$z_{2x} = -\left(A_{11}\widetilde{x}_0^2 + A_{12}\widetilde{x}_0\widetilde{a}_0 + A_{22}\widetilde{a}_0^2\right)\cos\mu - \left(B_{11}\widetilde{x}_0^2 + B_{12}\widetilde{x}_0\widetilde{a}_0 + B_{22}\widetilde{a}_0^2\right)\sin\mu,$$

$$z_{2a} = \left(A_{11}\widetilde{x}_0^2 + A_{12}\widetilde{x}_0\widetilde{a}_0 + A_{22}\widetilde{a}_0^2\right)\sin\mu - \left(B_{11}\widetilde{x}_0^2 + B_{12}\widetilde{x}_0\widetilde{a}_0 + B_{22}\widetilde{a}_0^2\right)\cos\mu,$$

$$z_{3x} = 0,$$

$$z_{3a} = k_s\beta^{\frac{3}{2}}\left(\widetilde{x}_0^2 \cos^2\mu + \widetilde{x}_0\widetilde{a}_0 \sin(2\mu) + \widetilde{a}_0^2 \sin^2\mu\right).$$

Since the one turn map in the new coordinates is a rotation up to the second order, we obtain the following equations

$$A_{11}\cos^2\mu - \frac{1}{2}A_{12}\sin(2\mu) + A_{22}\sin^2\mu - A_{11}\cos\mu - B_{11}\sin\mu = 0,$$

$$A_{11}\sin(2\mu) + A_{12}\cos(2\mu) - A_{22}\sin(2\mu) - A_{12}\cos\mu - B_{12}\sin\mu = 0,$$

$$A_{11}\sin^2\mu + \frac{1}{2}A_{12}\sin(2\mu) + A_{22}\cos^2\mu - A_{22}\cos\mu - B_{22}\sin\mu = 0,$$

$$B_{11}\cos^2\mu - \frac{1}{2}B_{12}\sin(2\mu) + B_{22}\sin^2\mu + A_{11}\sin\mu - B_{11}\cos\mu = -k_s\beta^{\frac{3}{2}}\cos^2\mu,$$

$$B_{11}\sin(2\mu) + B_{12}\cos(2\mu) - B_{22}\sin(2\mu) + A_{12}\sin\mu - B_{12}\cos\mu = -k_s\beta^{\frac{3}{2}}\sin(2\mu),$$

$$B_{11}\sin^2\mu + \frac{1}{2}B_{12}\sin(2\mu) + B_{22}\cos^2\mu + A_{22}\sin\mu - B_{22}\cos\mu = -k_s\beta^{\frac{3}{2}}\sin^2\mu.$$

After tedious but straightforward algebraic and trigonometric manipulations, the solution is obtained, which is

$$A_{11} = -k_s\beta^{\frac{3}{2}}\frac{\cos(\mu/2)\cos\mu}{2\sin(3\mu/2)}, \quad A_{12} = 0, \quad A_{22} = -k_s\beta^{\frac{3}{2}}\frac{\cos^3(\mu/2)}{\sin(3\mu/2)},$$

$$B_{11} = -\frac{1}{2}k_s\beta^{\frac{3}{2}}, \quad B_{12} = k_s\beta^{\frac{3}{2}}\frac{\cos(\mu/2)\cos\mu}{\sin(3\mu/2)}, \quad B_{22} = 0.$$

As a result, the transformation defines a coordinate system in which the motion is a rotation up to the second order. In other words, we have

$$\widetilde{x}^2 + \widetilde{a}^2 = \epsilon,$$

which take the form

$$\left(\widetilde{x} + A_{11}\widetilde{x}^2 + A_{22}\widetilde{a}^2\right)^2 + \left(\widetilde{a} + B_{11}\widetilde{x}^2 + B_{12}\widetilde{x}\widetilde{a}\right)^2 = \epsilon$$

in the old coordinates. Keeping only the terms up to the first order of $k_s$, we have

$$\widetilde{x}^2 + \widetilde{a}^2 + 2A_{11}\widetilde{x}^3 + 2B_{11}\widetilde{x}^2\widetilde{a} + 2\left(A_{22} + B_{12}\right)\widetilde{x}\widetilde{a}^2 = \epsilon.$$

Since

$$A_{22} + B_{12} = k_s\beta^{\frac{3}{2}}\left[-\frac{\cos^3\left(\mu/2\right)}{\sin\left(3\mu/2\right)} + \frac{\cos\left(\mu/2\right)\cos\mu}{\sin\left(3\mu/2\right)}\right]$$

$$= k_s\beta^{\frac{3}{2}}\frac{\cos\left(\mu/2\right)\left[\cos\mu - \cos^2\left(\mu/2\right)\right]}{\sin\left(3\mu/2\right)}$$

$$= -k_s\beta^{\frac{3}{2}}\frac{\cos\left(\mu/2\right)\sin^2\left(\mu/2\right)}{\sin\left(3\mu/2\right)} = -k_s\beta^{\frac{3}{2}}\frac{\sin\left(\mu/2\right)\sin\mu}{2\sin\left(3\mu/2\right)},$$

the perturbed invariant is

$$\widetilde{x}^2 + \widetilde{a}^2 - k_s\beta^{\frac{3}{2}}\left[\frac{\cos\left(\mu/2\right)\cos\mu}{\sin\left(3\mu/2\right)}\widetilde{x}^3 + \widetilde{x}^2\widetilde{a} + \frac{\sin\left(\mu/2\right)\sin\mu}{\sin\left(3\mu/2\right)}\widetilde{x}\widetilde{a}^2\right] = \epsilon,$$

or

$$\widetilde{x}^2 + \widetilde{a}^2 - \frac{k_s\beta^{\frac{3}{2}}}{\sin\left(3\mu/2\right)}\widetilde{x}\left[\widetilde{x}\cos\left(\mu/2\right) + \widetilde{a}\sin\left(\mu/2\right)\right]\left[\widetilde{x}\cos\mu + \widetilde{a}\sin\mu\right] = \epsilon.$$

Apparently the invariant diverge when $\mu \to 2\pi k/3$. The invariant in the above equation is shown in Fig. 11.3, which shows the presence of the third–integer resonance. As a comparison, the true invariant obtained through tracking is shown in Fig. 11.4. It is clear that the invariant obtained from the first order perturbation theory agrees with the exact invariant qualitatively but not quantitatively.

Like the half–integer resonance, we can go one step further to include the terms that are proportional to $k_s^2$. Again, we can attempt to find another coordinate system in which the motion is a rotation up to $k_s^2$. But first of all, we have to obtain the third order one turn map in the coordinates of $(\widetilde{x}, \widetilde{a})$. From eq. (11.8), we have, to the third order,

$$\begin{pmatrix}\widetilde{\widetilde{x}}_1 \\ \widetilde{\widetilde{a}}_1\end{pmatrix} = \mathcal{A} \circ \begin{pmatrix}x \\ a + k_s\beta^{\frac{3}{2}}x^2\end{pmatrix} \circ \begin{pmatrix}x\cos\mu + a\sin\mu \\ -x\sin\mu + a\cos\mu\end{pmatrix} \circ \mathcal{A}^{-1} \circ \begin{pmatrix}\widetilde{\widetilde{x}}_0 \\ \widetilde{\widetilde{a}}_0\end{pmatrix}$$

$$= \begin{pmatrix}x + A_{11}x^2 + A_{22}a^2 \\ a + B_{11}x^2 + B_{12}xa\end{pmatrix} \circ \begin{pmatrix}x\cos\mu + a\sin\mu \\ -x\sin\mu + a\cos\mu + k_s\beta^{\frac{3}{2}}\left(x\cos\mu + a\sin\mu\right)^2\end{pmatrix}$$

$$\circ \left[\begin{pmatrix}\widetilde{\widetilde{x}}_0 - A_{11}\widetilde{\widetilde{x}}_0^2 - A_{22}\widetilde{\widetilde{a}}_0^2 \\ \widetilde{\widetilde{a}}_0 - B_{11}\widetilde{\widetilde{x}}_0^2 - B_{12}\widetilde{\widetilde{x}}_0\widetilde{\widetilde{a}}_0\end{pmatrix} + \begin{pmatrix}2A_{11}^2\widetilde{\widetilde{x}}_0^3 + 2A_{22}B_{11}\widetilde{\widetilde{x}}_0^2\widetilde{\widetilde{a}}_0 + A_{22}B_{12}\widetilde{\widetilde{x}}_0\widetilde{\widetilde{a}}_0^2 \\ 2A_{11}^2\widetilde{\widetilde{x}}_0^2\widetilde{\widetilde{a}}_0 + 2A_{22}B_{11}\widetilde{\widetilde{x}}_0\widetilde{\widetilde{a}}_0^2 + A_{22}B_{12}\widetilde{\widetilde{a}}_0^3\end{pmatrix}\right].$$

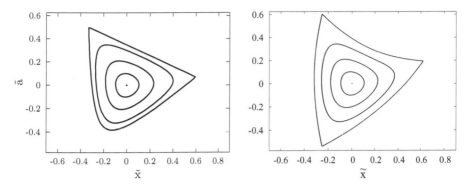

**FIGURE 11.3**: Invariant obtained through first order perturbation theory (left) and tracking (right), with $k_s\beta^{\frac{3}{2}} = 1$ and $\nu = 130/360$.

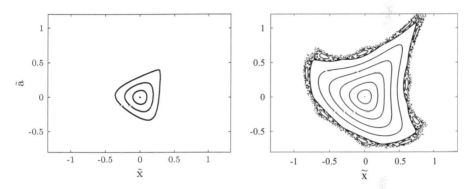

**FIGURE 11.4**: Invariant obtained through first order perturbation theory (left) and tracking (right), with $k_s\beta^{\frac{3}{2}} = 1$ and $\nu = 110/360$.

The last line is the function $\mathcal{A}^{-1}$ to the third order. It is straightforward to verify that it is true using the relation $B_{12} = -2A_{11}$. Yet $\mathcal{A}$ and $\mathcal{A}^{-1}$ are not symplectic to the third order, which is easily verified using eq. (5.6). The details of obtaining the symplectic version of $\mathcal{A}$ and $\mathcal{A}^{-1}$ are beyond the scope of this book. The result is actually very simple, which is

$$\mathcal{A} = \begin{pmatrix} x + A_{11}x^2 + A_{22}a^2 + A_{11}^2\tilde{x}_0^3 + A_{22}B_{11}\tilde{x}_0^2\tilde{a}_0 - A_{11}A_{22}\tilde{x}_0\tilde{a}_0^2 \\ a + B_{11}x^2 - 2A_{11}xa + A_{11}^2\tilde{x}_0^2\tilde{a}_0 + A_{22}B_{11}\tilde{x}_0\tilde{a}_0^2 - A_{11}A_{22}\tilde{a}_0^3 \end{pmatrix},$$

$$\mathcal{A}^{-1} = \begin{pmatrix} x - A_{11}x^2 - A_{22}a^2 + A_{11}^2\tilde{x}_0^3 + A_{22}B_{11}\tilde{x}_0^2\tilde{a}_0 - A_{11}A_{22}\tilde{x}_0\tilde{a}_0^2 \\ a - B_{11}x^2 + 2A_{11}xa + A_{11}^2\tilde{x}_0^2\tilde{a}_0 + A_{22}B_{11}\tilde{x}_0\tilde{a}_0^2 - A_{11}A_{22}\tilde{a}_0^3 \end{pmatrix}.$$

Again, it is straightforward to verify the symplecticity of the above map using eq. (5.6). Note that the third order part in $\mathcal{A}$ and $\mathcal{A}^{-1}$ is half of that in the non-symplectic version of $\mathcal{A}^{-1}$.

After a straightforward yet rather lengthy derivation, the final result is

$$
\begin{pmatrix} \widetilde{\widetilde{x}}_1 \\ \widetilde{\widetilde{a}}_1 \end{pmatrix} = \begin{pmatrix} \widetilde{\widetilde{x}}_0 \cos\mu + \widetilde{\widetilde{a}}_0 \sin\mu \\ -\widetilde{\widetilde{x}}_0 \sin\mu + \widetilde{\widetilde{a}}_0 \cos\mu \end{pmatrix} + \begin{pmatrix} A_{111}\widetilde{\widetilde{x}}_0^3 + A_{112}\widetilde{\widetilde{x}}_0^2\widetilde{\widetilde{a}}_0 + A_{122}\widetilde{\widetilde{x}}_0\widetilde{\widetilde{a}}_0^2 + A_{222}\widetilde{\widetilde{a}}_0^3 \\ B_{111}\widetilde{\widetilde{x}}_0^3 + B_{112}\widetilde{\widetilde{x}}_0^2\widetilde{\widetilde{a}}_0 + B_{122}\widetilde{\widetilde{x}}_0\widetilde{\widetilde{a}}_0^2 + B_{222}\widetilde{\widetilde{a}}_0^3 \end{pmatrix},
$$

where

$$
A_{111} = -k_s^2\beta^3 \frac{\cos^3(\mu/2)\sin\mu\cos^2\mu}{\sin(3\mu/2)},
$$

$$
A_{112} = k_s^2\beta^3 \frac{\cos^3(\mu/2)\cos\mu}{2\sin(3\mu/2)}\left[3\cos(2\mu)-1\right],
$$

$$
A_{122} = k_s^2\beta^3 \frac{\cos^4(\mu/2)\sin(\mu/2)}{\sin(3\mu/2)}\left[1+3\cos(2\mu)\right],
$$

$$
A_{222} = k_s^2\beta^3 \frac{4\cos^5(\mu/2)\sin^2(\mu/2)\cos\mu}{\sin(3\mu/2)}, \tag{11.9}
$$

and

$$
B_{111} = k_s^2\beta^3 \left\{ \frac{4\cos^5(\mu/2)\sin^2(\mu/2)\cos\mu}{\sin(3\mu/2)} + \frac{\cos(\mu/2)\cos^4\mu}{\sin(3\mu/2)} \right\},
$$

$$
B_{112} = k_s^2\beta^3 \left\{ -\frac{\cos^4(\mu/2)\sin(\mu/2)}{\sin(3\mu/2)}\left[1+3\cos(2\mu)\right] \right.
$$
$$
\left. +6\frac{\sin(\mu/2)\cos^2(\mu/2)\cos^3\mu}{\sin(3\mu/2)} \right\},
$$

$$
B_{122} = k_s^2\beta^3 \left\{ \frac{\cos^3(\mu/2)\cos\mu}{2\sin(3\mu/2)}\left[3\cos(2\mu)-1\right] \right.
$$
$$
\left. +12\frac{\sin^2(\mu/2)\cos^3(\mu/2)\cos^2\mu}{\sin(3\mu/2)} \right\},
$$

$$
B_{222} = k_s^2\beta^3 \left\{ \frac{\cos^3(\mu/2)\sin\mu\cos^2\mu}{\sin(3\mu/2)} + \frac{\cos(\mu/2)\cos\mu\sin^3\mu}{\sin(3\mu/2)} \right\}. \tag{11.10}
$$

The next step is to find a second order transformation (in terms of $k_s$) that the third order map in the newest coordinates is a rotation. Unlike the first order transformation, not all nonlinear terms in the map can be removed even if $\mu$ does not satisfy any resonance condition (i.e., $\nu$ is irrational.). It is much easier to illustrate this in the eigenspace of the linear map, which is complex. Let us denote $(s^+, s^-)$ as the complex coordinates which are related to the real coordinates by the relations

$$
\begin{pmatrix} s^- \\ s^+ \end{pmatrix} = \frac{1}{\sqrt{2}} \begin{pmatrix} 1 & i \\ 1 & -i \end{pmatrix} \begin{pmatrix} \widetilde{\widetilde{x}} \\ \widetilde{\widetilde{a}} \end{pmatrix}. \tag{11.11}
$$

The inverse is

$$\begin{pmatrix} \widetilde{\widetilde{x}} \\ \widetilde{\widetilde{a}} \end{pmatrix} = \frac{1}{\sqrt{2}} \begin{pmatrix} 1 & 1 \\ -i & i \end{pmatrix} \begin{pmatrix} s^- \\ s^+ \end{pmatrix}. \tag{11.12}$$

It is obvious, in this case, that the map in the complex coordinates is symplectic as long as that in the real coordinates is, since the determinant of the Jacobian in the complex coordinates equals that in the real coordinates, which equals to 1. The linear one turn map in the complex coordinates is

$$\begin{pmatrix} s_1^- \\ s_1^+ \end{pmatrix} = \frac{1}{2} \begin{pmatrix} 1 & i \\ 1 & -i \end{pmatrix} \hat{R}(\mu) \begin{pmatrix} 1 & 1 \\ -i & i \end{pmatrix} \begin{pmatrix} s_0^- \\ s_0^+ \end{pmatrix} = \begin{pmatrix} e^{-i\mu} & 0 \\ 0 & e^{i\mu} \end{pmatrix} \begin{pmatrix} s_0^- \\ s_0^+ \end{pmatrix}. \tag{11.13}$$

For the time being, let us consider the generic case that the map $\mathcal{M}$ contains a linear part $\mathcal{R}$ and a nonlinear part $\mathcal{S}$. In the complex coordinates, $\mathcal{R}$ is shown above. Furthermore, let us assume that the lowest order terms in $\mathcal{S}$ are of that of $m$. Note that if $m > 2$, $\mathcal{M}$ is the map that has been transformed through the nonlinear normal form transformations at least once. Now let us consider $\mathcal{M}$ only up to the $m$th order, i.e.,

$$\mathcal{M}_m = \mathcal{R} + \mathcal{S}_m.$$

Define the coordinate transformation as

$$\mathcal{A}_m = \mathcal{I} + \mathcal{T}_m,$$

where $\mathcal{I}$ is the unity map and $\mathcal{T}_m$ contains terms of the $m$th order only. Hence, to the $m$th order,

$$\mathcal{A}_m^{-1} = \mathcal{I} - \mathcal{T}_m.$$

The normalized map, to the $m$th order, is

$$\mathcal{N}_m =_m \mathcal{A}_m \circ \mathcal{M}_m \circ \mathcal{A}_m^{-1} =_m (\mathcal{I} + \mathcal{T}_m) \circ (\mathcal{R} + \mathcal{S}_m) \circ (\mathcal{I} - \mathcal{T}_m)$$
$$=_m (\mathcal{I} + \mathcal{T}_m) \circ (\mathcal{R} + \mathcal{S}_m \quad \mathcal{R} \circ \mathcal{T}_m) -_m \mathcal{R} + \mathcal{S}_m - (\mathcal{T}_m \circ \mathcal{R} - \mathcal{R} \circ \mathcal{T}_m).$$

The goal is to cancel as many terms in $\mathcal{S}_m$ as possible. Let us evaluate the map $\mathcal{T}_m \circ \mathcal{R} - \mathcal{R} \circ \mathcal{T}_m$, which is

$$\mathcal{T}_m \circ \mathcal{R} - \mathcal{R} \circ \mathcal{T}_m$$
$$= \sum_{k=0}^{m} \left[ \begin{pmatrix} T_{mk}^- (s^-)^k (s^+)^{m-k} \\ T_{mk}^+ (s^-)^k (s^+)^{m-k} \end{pmatrix} \circ \begin{pmatrix} e^{-i\mu} s_0^- \\ e^{i\mu} s_0^+ \end{pmatrix} - \begin{pmatrix} e^{-i\mu} s^- \\ e^{i\mu} s^+ \end{pmatrix} \circ \begin{pmatrix} T_{mk}^- \left(s_0^-\right)^k \left(s_0^+\right)^{m-k} \\ T_{mk}^+ \left(s_0^-\right)^k \left(s_0^+\right)^{m-k} \end{pmatrix} \right]$$
$$= \sum_{k=0}^{m} \begin{pmatrix} T_{mk}^- \left(s_0^-\right)^k \left(s_0^+\right)^{m-k} \left[ e^{i\mu(m-2k)} - e^{-i\mu} \right] \\ T_{mk}^+ \left(s_0^-\right)^k \left(s_0^+\right)^{m-k} \left[ e^{i\mu(m-2k)} - e^{i\mu} \right] \end{pmatrix}.$$

Apparently terms in $\mathcal{S}_m$ cannot be removed if the corresponding terms in the map $\mathcal{T}_m \circ \mathcal{R} - \mathcal{R} \circ \mathcal{T}_m$ are zero, and they are zero if

$$e^{i\mu(m-2k)} - e^{\pm i\mu} = 0.$$

That is

$$e^{i\mu(m-2k\mp1)} = 1,$$

which is fulfilled when

$$\mu(m - 2k \mp 1) = 2\pi n,$$

where $n$ is an integer. Apparently, the solutions of this equation can be divided into two classes: one that is independent of $\mu$, which is $k = (m \mp 1)/2$, and the other that depends on $\mu$. The solutions that depend on $\mu$ are called the resonance conditions, which can be avoided with the choice of $\mu$. The ones that are independent of $\mu$ provide the terms that cannot be removed from the map regardless of the choice of the tune. First, let us consider the case of $m = 2$. Since both $m$ and $2k$ are even, under no circumstance $m - 2k \mp 1 = 0$. This is consistent with the fact that a solution was found above that transforms the second order map into a linear map. For the $\mu$ dependent solutions, it is straightforward to verify that all six solutions are included in one simple relation, which is $3\mu = 2\pi n$. Note that this is none other than the condition of the third–integer resonance. Second, we consider the case of $m = 3$. There are two solutions that are independent of $\mu$, which are $k = 2$ for the top row and $k = 1$ for the bottom row. Again, all eight solutions that are dependent on $\mu$ are included in the expression $4\mu = 2\pi n$, which is the condition of the fourth–integer resonance. In case of the sextupole, it drives the third–integer resonance to the first order of $k_s$ and the fourth–integer resonance to the second order of $k_s$. Since we usually set the tune away from the third and the fourth–integer resonances, we can obtain a third order map that takes the form of

$$\mathcal{N}_3 = \begin{pmatrix} e^{-i\mu}s_0^- + S_{32}^- \left(s_0^-\right)^2 \left(s_0^+\right) \\ e^{i\mu}\ s_0^+ + S_{31}^+ \left(s_0^-\right) \left(s_0^+\right)^2 \end{pmatrix}.$$

Let us focus on the new map, since we will not try to obtain the second order distortion of the invariant. Going one step further, we have

$$\mathcal{N}_3 = \begin{pmatrix} \left[e^{-i\mu} + S_{32}^- s_0^- s_0^+\right] s_0^- \\ \left[e^{i\mu}\ + S_{31}^+ s_0^- s_0^+\right] s_0^+ \end{pmatrix}.$$

Transforming back to the real coordinates, we obtain

$$\mathcal{N}_3 = \begin{pmatrix} \widetilde{x}_0 \cos\mu + \widetilde{a}_0 \sin\mu \\ -\widetilde{x}_0 \sin\mu + \widetilde{a}_0 \cos\mu \end{pmatrix} + \frac{\widetilde{x}_0^2 + \widetilde{a}_0^2}{4} \begin{pmatrix} \left(S_{32}^- + S_{31}^+\right) \widetilde{x}_0 + i\left(S_{32}^- - S_{31}^+\right) \widetilde{a}_0 \\ -i\left(S_{32}^- - S_{31}^+\right) \widetilde{x}_0 + \left(S_{32}^- + S_{31}^+\right) \widetilde{a}_0 \end{pmatrix}.$$
$$(11.14)$$

As a result, we conclude that $S_{32}^- + S_{31}^+$ is real and $S_{32}^- - S_{31}^+$ is purely imaginary. In order to make clear the meaning of the third order terms in $\mathcal{N}_3$, let us first show that $\mathcal{N}_3$ is symplectic to the third order. First recall that

$$\mathcal{N}_3 =_3 \mathcal{A}_3 \circ \mathcal{A}_2 \circ \mathcal{M}_2 \circ \mathcal{A}_2^{-1} \circ \mathcal{A}_3^{-1}.$$

Defining $\mathcal{A}_{23} =_3 \mathcal{A}_3 \circ \mathcal{A}_2$, we have

$$\mathcal{N}_3 =_3 \mathcal{A}_{23} \circ \mathcal{M}_2 \circ \mathcal{A}_{23}^{-1}.$$

When carrying out the transformations, we usually make sure that $\mathcal{A}_{23}$ is symplectic up to the third order. Therefore, we have

$$\hat{N}_3 \hat{J} \hat{N}_3^T =_2 \hat{A}_{23} \hat{M}_2 \hat{A}_{23}^{-1} \hat{J} \left( \hat{A}_{23} \hat{M}_2 \hat{A}_{23}^{-1} \right)^T =_2 \hat{A}_{23} \hat{M}_2 \hat{A}_{23}^{-1} \hat{J} \left( \hat{A}_{23}^{-1} \right)^T \hat{M}_2^T \hat{A}_{23}^T$$

$$=_2 \hat{A}_{23} \hat{M}_2 \hat{J} \hat{M}_2^T \hat{A}_{23}^T =_2 \hat{A}_{23} \hat{J} \hat{A}_{23}^T =_2 \hat{J}.$$

Now that we have shown that $\mathcal{N}_3$ is symplectic up to the third order, we can find out the relations between the terms. The Jacobian of $\mathcal{N}_3$ can be written as

$$\hat{N}_3 = \hat{R} + \hat{S}_3,$$

where

$$\hat{R} = \begin{pmatrix} e^{-i\mu} & 0 \\ 0 & e^{i\mu} \end{pmatrix} \quad \text{and} \quad \hat{S}_3 = \begin{pmatrix} S_{32}^- s_0^- s_0^+ & S_{32}^- \left(s_0^-\right)^2 \\ S_{31}^+ \left(s_0^+\right)^2 & S_{31}^+ s_0^- s_0^+ \end{pmatrix}.$$

The symplectic condition becomes

$$\left( \hat{R} + \hat{S}_3 \right) \hat{J} \left( \hat{R}^T + \hat{S}_3^T \right) =_2 \hat{J},$$

which leads to the relation

$$\hat{S}_3 \hat{J} \hat{R}^T + \hat{R} \hat{J} \hat{S}_3^T = 0.$$

Plugging in the matrices $\hat{R}$ and $\hat{S}_3$, we have

$$\begin{pmatrix} S_{32}^- s_0^- s_0^+ & S_{32}^- \left(s_0^-\right)^2 \\ S_{31}^+ \left(s_0^+\right)^2 & S_{31}^+ s_0^- s_0^+ \end{pmatrix} \begin{pmatrix} 0 & 1 \\ -1 & 0 \end{pmatrix} \begin{pmatrix} e^{-i\mu} & 0 \\ 0 & e^{i\mu} \end{pmatrix}$$

$$+ \begin{pmatrix} e^{-i\mu} & 0 \\ 0 & e^{i\mu} \end{pmatrix} \begin{pmatrix} 0 & 1 \\ -1 & 0 \end{pmatrix} \begin{pmatrix} S_{32}^- s_0^- s_0^+ & S_{31}^+ \left(s_0^+\right)^2 \\ S_{32}^- \left(s_0^-\right)^2 & S_{31}^+ s_0^- s_0^+ \end{pmatrix} = 0,$$

which can be simplified to

$$\left( S_{32}^- e^{i\mu} + S_{31}^+ e^{-i\mu} \right) s_0^- s_0^+ \begin{pmatrix} 0 & 1 \\ -1 & 0 \end{pmatrix} = 0.$$

Defining

$$T_S = i S_{32}^- e^{i\mu},$$

we obtain

$$S_{32}^- = -i T_S e^{-i\mu}, \quad S_{31}^+ = i T_S e^{i\mu}.$$

Furthermore, we obtain that

$$S_{32}^- + S_{31}^+ = -2T_S \sin\mu, \quad S_{32}^- - S_{31}^+ = -2iT_S \cos\mu.$$

Hence we arrive at the conclusion that $T_S$ is real. Therefore, we have

$$\mathcal{N}_3 = \begin{pmatrix} e^{-i\mu}\left[1 - iT_S s_0^- s_0^+\right] s_0^- \\ e^{i\mu}\left[1 + iT_S s_0^- s_0^+\right] s_0^+ \end{pmatrix}.$$

To the first order of $T_S$, $\mathcal{N}_3$ can be expressed as

$$\mathcal{N}_3 = \begin{pmatrix} \exp\left[-i\left(\mu + T_S s_0^- s_0^+\right)\right] s_0^- \\ \exp\left[\phantom{-}i\left(\mu + T_S s_0^- s_0^+\right)\right] s_0^+ \end{pmatrix}.$$

It is clear by far that the remaining terms in the normalized map $\mathcal{N}_3$ contribute to the change of the tune only. It is worth noting that the change of the tune is a function of the invariant, which is sometimes called the tune shift with amplitude. Computationally, the above described procedure can be easily carried out using the Differential Algebraic (DA) technique, which is valid for any given order.

Now the tune shift with amplitude can be determined through the relations between the coefficients in the real and the complex coordinates. Repeating eq. (11.13) to the third order, using eqs. (11.11) and (11.12), we have

$$\begin{pmatrix} s_1^- \\ s_1^+ \end{pmatrix} = \frac{1}{2}\begin{pmatrix} 1 & i \\ 1 & -i \end{pmatrix}\begin{pmatrix} \cos\mu & \sin\mu \\ -\sin\mu & \cos\mu \end{pmatrix}\begin{pmatrix} 1 & 1 \\ -i & i \end{pmatrix}\begin{pmatrix} s_0^- \\ s_0^+ \end{pmatrix}$$

$$+ \frac{1}{4}\begin{pmatrix} 1 & i \\ 1 & -i \end{pmatrix}\begin{pmatrix} A_{111}\left(s_0^- + s_0^+\right)^3 + A_{112}\left(s_0^- + s_0^+\right)^2\left(-is_0^- + is_0^+\right) \\ B_{111}\left(s_0^- + s_0^+\right)^3 + B_{112}\left(s_0^- + s_0^+\right)^2\left(-is_0^- + is_0^+\right) \end{pmatrix}$$

$$+ \frac{1}{4}\begin{pmatrix} 1 & i \\ 1 & -i \end{pmatrix}\begin{pmatrix} A_{122}\left(s_0^- + s_0^+\right)\left(-is_0^- + is_0^+\right)^2 + A_{222}\left(-is_0^- + is_0^+\right)^3 \\ B_{122}\left(s_0^- + s_0^+\right)\left(-is_0^- + is_0^+\right)^2 + B_{222}\left(-is_0^- + is_0^+\right)^3 \end{pmatrix}$$

$$= \begin{pmatrix} e^{-i\mu} & 0 \\ 0 & e^{i\mu} \end{pmatrix}\begin{pmatrix} s_0^- \\ s_0^+ \end{pmatrix}$$

$$+ \frac{1}{4}\begin{pmatrix} (A_{111}+iB_{111})\left(s_0^- + s_0^+\right)^3 + (A_{112}+iB_{112})\left(s_0^- + s_0^+\right)^2\left(-is_0^- + is_0^+\right) \\ (A_{111}-iB_{111})\left(s_0^- + s_0^+\right)^3 + (A_{112}-iB_{112})\left(s_0^- + s_0^+\right)^2\left(-is_0^- + is_0^+\right) \end{pmatrix}$$

$$+ \frac{1}{4}\begin{pmatrix} (A_{122}+iB_{122})\left(s_0^- + s_0^+\right)\left(-is_0^- + is_0^+\right)^2 + (A_{222}+iB_{222})\left(-is_0^- + is_0^+\right)^3 \\ (A_{122}-iB_{122})\left(s_0^- + s_0^+\right)\left(-is_0^- + is_0^+\right)^2 + (A_{222}-iB_{222})\left(-is_0^- + is_0^+\right)^3 \end{pmatrix}.$$

It is straightforward to extract the coefficients $S_{32}^-$ and $S_{31}^+$, which are

$$S_{32}^- = \frac{1}{4}\left[3\left(A_{111}+iB_{111}\right) - i\left(A_{112}+iB_{112}\right) + \left(A_{122}+iB_{122}\right) - 3i\left(A_{222}+iB_{222}\right)\right],$$

$$S_{31}^+ = \frac{1}{4}\left[3\left(A_{111}-iB_{111}\right) + i\left(A_{112}-iB_{112}\right) + \left(A_{122}-iB_{122}\right) + 3i\left(A_{222}-iB_{222}\right)\right].$$

Furthermore, we have

$$\frac{1}{4}\left(S_{32}^- + S_{31}^+\right) = \frac{1}{8}\left(\ 3A_{111} + B_{112} + A_{122} + 3B_{222}\right),$$

$$\frac{i}{4}\left(S_{32}^- - S_{31}^+\right) = \frac{1}{8}\left(-3B_{111} + A_{112} - B_{122} + 3A_{222}\right).$$

From eqs. (11.9) and (11.10), we obtain

$$\frac{1}{4}\left(S_{32}^- + S_{31}^+\right) = \frac{3}{8}k_s^2\beta^3 \frac{\cos\left(\mu/2\right)\sin\mu\cos\mu}{\sin\left(3\mu/2\right)},$$

$$\frac{i}{4}\left(S_{32}^- - S_{31}^+\right) = -\frac{3}{8}k_s^2\beta^3 \frac{\cos\left(\mu/2\right)\cos^2\mu}{\sin\left(3\mu/2\right)}.$$

As a result, eq. (11.14) becomes

$$\mathcal{N}_3 = \begin{pmatrix} \widetilde{\widetilde{x}}_0 \cos\mu + \widetilde{\widetilde{a}}_0 \sin\mu \\ -\widetilde{\widetilde{x}}_0 \sin\mu + \widetilde{\widetilde{a}}_0 \cos\mu \end{pmatrix}$$

$$+ \frac{3}{8}k_s^2\beta^3 \begin{pmatrix} \cos\left(\mu/2\right)\sin\mu\cos\mu/\sin\left(3\mu/2\right) \\ \cos\left(\mu/2\right)\cos^2\mu/\sin\left(3\mu/2\right) \end{pmatrix} \left(\widetilde{\widetilde{x}}_0^2 + \widetilde{\widetilde{a}}_0^2\right) \widetilde{\widetilde{x}}_0$$

$$+ \frac{3}{8}k_s^2\beta^3 \begin{pmatrix} -\cos\left(\mu/2\right)\cos^2\mu/\sin\left(3\mu/2\right) \\ \cos\left(\mu/2\right)\sin\mu\cos\mu/\sin\left(3\mu/2\right) \end{pmatrix} \left(\widetilde{\widetilde{x}}_0^2 + \widetilde{\widetilde{a}}_0^2\right) \widetilde{\widetilde{a}}_0$$

$$= \begin{pmatrix} \cos\mu & \sin\mu \\ -\sin\mu & \cos\mu \end{pmatrix} \begin{pmatrix} \widetilde{\widetilde{x}}_0 \\ \widetilde{\widetilde{a}}_0 \end{pmatrix}$$

$$- \frac{3}{8}k_s^2\beta^3 \frac{\cos\left(\mu/2\right)\cos\mu}{\sin\left(3\mu/2\right)}\left(\widetilde{\widetilde{x}}_0^2 + \widetilde{\widetilde{a}}_0^2\right)\begin{pmatrix} -\sin\mu & \cos\mu \\ -\cos\mu & -\sin\mu \end{pmatrix}\begin{pmatrix} \widetilde{\widetilde{x}}_0 \\ \widetilde{\widetilde{a}}_0 \end{pmatrix}$$

$$=_3 \begin{pmatrix} \cos\left(\mu + \Delta\mu\right) & \sin\left(\mu + \Delta\mu\right) \\ -\sin\left(\mu + \Delta\mu\right) & \cos\left(\mu + \Delta\mu\right) \end{pmatrix}\begin{pmatrix} \widetilde{\widetilde{x}}_0 \\ \widetilde{\widetilde{a}}_0 \end{pmatrix},$$

where $\Delta\mu = -(3/8) \cdot k_s^2\beta^3 \cdot (\cos\left(\mu/2\right)\cos\mu/\sin\left(3\mu/2\right)) \cdot (\widetilde{\widetilde{x}}_0^2 + \widetilde{\widetilde{a}}_0^2)$. Note that $\Delta\mu$ is proportional to $\widetilde{\widetilde{x}}_0^2 + \widetilde{\widetilde{a}}_0^2$, which is an invariant of motion. Note that the distortion of the invariant of motion is proportional to $k$. Hence the tune shift is small compared to the distortion of the invariant. The result is that the third–integer resonance usually leads to arbitrary large distortion of the invariant. For higher order resonances, the distortion is either of the same order or smaller than the tune shift. The resonances are, therefore, confined in the phase space.

# References

[1] L. W. Alvarez. Linear accelerator. US Patent 2,545,595, 1951. Filed 1947.

[2] J. Arthur, P. Anfinrud, P. Audebert, et al. Linac Coherent Light Source (LCLS) conceptual design report. Technical Report SLAC-R-593, SLAC National Accelerator Laboratory, 2002.

[3] U. Bechstedt, J. Dietrich, R. Maier, et al. The cooler synchrotron COSY in Jülich. *Nuclear Instruments and Methods*, B113(1-4):26–29, 1996.

[4] M. Berz. Differential algebraic description of beam dynamics to very high orders. Technical Report SSC-152, also ON: DE90013777 and TRN: 90-023092, Lawrence Berkeley National Laboratory, SSC Central Design Group, 1988. OSTI ID: 6876262. Also *Particle Accelerators*, 24:109, 1989.

[5] M. Berz. *Modern Map Methods in Particle Beam Physics*. Academic Press, San Diego, 1999.

[6] M. Berz, H. C. Hofmann, and H. Wollnik. COSY 5.0, the fifth order code for corpuscular optical systems. *Nuclear Instruments and Methods*, A258:402–406, 1987.

[7] M. Berz and K. Makino. COSY INFINITY. http://cosyinfinity.org, (accessed August 2014).

[8] I. G. Brown, editor. *The Physics and Technology of Ion Sources*. Wiley-VCH, Weinheim, second edition, 2004.

[9] K. L. Brown, F. Rothacker, D. C. Carey, and C. Iselin. TRANSPORT a computer program for designing charged particle beam transport systems. Technical Report SLAC-91 Rev. 3, also CERN-80-04 and NAL-91, Stanford Linear Accelerator Center, Fermi National Accelerator Laboratory, CERN, 1983.

[10] P. J. Bryant and K. Johnsen. *The Principles of Circular Accelerators and Storage Rings*. Cambridge University Press, Cambridge, 1993.

[11] D. C. Carey. *The Optics of Charged Particle Beams*. Harwood Academic, New York, 1987.

[12] A. W. Chao. *Physics of Collective Beam Instabilities in High Energy Accelerators*. Wiley, New York, 1993.

[13] A. W. Chao and M. Tigner, editors. *Handbook of Accelerator Physics and Engineering*. World Scientific, New Jersey, second edition, 1999.

[14] J. D. Cockcroft and E. T. S. Walton. Experiments with high velocity positive ions. – (I) Further developments in the method of obtaining high velocity positive ions. *Proceedings of the Royal Society of London, A*, 136:619–630, 1932.

[15] M. Conte and W. W. MacKay. *An Introduction to the Physics of Particle Accelerators*. World Scientific, New Jersey, second edition, 2008.

[16] E. D. Courant and H. S. Snyder. Theory of the alternating-gradient synchrotron. *Annals of Physics*, 3:1, 1958.

[17] K. R. Crandall, R. H. Stokes, and T. P. Wangler. RF quadrupole beam dynamics design studies. In R. L. Witkover, editor, *Proceedings of the 1979 Linac Accelerator Conference*, pages 205–216. Brookhaven National Laboratory, 1979. BNL-51134.

[18] R. J. Van de Graaff. Electrostatic generator. US Patent 1,991,236, 1935. Filed 1931.

[19] R. J. Van de Graaff. Tandem electrostatic accelerators. *Nuclear Instruments and Methods*, 8:195–202, 1960.

[20] R. J. Van de Graaff, K. T. Compton, and L. C. Van Atta. The electrostatic production of high voltage for nuclear investigations. *Physical Review*, 43(3):149–157, 1933.

[21] L. Deniau, F. Schmidt, C. Iselin, et al. MAD – Methodical Accelerator Design.
http://mad.web.cern.ch/mad/, (accessed August 2014).

[22] A. J. Dragt. MaryLie code and MaryLie manual information and download page.
http://www.physics.umd.edu/dsat/dsatmarylie.html, (accessed August 2014).

[23] A. J. Dragt, L. M. Healy, F. Neri, and R. Ryne. MARYLIE 3.0 – a program for nonlinear analysis of accelerators and beamlines. *IEEE Transactions on Nuclear Science*, NS-3,5:2311, 1985.

[24] D. A. Edwards and M. J. Syphers. *An Introduction to the Physics of High Energy Accelerators*. Wiley, New York, 1993.

[25] D. A. Edwards and L. C. Teng. Parametrization of linear coupled motion in periodic systems. *IEEE Transactions Nuclear Science*, 20(3):885–888, 1973.

[26] D. Einfeld, J. Schaper, and M. Plesko. Design of a diffraction limited light source (DIFL). In *Proceedings of the 1995 Particle Accelerator Conference*, volume 1, pages 177–179. IEEE, 1996.

[27] E. Forest. *Beam Dynamics: A New Attitude and Framework*. Harwood Academic Publishers, Amsterdam, 1998.

[28] R. Geller. New high intensity ion source with very low extraction voltage. *Applied Physics Letters*, 16(10):401–404, 1970.

[29] H. Goldstein. *Classical Mechanics*. Addison-Wesley, Reading, MA, second edition, 1980.

[30] P. W. Hawkes and E. Kasper. *Principles of Electron Optics*, volume 1–3. Academic Press, London, 1996.

[31] J. Ishikawa. Negative ion sources. In I. G. Brown, editor, *The Physics and Technology of Ion Sources*, pages 285–310. Wiley-VCH, Weinheim, second edition, 2005. http://onlinelibrary.wiley.com/doi/10.1002/3527603956.ch14/summary.

[32] G. H. Jansen. *Coulomb Interactions in Particle Beams*. Advances in Electronics and Electron Physics: Supplement 21. Academic Press, New York, 1990.

[33] D. Johnson. Private communication for the lattice data of the FODO cell of the Fermilab Main Injector at Fermi National Accelerator Laboratory.

[34] L. W. Jones and K. M. Terwilliger. A small model fixed field alternating gradient radial sector accelerator. In E. Regenstreif, editor, *Proceedings of CERN Symposium on High Energy Accelerators and Pion Physics*, pages 359–365. CERN, European Organization for Nuclear Research, 1956. CERN 56-25, Volume 1.

[35] S. P. Kapitza. *The Microtron*. Accelerators & Storage Rings, 1. Harwood Academic, London, 1978. Originally published by I. Nauka, Moscow, 1969.

[36] M. Kauderer. *Symplectic Matrices: First Order Systems and Special Relativity*. World Scientific Publishing Co., Singapore, 1994.

[37] D. W. Kerst. The acceleration of electrons by magnetic induction. *Physical Review*, 60:47–53, 1941. https://journals.aps.org/pr/abstract/10.1103/PhysRev.60.47.

[38] D. W. Kerst. Magnetic induction accelerator. US Patent 2,335,014, 1943. Filed 1942.

[39] O. L. Krivanek, N. Dellby, and M. F. Murfitt. Aberration correction in electron microscopy. In J. Orloff, editor, *Handbook of Charged Particle Optics*, pages 601–640. CRC Press, Taylor & Francis Group, London, second edition, 2009.

[40] E. O. Lawrence. Method and apparatus for the acceleration of ions. US Patent 1,948,384, 1934. Filed 1932.

[41] J. D. Lawson. *The Physics of Charged-Particle Beams*. Clarendon Press, Oxford, 1988.

[42] LBNL. 1–2GeV synchrotron radiation source, conceptual design report. Technical Report LBNL PUB-5172 Rev., Lawrence Berkeley National Laboratory, 1986.

[43] S. Y. Lee. *Accelerator Physics*. World Scientific, New Jersey, second edition, 2004.

[44] M. P. Level, P. C. Marin, P. Nghiem, E. M. Sommer, and H. Zyngier. Progress report on Super-ACO. In E. R. Lindstrom and L. S. Taylor, editors, *Proceedings of the 1987 IEEE Particle Accelerator Conference*, volume 1, pages 470–472. National Bureau of Standards, Los Alamos National Laboratory, 1987. OSTI ID: 5125784, CONF-870302-Vol.1.

[45] M. S. Livingston and J. P. Blewett. *Particle Accelerators*. McGraw-Hill, New York, 1962.

[46] E. J. Lofgren. Bevatron operational experiences. In E. Regenstreif, editor, *Proceedings of CERN Symposium on High Energy Accelerators and Pion Physics*, pages 496–503. CERN, European Organization for Nuclear Research, 1956. CERN 56-25, Volume 1.

[47] J. M. J. Madey. Stimulated emission of bremsstrahlung in a periodic magnetic field. *Journal of Applied Physics*, 42(5):1906–1913, 1971.

[48] K. Makino. *Rigorous Analysis of Nonlinear Motion in Particle Accelerators*. PhD thesis, Michigan State University, 1998.

[49] K. Makino and M. Berz. Perturbative equations of motion and differential operators in nonplanar curvilinear coordinates. *International Journal of Applied Mathematics*, 3,4:421–440, 2000.

[50] K. Makino and M. Berz. COSY INFINITY version 9. *Nuclear Instruments and Methods*, A558:346–350, 2006.

[51] T. Matsuo, H. Matsuda, Y. Fujita, and H. Wollnik. Computer program TRIO for third order calculation of ion trajectory. *Mass Spectroscopy*, 24:19–61, 1976.

[52] L. Michelotti. *Intermediate Classical Dynamics with Applications to Beam Physics*. Wiley, New York, 1995.

[53] M. Nishiguchi and M. Toyoda. Computer program TRIO 2.0 for calculation and visualization of ion trajectories. *Physics Procedia*, 1:325–332, 2008.

[54] H. Nishimura. Private communication for the lattice data of the Booster to Storage Ring beam transfer line at the Advanced Light Source at Lawrence Berkeley National Laboratory.

[55] J. Orloff, editor. *Handbook of Charged Particle Optics*. CRC Press, Taylor & Francis Group, London, second edition, 2009.

[56] J. Picht. Beiträge zur Theorie der geometrischen Elektronenoptik. *Annalen der Physik*, 407(8):926–964, 1932.

[57] J. R. Pierce. Rectilinear electron flow in beams. *Journal of Applied Physics*, 11:548–554, 1940.

[58] M. Reiser. *Theory and Design of Charged Particle Beams*. Wiley-VCH, Weinheim, second edition, 2008.

[59] H. Rose. Historical aspects of aberration correction. *Journal of Electron Microscopy*, 58(3):77–85, 2009.

[60] H. Rose. *Geometrical Charged-Particle Optics*. Springer-Verlag, Berlin, second edition, 2012.

[61] Y. Sasaki and S. Maruse. Über die Arbeitsweise und die elektronenoptischen Eigenschaften der Spitzenkathode. In G. Möllenstedt, H. Niehrs, and E. Ruska, editors, *Physikalisch-Technischer Teil, Band 1*, pages 9–13. Springer-Verlag, 1960. Fourth International Conference on Electron Microscopy, Berlin 1958.

[62] O. Scherzer. Über einige Fehler von Elektronenlinsen. *Zeitschrift für Physik*, 101(9–10):593–603, 1936.

[63] A. Septier, editor. *Applied Charged Particle Optics*. Academic Press, New York, 1980, 1983. Part A, B, C.

[64] D. H. Sloan and E. O. Lawrence. Production of heavy high speed ions without the use of high voltages. *Physical Review*, 38:2021–2032, 1931. https://journals.aps.org/pr/abstract/10.1103/PhysRev.38.2021.

[65] C. Steier. Private communication for the lattice data of the Advanced Light Source at Lawrence Berkeley National Laboratory.

[66] V. Suller. Private communication for the lattice data of the storage at Center for Advanced Microstructures and Devices at Louisiana State University.

[67] M. Szilagyi. *Electron and Ion Optics*. Plenum Press, New York, 1988.

[68] R. M. Tromp, J. B. Hannon, A. W. Ellis, W. Wan, A. Berghaus, and O. Schaff. A new aberration-corrected, energy-filtered LEEM/PEEM instrument. I. Principles and design. *Ultramicroscopy*, 110:852–861, 2010.

[69] V. Veksler. A new method of acceleration of relativistic particles. *Journal of Physics (Moscow)*, 9(3):153, 1945.

[70] T. P. Wangler. *RF Linear Accelerators*. Wiley-VCH, Weinheim, second edition, 2008.

[71] H. Weick. GICOSY – based on COSY 5.0 with additions done later in Giessen. http://web-docs.gsi.de/ weick/gicosy/, (accessed August 2014).

[72] H. Weick. GIOS – General Ion Optical System. http://web-docs.gsi.de/ weick/gios/, (accessed August 2014).

[73] R. Wideröe. Über ein neues Prinzip zur Herstellung hoher Spannungen. *Archiv für Elektrotechnik*, 21:387, 1928.

[74] H. Wiedemann. *Particle Accelerator Physics*. Springer, Berlin, third edition, 2007.

[75] K. Wille. *The Physics of Particle Accelerators: An Introduction*. Oxford University Press, New York, 2000. In English, orignal in German, 1996.

[76] E. J. N. Wilson. *An Introduction to Particle Accelerators*. Oxford University Press, New York, 2001.

[77] B. Wolf, editor. *Handbook of Ion Sources*. CRC Press, New York, 1995.

[78] H. Wollnik. *Optics of Charged Particles*. Academic Press, Orlando, 1987.

[79] H. Wollnik and M. Berz. Relations between the elements of transfer matrices due to the condition of symplecticity. *Nuclear Instruments and Methods*, A238:127–140, 1985.

[80] H. Wollnik, B. Hartmann, and M. Berz. Principles of GIOS and COSY. *AIP Conference Proceedings*, 177:74, 1988.

[81] M. Yavor. *Optics of Charged Particle Analyzers*, volume 157 of *Advances in Imaging and Electron Physics*. Academic Press, San Diego, 2009.

# *Index*

Printed and bound by CPI Group (UK) Ltd, Croydon, CR0 4YY

21/10/2024

01777083-0011